FRESHWATER ALGAE
their microscopic world explored

"THE SCRAP COLLECTION"

PROFESSOR F. E. FRITSCH - 1922

Drawn by F. M. Haines

FRESHWATER ALGAE
their microscopic world explored

Hilda Canter-Lund
PhD, DSc, FRPS

John W. G. Lund
CBE, DSc, DSc (hc), CBiol, FIBiol, FIWEM, FRS

Honorary Research Fellows of the
Freshwater Biological Association

Biopress Limited

© Biopress Ltd., 1995

All rights reserved. No part of this publication may be reproduced, stored in a retrieval system, or transmitted, in any form or by any means, electronic, mechanical, photocopying, recording or otherwise, without the prior permission of the copyright owner.

ISBN 0-948737-25-5

PUBLISHED BY:

Biopress Ltd.
The Orchard
Clanage Road
Bristol
BS3 2JX
England

First published 1995

Reprinted 1996

Reprinted 1998

British Library Cataloguing in Publication Data

A catalogue record for this book is available from the British Library

© Text – J. W. G. Lund (1995)

© Photographs – H. Canter-Lund (1995)

except © J. W. G. Lund, Nos 1–4, 7–10, 127–129, 166, 204, 258, 342, 421, 438, 447, 448, 450, 451, 462–464; T. Furnass, Nos 5, 6, 93, 118, 119, 122, 123, 343, 345, 362, 403, 435, 436, 440, 449; and J. Clayton, No. 11.

Colour reproduction by Tenon & Polert Ltd., Hong Kong
Printed in Hong Kong by Dah Hua Ltd.

Dedication

To our teachers and friends,
Professor C. T. Ingold and the late
Professors F. E. Fritsch and
W. H. Pearsall for their inspiration
and kindness.

The frontispiece is a cartoon by the late Professor F. M. Haines of Professor Fritsch working some 70 years ago on his now famous collection of illustrations of freshwater algae. Its size at the time was much exaggerated in the drawing. In reality it was minute compared to how it has grown over the years. Since Fritsch's death the Collection has been maintained and expanded by the Freshwater Biological Association at its Windermere laboratory.

Contents

Foreword	ix
Acknowledgements	xi
Preface	xiii
Photography	xiv
Measurements and Abbreviations	xv
Arrangement of Illustrations	xv

Chapter One	**Introduction**	1
	What are algae?	1
	The size of algae	2
	The colour of algae	2
	The structure and reproduction of algae	3
	The naming of algae	4
	The occurrence of algae	8
	The ecology of algae with special reference to eutrophication and plankton	10
	Eutrophication	12
	Plankton – Habitat and seasonal cycle of the phytoplankton	14
Chapter Two	**Chlorophytes** (green algae)	21
	Unicellular and colonial forms	21
	Conjugate algae	38
	Desmids	39
	The *Spirogyra* group	56
	The *Oedogonium* group	62
	The *Ulothrix* group	66
	The *Chaetophora* group	68
	The *Cladophora* group	72
	The *Enteromorpha* group	78
	Free-swimming forms	80
Chapter Three	**Euglenophytes**	94
Chapter Four	**Xanthophytes** (Tribophytes)	108
Chapter Five	**Bacillariophytes** (diatoms)	118
Chapter Six	**Chrysophytes**	148
Chapter Seven	**Haptophytes** (Prymnesiophytes), **Raphidiophytes** and **Cryptophytes**	166
Chapter Eight	**Dinophytes**	172
Chapter Nine	**Rhodophytes**	182
Chapter Ten	**Cyanophytes** (Cyanobacteria: blue-green algae)	194

Chapter Eleven	**Symbiosis**	238
	Cyanophytes and plants	238
	Chlorella and invertebrates	242
	Lichens	246
	Cyanelles	252
Chapter Twelve	**Animals feeding on algae**	254
	Protozoans	254
	Some non-protozoan animals	274
	Rotifers	274
Chapter Thirteen	**Fungi living on and in algae**	279
	Chytrids	281
	Some aspects of the ecology of fungal parasitism	300
	Hyphochytrids	310
	Biflagellate fungi	310
	Hyperparasites	320
	Fungal parasitism and man's activites	323

Postscript	325
Glossary	327
Suggestions for further reading	344
Index of Algal Genera	347
Index of Animal Genera	351
Index of Fungal Genera	352
Index of Algal Genera associated with Animals or Fungi	353
General Index	356

Foreword

I do not think that, at any time in the long history of the Freshwater Biological Association, a Director could look forward so much to the appearance of a book. I am delighted, particularly as a microbiologist who has been brought up in the broadest sense of that word, to see Hilda's exquisite photographs and John's words translated into a volume that we shall all enjoy. This appreciation will extend beyond that felt by scientists, because I know that those involved in arts and crafts, for example, have found these photographs to be an inspiration.

Professor J. Gwynfryn Jones

Director
The Freshwater Biological Association
The Ferry House
Far Sawrey
Ambleside
Cumbria LA22 0LP
England

Acknowledgements

We are indebted to the Council of the Freshwater Biological Association for financial help and particularly to its secretary and director, Professor J. Gwynfryn Jones for his interest and support.

To our colleague Mr Trevor R. Furnass we owe special thanks, both for his photographic help and artistic advice, and for his kindness and encouragement during periods of difficulty, despite the many calls on his time. A list of photographs he has contributed is given later.

We thank Professor Frank E. Round and members of Biopress Limited for making publication finally possible. We are particularly grateful to Mrs Gillian Lockett for her expertise, unfailing attention to detail and warm support which have eased away various vicissitudes during the production of the final typescript. During later stages of preparation we were fortunate also to have the help of Mr Anthony Lockett.

Early drafts of the book were typed by Mr D. P. Kennedy in his spare time (such as he had!), an act of kindness which was of great help to us.

We thank all colleagues and friends who have supplied us with samples of algae and with whom we have had useful discussions. We have had a long, happy and fruitful association with the FBA. Among the many facilities from which we also benefitted, we would like to mention the splendid library, and the services it provides for the staff and members of the Association.

From abroad, we thank Professor W. T. Edmondson for samples from Lake Washington and Professor P. A. Tyler for desmid material from N. Australia. We also express our thanks to the following for valuable additions to the book.

Mr K. B. Clarke for lending us slides of diatoms taken from the Thomas Brightwell Collection, property of the Norfolk and Norwich Naturalists' Society (fig. 259).

Mr John W. Clayton, Biologist, Tweed Purification Board, for the photograph of a *Botryococcus* waterbloom (fig. 11).

Professor G. E. Fogg for allowing us to reproduce his beautiful paintings (figs 435, 436).

Mr Klaus Kemp, who specially prepared for us the most elegant arrangement of diatoms (fig. 260).

The picture of *Botrydium* (fig. 198) is reproduced from a photograph given to the second author by the late Mr E. A. Ellis.

The late Professor I. Manton took the picture reproduced in fig. 305.

H. C-L acknowledges with much gratitude a grant from the Royal Society, London for the provision of the microscope used to take the photomicrographs shown in this book. Other photographs have been provided as follows:

Mr T. I. Furnass, A.R.P.S., nos: 5, 6, 93, 118, 119, 122, 123, 343, 345, 362, 403, 435, 436, 440, 449.

Dr J. W. G. Lund, nos: 1–4, 7–10, 127–129, 166, 204, 258, 342, 421, 438, 447, 448, 450, 451, 462–464.

Preface

The purpose of this book is to take people into a world of great beauty and fascination which they normally never see but which is all around them whenever they go outdoors.

We also hope that those of a more artistic turn of mind or interested in design will obtain pleasure and perhaps ideas from the pictures.

The book contains 387 colour and 253 monochrome photographs with commentaries on them. It is written in simple language but with every effort to avoid "simple" becoming scientifically incorrect. It is intended to be understandable by everyone. Technical terms are reduced to a minimum. They are explained when first encountered and again in the glossary. Some explanations in the glossary are broader than in the text. For example, they may enlarge on the use of a word or refer to recent systems of classification which depend predominantly on what has been revealed by electron microscopy. It should be borne in mind that technical words do not always have the same meaning in all branches of science or biology.

In no sense is the organization of the book in accord with the normal textbook tradition. Textbook pictures illustrate the text, here the text is a commentary based on the pictures. It follows, therefore, that certain features such as the structural methods of growth and the reproductive processes which occur in the algae shown are not mentioned as they would be in a textbook.

The word "freshwater" in the title is not wholly correct, for some of the algae shown live in salty water. A more correct title would be Algae of Inland Waters or Non-marine Algae. Inland waters, however, can be salty, even more salty than the sea and any freshwater will become more and more salty as it dries up. Floristic works on "freshwater" algae often also are in fact algal floras of inland waters, so we use the word freshwater in much the same sense as others. We also follow others in including terrestrial algae. In any case, terrestrial algae depend on rain or mist for growth and so on freshwater.

We hope that those of our readers who have some knowledge of biology, botany or algae will not be offended by explanations of things long familiar to them. As a small compensation, they probably will learn some things about algae and their "enemies" which they did not know before.

The reader may be irritated by the frequency with which such words as usually, commonly, generally, often or rarely occur. Unfortunately, nature does not conform to our desire for neat, clear-cut categorisation. Even as there so often is the "odd man out" in a group of people, so, in this case, almost invariably there are algae which diverge in some way from the general mass in the group to which they belong. If, for example, a group of algae is characterized by some form of structure or reproduction, there are sure to be species which diverge to a smaller or greater extent from the norm but undoubtedly do belong to the group concerned. This situation is especially irritating for those who classify algae. It is the reason why the adverbs referred to above are used so frequently.

Nearly all the algae depicted are to be found all over the world and commonly so in the kinds of habitat they normally live in.

We show pictures of algae from the major groups which we have called —phytes, including the Cyanobacteria. Thus we have Chlorophytes (green algae), Euglenophytes, Xanthophytes, Bacillariophytes (here the word diatom is used in the text), Chrysophytes, Haptophytes, Raphidiophytes, Cryptophytes,

Dinophytes, Rhodophytes (red algae) and Cyanophytes (blue-green algae). For the derivation and meaning of words containing the syllables –phyte or –phyc, see the glossary under these syllables.

An introduction to algae would not be complete without a word about those organisms which depend upon them for food and so may be regarded as their "enemies". We have illustrated grazing by crustacea, rotifers and protozoa, and parasitism by aquatic fungi. In the latter instance, not only are pictures of a general nature shown, such as fungi growing on or in algae, but also some which will help the reader gain a greater understanding of more specific stages which occur in their life histories. Other parasites of algae not shown here include bacteria and viruses.

A species list is not given for three reasons. First, the specific identity of some of the organisms shown is not known. The pictures were not taken with such identifications in view. Moreover, correct names of several of the algae shown are the subject of argument among specialists, notably in regard to Cyanophytes. Second, it would be incorrect to give the impression that most of the algae shown can be identified from a single photograph. In many cases, identification by light microscopy or even with the aid of electron microscopy as well is not possible. Third, even as this is not a textbook, so also it is neither a flora nor a forum for taxonomic arguments.

A few species names are given for reasons which will be obvious. Among the fungi and protozoa there are some which as yet are so little known that they are not even identifiable at the level of genus.

Those who start to identify algae are well advised to familiarize themselves with genera before entering the much more difficult task of identifying species. Another good piece of advice is to familiarize oneself with living specimens collected from nature before making other kinds of observations, though the latter may be essential later.

Photography

All the details of the organisms shown can be seen with a light microscope. There are also a few pictures of habitats, waterblooms, etc. The vast majority of the photographs are from live, natural populations, that is, the algae are as they appear when collected or seen in the mass.

The photographs were taken using films (rated from 25-200 ASA) and electronic flash coupled to a Zeiss Photomicroscope I. A number of different types of objectives and condensers have been employed which in turn confer different qualities of illumination to the specimens under observation. In bright-field, organisms are viewed without any alteration, whereas phase-contrast and negative phase-contrast are mostly used for enhancing structures which are colourless. Differential interference contrast (Nomarski) imparts a 3D effect to specimens. It can also be used in the purely aesthetic sense to produce false coloured pictures and backgrounds. Dark-field illumination on the other hand always gives a jet-black background. In the legends to the photographs the type of illumination is only designated (in an abbreviated form) when it is other than bright-field (i.e. phase-contrast (P); negative phase-contrast (Pn); Nomarski (N) and dark-field (D). Dyes have been used occasionally to highlight certain structures and specimens have been mounted in Indian ink (I) to show the presence of mucilage which otherwise would have remained invisible.

Measurements: Abbreviations

Most freshwater algae are so small that their linear, square and cubic dimensions are given in micrometres (μm, μm^2, μm^3), the old name for which was microns. One micrometre = 0.000039 inches (in). The other metric units used are millimetre (mm), centimetre (cm), metre (m), kilometre (km) and ton (tonne). One metric ton equals 1.10 American short and 0.98 British long tons.

In the legends to the pictures of illustrations, the size of one or more algae depicted is given or that of cells or parts thereof (e.g. the length or breadth or both, lengths of spines, stalks, flagella etc). l. = length; br. = breadth or width; diam. = diameter.

In the case of the monochrome photographs, sizes can be calculated from the magnification given for each picture.

Arrangement of Illustrations

Colour pictures always occupy and fill righthand pages. Monochrome pictures are distributed throughout the text, mainly on lefthand pages and opposite a page of colour pictures.

Each single picture or the position and arrangement of a group of pictures are depicted in the form of numbered boxes, e.g.

1

or

1	2
3	4

The boxes are placed in the righthand margin of the lefthand page, except in rare instances where there are monochrome pictures on a righthand page. The legends to the pictures are below the boxes.

The sequence of numbering is continuous throughout the book. The numbers pertaining to colour pictures are printed in **bold type**.

Both boxes and legends of monochrome pictures are put opposite or as close as possible to the pictures concerned, those of colour pictures are put wherever most convenient. Because of this arrangement, the placing of the boxes in the margin of a lefthand page may not be in strict numerical order. Examples are:- p. 30, fig. **31** precedes figs 29, 30 and p. 32, figs **35-37** precede figs 32-34. However, the combination of the diagrams indicating the arrangement of the pictures and the differences in type for monochrome and colour should assure that there is no difficulty in matching the figure numbers to their relevant illustrations.

Chapter One

Introduction

What are algae?

A book about algae should start by answering the above question. A correct, but hardly satisfactory answer is that algae are a diverse group of organisms once thought to belong to a single class of plants. That algae cover a wide variety of organisms will be obvious when the pictures have been seen and the commentaries on them read. Hence, it could be easier to answer the question at the end of the book. All the matters mentioned in this Introduction indeed are repeated or enlarged upon when certain kinds of algae or their special types of cells are first encountered in the pictures.

Originally, algae were considered to be simple or lowly plants lacking the leaf, stem, root and reproductive systems of Higher Plants such as mosses, ferns, conifers and flowering plants. However, early on it was realised that some algae have animal-like characteristics so that they were included in both the plant and animal kingdoms, indeed they still are.

Though the animals and plants familiar to all obviously are very different kinds of organisms, at the algal level there is no distinction between the plant and animal kingdoms. Most algae are plants but some are animals and in between there is a considerable number of algae with both plant and animal characteristics. Some of these algae are more like plants than animals, others more like animals than plants and there is every gradation between the two.

Today we know that living organisms can be divided into two distinct groups called prokaryotes and eukaryotes. Prokaryotes include bacteria of all kinds, eukaryotes include all non-bacterial organisms. One group of algae, the Cyanophytes or blue-green algae, are now known to belong to the prokaryotes, in fact they are bacteria. In the scientific literature they are generally but not always referred to as Cyanobacteria. In the media they continue to be called blue-green algae.

Since Cyanophytes superficially look like algae and are photosynthetic, it was reasonable in the past to consider them to be plants. Nevertheless, even long ago there were some scientists who suggested a relationship between blue-green algae and bacteria. The reason why the similarities between bacteria and blue-green algae (Cyanophytes) were not evident in the past is that before the days of electron microscopes the so-called submicroscopic structural features common to both could not be seen. In addition, certain biochemical features (e.g. nitrogen fixation) had not been elucidated.

The only features common to all algal groups (including Cyanophytes) are that their photosynthetic representatives contain the pigment chlorophyll *a* and release oxygen in the process of photosynthesis. In these features they are like Higher Plants.

Included within the algal groups are some species which neither contain chlorophyll *a* nor photosynthesize. Nevertheless they possess structural or reproductive features which clearly relate them to the photosynthetic forms.

The size of algae

Most freshwater algae are of microscopic size, that is they themselves cannot be seen or their general structure cannot be seen by unaided vision. The stoneworts (Charophytes), not shown here, are an exception in that most are bushy plants 20 or more cm in height. Few algae are so small that they cannot be seen under an "ordinary" microscope (light microscope), though little more than the shape of the very small species may be visible. The smallest inland algae consist of single cells about one thousandth of a millimetre in diameter. The largest known alga, a seaweed, can reach about 60 metres in length, that is as long as a very tall tree is high.

The photomicrographs (pictures taken under a microscope) in this book were taken using an "ordinary" or light microscope, that is one depending on light rays to produce a magnified image of an object. Magnifications of about 1500 times are obtainable with a light microscope whereas with an electron microscope vastly greater magnifications (hundreds of thousands) are possible. Live algae cannot be examined by electron microscopy. The algae have to be killed or "fixed" in such a way that as much as possible of their structure is so preserved that it is not harmed by the conditions within an electron microscope. The minutest alga or part thereof can then be studied. For such minute details revealed by electron microscopy the words fine structure and ultrastructure are used.

The colour of algae

Even as algae vary greatly in size, so they do in colour. Seaweeds, the large algae of the seashore, are various shades of green, brown and red. Microscopic algae of the sea and inland waters also show almost every shade between and mixture of these colours plus forms which are blue-green and, in a few cases, blue. The colour common to many members of a group can be the basis of its name.

Thus, green coloured seaweeds belong to the Chlorophyte (Greek: *chloros* – green) group, and red ones to the Rhodophytes (Greek: *rhodon* – a rose, hence red).

Whereas marine and freshwater Chlorophytes usually are green in colour, freshwater Rhodophytes rarely are red. The brown seaweeds belong to a group of algae with very few freshwater representatives but many microscopic marine and freshwater algae, belonging to other major groups, are brown. Therefore, colour alone is not necessarily a safe guide to the group to which an alga belongs.

The structure and reproduction of algae

The structure of algae is a large subject but a few introductory facts will be helpful. The bodies of many algae consist of a single cell, such algae are unicellular. In some cases a varied, indeterminate number of cells live in an irregular and unorganised mass. In contrast to this formless aggregation of unicells, other algae live in organised groups. The degree of organization can vary from low to very high. Such structured groupings of algal cells are called colonies. Since these colonies basically are groups of individual cells, should a cell become separated from the colony, usually it can continue to grow and reproduce resulting in the formation of a new colony.

Cells can be arranged in thread-like chains, called filaments. Filaments can be solitary or in groups and branched or unbranched. In a few genera, filaments consist of a single, tubular cell. Only under special circumstances (e.g. reproduction) are these tubes divided up into one or more parts.

Lastly, cells can be united together in two or three planes to form tissue-like bodies reminiscent of the soft internal tissue of Higher Plants called parenchyma.

Unicellular, colonial and filamentous algae can be motile or non-motile. Motility is conferred in various ways from swimming to creeping, pushing or by flotation devices. Many non-motile algae reproduce by motile cells.

Reproduction, the production of new individuals, can take place in many ways. The separation of a cell from a colony mentioned above is reproduction because that cell can produce a new colony. In the same way, fragmentation of an algal filament is reproduction when the result is an increase in the number of filaments present. Cell division leading to an increase in size of an alga is no more reproduction than is the growth of a baby. However, in some algae, when a cell divides into two both parts separate, so that here cell division is also reproduction.

Many algae produce special reproductive bodies, for most of which we use the common word spore. Spores usually are produced in cells of a special kind or in ordinary cells which have become changed for the purpose of producing them. A spore-producing cell often is called a sporangium.

Spores are of many kinds and so have many technical names, the majority of which need not concern us. The ones mentioned later in the text are zoospore, gamete, zygote and resting spore or cyst. Zoospores are free-swimming spores. Gametes are sexual cells which fuse in pairs. Gametes may all be much the same in size and behaviour or so different in one or both features that they can be considered as male or female. Egg cells are female gametes which are much larger than their male sexual partners which are then called sperms.

The product of sexual fusion is a zygote. Zygotes usually do not germinate quickly and so also act as resting spores. A resting spore or cyst, the meaning of the two words overlap, is a special kind of reproductive body the production of which may or may not involve sex. Whether sexually formed or not, as the words resting spore suggest, such a spore often enables an alga to, as it were, lie dormant. Thus, they very often are produced when conditions are unfavourable for further growth and are resistant to these unfavourable conditions (e.g. drought, heat, cold, lack of nutrients). They germinate after months or years, when conditions may well be favourable for renewed growth.

As with ordinary spores there are many kinds of resting spore or cyst, several of which are shown but not given their various technical names. An exception is the digestion cyst of certain animals feeding on algae, in which, as the name suggests, the food they have taken in is digested and out of which a further animal or animals will come when digestion is completed.

The naming of algae

A few words about the names of algae will be helpful to those not familiar with the subject of nomenclature.

Briefly, it can be said that the classification and nomenclature of algae follows plant, animal or bacterial systems depending on whether you consider a given alga to be a plant, animal or bacterium. The commonest system treats all algae as plants. Organisms which can be seen easily by eye and are common usually have common names. Such names are very likely to differ from country to country or even possess regional variations. This is not so with their scientific names. To give an example, the tree in English called oak is chêne in French, Eiche in German and dub in Russian but in the universal language of science all oaks belong to a genus called *Quercus* (printed in scientific works in italic type). This group or genus contains many different kinds or species of oak.

Like the genera of other organisms, those of algae may consist of a single species or more than one species, in a few cases even over 1000 species. With very few exceptions (e.g. *Hydrodictyon reticulatum*, p. 32) we only give the name of the genus. The names of genera used are those to be found in the present generation of textbooks and floras. However, we are in a period when many changes in naming have been proposed. When a genus mentioned has had its name changed recently or has been broken up into two or more genera and such changes are likely to become permanent we mention them. We do this because they will appear in future textbooks and floras.

The scientific names of algae and other organisms are in Latin (often derived from Greek words) or the latinised form of a non-Latin and non-Greek word. An example of the latter type of name is the common garden plant, the dahlia (*Dahlia*) which is the latinised name of a person (Dahl). There is a plant called *Linnea* which is named after the Swedish inventor of modern scientific nomenclature, Linnaeus who latinised his own name of Carl von Linné.

Most algae do not have a common name, only a scientific one either because they are microscopic or, even if easily seen without magnification, are not so familiar to the layman that they have acquired common names. One of the few freshwater algae large enough and of sufficiently striking appearance to have a common name in English is a species of the genus *Hydrodictyon*. The species shown (figs 32-34, 38), *Hydrodictyon reticulatum*, is called the water net. The word *Hydrodictyon* is derived from two Greek words, appropriately meaning water and net. The species name *reticulatum* means reticulate, that is net-like and so repeats part of the name of the genus.

As with all plants and animals, algae are classified in an hierarchical system. From the top downwards, each rank in this graded system includes every other rank below it. The top category in this hierarchy is the division (e.g. Chlorophyta) and below in succession come the class (e.g. Chlorophyceae), order (e.g. Chlorococcales), family (e.g. Chlorococcaceae), genus (e.g. *Hydrodictyon*) and species (*Hydrodictyon reticulatum*). At the level of species there may be varieties or forms. Various intermediate ranks are possible (e.g. subgenus). Unfortunately, there is no universal agreement as to the category to which an organism must belong. One specialist may consider a group of organisms is worthy of the rank of order, another that it is only worth the name of family and so on. Further, one specialist may hold to a certain classification of the living world which is not supported by another.

Taxonomy is the science on which classification is based and its practitioners are called taxonomists. They try to be objective when deciding how to build a system of classification and into which taxonomic category (taxon) an organism or group of organisms should be placed. However, these are human conceptions and so are subjective to some degree.

There are two main methods on which taxonomy and classification are based. In practice, the two methods overlap. Characteristic features of organisms can be compared without any consideration of the importance which should be given to any of them. Alternatively, more weight can be given to some characteristics than to others. For example, the presence or absence of wings could be used as a distinctive feature. Of course, this is an over-simplified example and nobody would compare a single presence or absence alone; as many characters as possible would be compared. For example, flies have wings but they have many other features distinguishing them from other winged organisms and the structure of their wings itself is different. Bats have wings and look more like birds than do flies but other features show clearly that they are mammals and not birds, just as whales are mammals, not fishes.

Whatever taxonomic method is used, there is an overall desire to produce a natural classification, that is one based on true relationships, not mere similarities. True relationships are those arising from the evolutionary histories of the organisms concerned. Thus, bats have diverged in the course of evolution from ancestors common to them and to birds and a natural classification puts them in our group, the mammals.

Of course, relationships often are not yet clear and are a cause of controversy, sometimes heated controversy. Moreover, biologists can be divided into lumpers and splitters. Lumpers will tend to have broader categories than splitters, hence their genera will tend to have more species than splitters would accept. So far as algae are concerned, the splitters are in ascendancy today. One can see this by the way attempts are made to break up genera with large numbers of species into several separate genera. Examples of such genera are *Cosmarium* (p. 40) and *Navicula* (p. 237), each a genus with some 2000 named species. In the first case, attempts to split the genus have failed, in the second case some success has already been achieved. However, the present ascendancy of splitters has led to the breakup of genera with many less species, even only a few species. The result has been a great increase in the number of genera in recent times. No doubt, in some cases splitters are correct and in others lumpers are. We too have our views or prejudices. So far as *Navicula* is concerned, it seems probable that the present splitting up of the genus is justified by the more comprehensive study of the species concerned than existed in the past. On the other hand, so far from splitting up the genus *Cosmarium*, an alternative case could justifiably be made for uniting it with five other genera of the group to which it belongs. Such a merged genus would contain about 4000 species and so be an anathema to splitters. Though holding a lumper's view here, we doubt if such a generic reorganisation

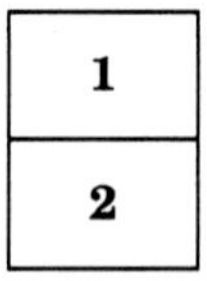

Figs 1, 2. Green algae, predominantly *Apatococcus*, on wooden fencing and wall.

of desmids would assist in the identification of species and justify all the resulting reorganisation of species names involved. Hence, our prejudice is that the present, long used, unnatural classification of desmids might as well remain.

The occurrence of algae

Wherever you are, in town or country, algae are around you or under your feet, for example on palings (fig. 1), walls (fig. 2) and trees. They live between paving stones, on steps, on the roofs and in the gutters of houses and in or on garden and field soil (fig. 3). Just as paving stones may be slippery because of the algae on them, so may rocks and rock faces in the mountains and in flowing waters. Wherever there is moisture and light there will be algae.

The greatest variety of algae is found in the permanent or semi-permanent waters of lakes, reservoirs, ponds, bogs, rivers, streams and canals. Those in marginal or shallow water often are visible as floating masses, tangles around waterweeds or other underwater objects or as fine threads on or floating out from rocks, stones or other surfaces (e.g. figs 4-6, 93, 118, 119, 127, 128). Sometimes accumulations of microscopic algae can be seen on the surface of the water. When such algae are so numerous that they form a coloured coating to the surface they produce what is called a waterbloom (figs 7, 10, 11). Very small temporary waterbodies such as puddles, rock hollows, bird baths (fig. 9), and even the footprints of cattle can contain algae, sometimes in great numbers. Such waterbodies often dry up and may remain dry for a considerable or long time, notably in hot countries with well marked dry and wet seasons. Live algae may still be present though unable to grow until the next rain comes. Such air-dry algae may be dislodged by the wind and blown away. As a result, the air contains algae but they do not grow or multiply in it. They have even been found high up in the atmosphere.

In winter in cool or cold climates, lakes and pools are covered by ice and snow, often for prolonged periods. Though conditions for survival or growth may be severe, the populations of algae present in summer rarely are destroyed, nevertheless their numbers can be greatly reduced. Just as the algae of temporary waters (e.g. bird bath, fig. 9) have strategies for withstanding drying, so permanently aquatic algae have means of existing in winter. Some exist passively (i.e. not growing) until spring comes, others continue to grow, albeit relatively slowly. Figure 8 shows a bottle of water taken from a pool which had been covered by snow and ice for over eight weeks. Such a mass of algae was living just under the ice sheet that the water was deep green in colour.

The effect of the combination of low light, short winter days and the coldness of the water reaches its maximum in

3	4
5	6

Fig. 3. Soil algae, predominantly Chlorophytes and Xanthophytes and seedling garden plants.

Fig. 4. Mat of filamentous algae, predominantly *Spirogyra* on the surface of a pond.

Fig. 5. *Stigeoclonium* growing on a stone.

Fig. 6. Actively photosynthesizing filaments of *Cladophora*, the white specks are bubbles of oxygen.

the winter darkness of polar regions. Yet, in Antarctica, algae are present, sometimes in abundance, in lakes which are always under ice because in such a cold climate the ice cover is not wholly lost in summer. Some algae grow relatively fast in cold water, even close to freezing point. Such algae can begin their spring growth before an ice sheet is lost and their populations reach a maximum at a temperature of about 4°C.

About 30°C is the upper limit at which most algae can grow but some of those which live around hot springs can grow at a temperature of about 60°C. Hot springs are not the only places where algae might be expected to be unable to grow. Returning to cold places, algae can be found on permanent or semi-permanent snowfields ("red" or "green" snow) and a little way below the surface of rocks in parts of Antarctica where the air temperature never exceeds freezing point and is extremely low in winter. On the other hand, a rockface can be warmer in the continuous daylight of the Antarctic summer.

Deserts contain algae. They are least common in the windblown sandy deserts and most common in the pebbly, rocky or clayey deserts. They can be numerous in deserts which receive no rain for a year or more. This is because in the cold desert night dew may fall and trickle down the sides of rocks and pebbles. Hence, most algae are found under stones, particularly translucent ones (e.g. quartz), or close to their base. Here they are protected from the fiery midday sun and the moisture lasts longest. Parts of deserts near the sea may not receive rain but considerable mist.

Some algae, as we shall see, live inside animals or plants. Yet other algae live with fungi in the form of what one can call dual organisms, in this instance forming the group of organisms known commonly as lichens (p. 246).

The ecology of algae (with special reference to eutrophication and plankton).

In the previous section the subject of algal ecology, that is their interrelations with the environment and with one another has been mentioned. Like all other organisms, they are affected by the variations in the physical, chemical and biological conditions in space and time. Ecology is a vast subject and only a few aspects are touched upon here. Symbiosis, animals feeding on algae and fungi parasitizing algae occupy separate chapters and what is meant by such words as symbiosis, grazing and parasitism are explained in them. On the other hand, the words eutrophication, plankton and stratification, which are used to describe important aspects of lake ecology are referred to more than once and do not form the subjects of specific chapters. Therefore some general remarks concerning these words are given here.

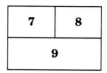

Fig. 7. Scum of *Euglena sanguinea* on a pond.

Fig. 8. Bottle of water, taken from beneath the snow covered ice sheet, containing a massive accumulation of *Chlamydomonas*.

Fig. 9. A dry bird bath discoloured by cysts of *Haematococcus*.

Eutrophication

Waters are said to be oligotrophic or eutrophic on the basis of their potential for supporting little or much growth of plants and so, in turn, of animals. These words, like so many other scientific terms, are derived from classical Greek. Oligotrophic (*oligos*, few or little: *trophe*, nourishment) waters are poor in plant nutrients and eutrophic (*eu*, well) waters are rich in plant nutrients. Clearly, there is every gradation between these two states. Further, in general, these words have no exact numerical values, any more than is indicated by the statement that a soil is of poor (oligotrophic) or of good (eutrophic) quality. Lakes and rivers receiving water from areas of thin (e.g. mountain soils) or poor soils (e.g. lowland heath soils) will be oligotrophic, whereas those receiving water from deep lowland or good agricultural soils will be eutrophic. However, what may be expected can be altered by substances added by man. These can be effective fertilisers. Add suitable fertilisers to an oligotrophic water and it will become a eutrophic water.

Algae need about 16 chemical elements for life. They are supplied in various forms, for example as the nitrate, phosphate, potash and lime so important for gardeners or farmers. A lack of any one of these essential elements can affect the rate at which an alga grows or, if there is sufficiently severe shortage, can stop growth. The major element controlling algal growth in most temperate waters is phosphorus in the form of phosphate. The next most important element is nitrogen in the form of nitrate and ammonium salts, except for some Cyanophytes which can use nitrogen gas (p. 210). In the tropics, it is nitrogen rather than phosphorus which seems to be the major element controlling growth in most waters.

Basically, eutrophication (enrichment) is not a bad thing. More nutrients means more plants and animals. However, as the saying goes, one can have too much of a good thing. Too many algae and other organisms can reduce water quality, block rivers and canals, add significantly to the cost of water supply, harm amenity and endanger plants and animals which we wish to conserve. Further, what lives eventually will die. The products of death and decay can be harmful, for example the process of decomposition uses up the oxygen in the water which is essential for the life of fish and other organisms. Eutrophication alters the structure of the communities of plants and animals present. Fish, which fishermen wish to catch are replaced by other less desirable kinds. Plants and animals which have rarity value or we wish to conserve are replaced by others. The main causes of undesirable eutrophication are the effluent from sewage works, agricultural fertilisers and farm wastes (e.g. from cattle

Fig. 10. Waterbloom, Cyanophyte scum; pale blue part near the shore, pigments released from dead (lysed) cells.

Fig. 11. Waterbloom of *Botryococcus*, from a photograph by John W. Clayton.

lots and piggeries). Hence, this artificial enrichment of water sometimes is called cultural or man-made eutrophication. It is not pollution in the ordinary sense of the term, that is the entry into water of dangerous or poisonous chemicals but its secondary effects can be similar to those of direct pollution. Eutrophication is considered further on pp. 76, 220.

Plankton – Habitat and seasonal cycle of the phytoplankton.

Plankton is the assemblage of organisms which carry out all or a major part of their growth and reproduction in the open waters of the sea, lakes, pools and rivers. How long they can remain dispersed in the water or that part of it which permits growth and reproduction depends on their own inherent properties and the movements of the water in which they are suspended.

The plant and plant-like organisms form the phytoplankton, (Greek: *phyton* – a plant), the animal or animal-like organisms the zooplankton, (Greek: *zoon* – animal) and the bacterial ones the bacterioplankton. The phytoplankton consists almost wholly of algae. Since algae contain forms covering every gradation between plants and animals and include bacterial forms (prokaryotes), the accepted constitution of the phytoplankton can be viewed as a convenience rather than a strictly scientific category. However, by phytoplankton, scientists almost invariably mean photosynthetic algae and so the word "plant-like" is acceptable. It is such algae which are considered here.

In a lake, pool or river, the sides and bottom form a habitat to which algae can be attached or on which they can reside. This habitat is called the benthos. It can extend outwards from the sides or upwards from the bottom into the water in the form of things sticking or growing out of it, for example aquatic plants. One can consider a very weedy pool as an almost wholly benthic habitat because there is very little free water for the development of plankton. An alga can be both planktonic and benthic, spending one stage of its life in the water and another on the bottom. This can happen even in large, deep lakes but clearly is unlikely in the middle of the Pacific Ocean where the sides and bottom are so far away.

It is self-evident that an algal plankton can only arise if the algae have time to multiply in the water, so that the gain in numbers in unit time exceeds the loss in numbers. Moreover, the rate of loss of algae from the plankton can affect both its quantity and quality, as is well seen in river plankton.

If a river is short and fast flowing then it cannot have an algal plankton because the water has reached the sea or, if it is a tributary, the main river, before any potentially planktonic cells present have multiplied significantly.

When a river becomes long enough or slow flowing, time is available for the development of phytoplankton. Even then, time can affect the quality of the phytoplankton. If the time of residence in the river is short, then the phytoplankton is likely to be dominated by small algae and the larger algae, so common in lake plankton, be rare or absent. In general, small algae multiply faster than large ones. Therefore a plankton of small algae is likely to be able to maintain itself when rates of loss of cells are such that they exceed the rates of growth of large algae. In Britain, a country without large continental rivers, the poverty of the phytoplankton in many rivers and the dominance of small algae in those rivers long or slow enough to possess a well-developed phytoplankton is very marked. It is also very noticeable that it is in the longest British rivers that large algae are most prominent but even then some only so in drought years.

In large continental rivers, such as those of America, Africa and Russia, there is ample time for plankton to develop. In addition, the larger the river system, the more likely that there are many lakes in it supplying the river with typical lake plankton. To these lakes in the river system have been added reservoirs and dams, some of which are very large. The construction of them can change the character of a long stretch of a river. For example, in the last century the Volga and Dnieper were rivers, today long stretches have become chains of reservoirs.

The plankton of lakes too is affected by the time factor, by how rapidly the water in it is replaced by new water entering it. A lake can be considered as a bulge in a river but usually the "bulge" is so large that the residence time of the water in it is ample for the development of plankton. However, there are lakes through which the flow is so great for some or most of the time that this is the major factor controlling the phytoplankton.

At this point, a word should be said about what we mean by small and large planktonic algae. There is no actual criterion but roughly by small we mean algae whose volume is less than 1000 μm^3. Such algae can be of very varied shape, for example spherical (e.g. 10 μm or less in diameter) or long and thin (e.g. 100 μm or less in length and less than 4 μm in width).

So far, only the horizontal time factor has been considered, that is how fast river water flows or how rapidly lake water is replenished. In lakes and to some extent in deep rivers there is a vertical time factor determining how long a plankton alga spends in the upper, well illuminated layers of the water. Light will both decrease with depth and will change in colour since different parts of the spectrum will be absorbed to different degrees. It should be pointed out here that light is what the human eye sees. This is not exactly the same as the waveband a plant uses for photosynthesis but near enough to use the word light here.

Clearly, if a lake is deep enough there will be too little light in its lower regions for photosynthesis and so for growth. At what depth these lower regions begin depends on the clarity of the water. Nevertheless, if the water is mixed throughout, an alga may not spend so much of its time in the dark that its growth ceases. However, lake water is not always completely mixed by water movements from the top to bottom. The degree of mixing and its periodicity have profound effects on plankton algae. The degree of mixing depends in most lakes on wind action and temperature. There is a limit to the depth to which wind alone can mix a water-body. Hence, in deep lakes and the oceans other factors play a major role in the sinking, rising and mixing of the water. In what follows, lakes of moderate depth (e.g. less than about 100 m) are considered. If there is no density barrier (e.g. thermal stratification, see below), wind potentially can mix the whole mass of water. So far as algal growth dependent on photosynthesis is concerned, sufficient light is unlikely to be present below 30 m. Though some algae can move, their power of movement is small compared to movements of the water which the wind can and so often does produce. However, if the density of the water increases with depth, then it can impede the downward mixing produced by the energy of the wind. The commonest cause of density differences with depth is temperature. Hence, this effect is called thermal stratification. How it arises can be explained by following the seasonal changes likely in an imaginary lake of moderate size in the north or south temperate region and one which is frozen in winter.

The first thing to remember is that the maximum density of water is at approximately 4°C, above or below 4°C it becomes progressively less dense. The actual temperature of maximum density is affected by pressure and by the salt content of the water, neither of which have much effect in most "fresh" waters. In winter under ice, the temperature and so density of the water will be less near the ice sheet than near the bottom because it is the surface water which cools and freezes. The difference in temperature from top to bottom may be quite small. Greater temperature difference and so stronger thermal stratification can exist in summer. On the other hand, the water under ice is unaffected by turbulent mixing produced by the wind. Under ice, light penetration may be very good or very poor but even if good the rate of growth of algae under the ice is retarded by the short days and cold water. Light penetration through clear ice is almost as good as through distilled water but is poor if the ice is cloudy and especially so if it is covered by snow. Snow reflects most of the light falling on it, hence its whiteness.

As spring approaches, the days lengthen, the sun rises higher in the sky and the snow and ice begin to melt. Penetration of light increases. Finally, ice cover is lost and the water is mixed to all depths by the wind. This situation continues while the water temperature rises to 4°C and some way above it. Nevertheless, the upper layers of the water are warming up, since they absorb the incident heat rays. So long as the heating and so reduction in density of the water is not too great, any incipient thermal stratification will soon be broken down by wind action. Eventually the amount of heat entering the upper layers of the water in unit time is too great for the wind to dissipate throughout the whole depth of the lake. By virtue of their density differences, the upper part of the water becomes separated from the lower colder part. The lake water now consists of three layers. The warm, well illuminated part is called the epilimnion and the cold darker and lower part, the hypolimnion. In between these two layers is a region in which temperature falls sharply with depth, this is the metalimnion or thermocline (the difference in the exact meanings of the two words is not important here). With the onset of autumn the surface waters cool, thermal stratification weakens, and the epilimnion and metalimnion deepen, so reducing the volume of the hypolimnion. Finally, the wind mixes the whole water mass so that its temperature is the same from top to bottom. Then, the reverse of summer stratification takes place. The water becomes colder from the top downwards, every quiet night more ice is formed and finally it covers the whole lake.

Obviously, there is an infinite variety of seasonal cycles of stratification, depending on how big and how deep a lake is, where it is situated and in what kind of climate (e.g. continental or oceanic). Tropical lakes, despite the absence of winter and summer, can have seasonal variations in thermal stratification because of marked seasonal changes in wind or rainfall, which will be accompanied by more cloudy and so cooler weather. However, often the pattern of thermal stratification in a tropical lake is irregular. It can be daily in shallow lakes. Despite the great input of heat by day, the loss by night can set up strong convection currents. The same situation can arise in sunny summer weather in small pools in temperate lands.

Let us consider an imaginary lake in the temperate zone but also in a continental region, with the seasonal cycle of thermal stratification described above. This lake has a maximum depth of 50 m and a mean one of 25 m. In summer, the epilimnion might be from 5-10 m in depth. It will of course fluctuate in relation to variations in the weather. The metalimnion could be from 10-15 m below the surface. What proportion of the water mass is in each layer will depend on the shape of the lake basin. The percentage of the water mass of our lake in the epilimnion, metalimnion and hypolimnion could be of the order of 30%,

10% and 60% respectively. These rough figures are only given in order to provide some idea of the likely proportion of the water available for the growth of phytoplankton. Light penetration at best is only likely to permit significant photosynthesis as far down as the upper part of the hypolimnion. Further, if there are large populations of algae in the epilimnion, they themselves will reduce the transparency of the water and so the depth to which photosynthesis is possible.

Planktonic algae must have some means of remaining in suspension for a long time, for example weeks or months, in order to produce their often large populations. When a water is mixed from top to bottom, their own motions are of little or no significance in determining where they are. The turbulent motions produced by a strong wind blowing on unstratified water can have an affect similar to shaking water containing algae in a bottle. They are distributed at random, nobody can predict where a given alga will be at a given moment. The less turbulent the water is, the more the chance increases that the position of an algae will be determined by its own motion relative to that of the water it is in, for example by its sinking rate. In our imaginary lake, algae may be mixed more or less at random in the upper epilimnion but the nearer they come to the metalimnion, the less turbulent the water. If an alga enters the metalimnion then it will probably sink through it and be lost to the hypolimnion if it has no form of mobility to bring it back into the epilimnion.

Most algae have a positive rate of sinking, that is their weight in water is greater than that of the water itself. The density of protoplasm is somewhat greater than that of water and the components of the cell wall or solids in the cell may increase the cell's overall density. However, there is a large minority of planktonic algae which either have a positive rate of rising or can swim. Some are so small that their rate of sinking is negligible. The majority of the planktonic algae with solitary spherical, globose, oval or ovoid cells are small (e.g. for spherical cells, less than 10 μm diameter; the smallest are about 1 μm diameter). When in groups (colonies, p. 3) they usually are embedded in mucilage which itself is extremely watery and which may well reduce the overall rate of sinking. All such cells have a low ratio of surface area to volume, a sphere the lowest possible. The greater the ratio of surface area to volume, the slower the rate of sinking will be, provided of course that the total volume remains the same. The fact that so many planktonic algae have cells or groups of cells with a high ratio of surface area to volume, compared to a sphere, is considered to be an adaptation to ensure a slow rate of sinking. (good examples are seen in Chapter 5, figs 239, 240, 243). A second potential advantage is that a greater surface area is available for absorption of light and nutrients. A third advantage is that they can be of such large size and awkward shape that many planktonic animals cannot ingest them. However, this advantage is reduced by the fact that some animals enter into such algae and "eat" them from the inside or attach themselves to cells and "suck" out their contents (concerning animal grazing see Chapter 12).

Sinking has the disadvantage that photosynthesis and so growth will get less and less as the bottom is approached. However, it is not always a disadvantage. Staying too long in the warm, well-illuminated epilimnion can lead to the growth of a population exceeding the rate of supply of nutrients needed to maintain or increase its numbers. This state of affairs regularly arises in the spring and after thermal stratification has started. If algae sink out of the epilimnion while still alive and reach water less depleted of nutrients, they can replenish their cells and, if sufficient light is available, renew growth. In our imaginary lake, the more probable result is that they will enter the dark hypolimnion. If they can live in the dark long enough, they have a good chance of returning to the lake water as a whole when thermal stratification breaks down in autumn or winter. Then, they can be the inoculum for the next year's vernal (spring) algal maximum in the epilimnion. Many algae can produce special kinds of cells which can resist unfavourable conditions for a long time. Such resting spores will, like seeds, germinate later, stimulated by some change in the environment (e.g. an increase in the temperature of the water or exposure to light after a period in the dark). The great population increase from early spring to summer is very commonly succeeded by a mass decrease from death and sedimentation. This in turn produces a more transparent water because the high numbers of algae previously reduced light penetration. This so-called clear-water phase can be dominated by small algae which in general are more successful in waters poor in nutrients (oligotrophic, p. 12) than are large algae. Small algae generally grow faster than large ones but their numbers can be severely depleted by the filter-feeding animals. Their sinking rate can be extremely slow but a number can produce resting cells which sink faster than the ordinary cells. During summer, a series of algal maxima can arise depending on how much new nutrient material is available, the reactions of the algae concerned to other environmental features and their rate of sinking or rising.

One group of algae, the Cyanophytes (Chapter 10), contains species which have gas-containing bodies in their cells and are especially common in warm water and so also in the tropics. Such algae mass in the epilimnion and if the weather is quiet enough produce surface scums (fig. 10). Advantageous though rising rather than sinking may be, ensuring entry into the warmest and best illuminated layers of the water, it also has its dangers. The disadvantages are analogous to those mentioned concerning eutrophication, namely that one can have too much of a good thing. Very high irradiation such as will reach a surface scum or the uppermost layers of the water can harm algae or the water may become too hot. Ultraviolet light is especially dangerous, being absorbed in the uppermost layers of the water. What caused the death of the part of the surface scum near the shore seen in fig. 10 is not known but this is just the situation in which ultraviolet light could be the cause. Clearly, if the destruction of the ozone layer continues, danger from ultraviolet light will increase. These algae containing gas chambers in their cells do have some ability to escape from the surface layers and some can even poise themselves in or near the metalimnion by alterations in the amount of gas in their cells.

Being able to swim would seem to be wholly advantageous, apart from the fact that swimming could utilise much energy which might be essential for other needs. however, it seems that this is not so. Therefore, it is unclear why free-swimming algae are not the dominant planktonic forms. Under ice, particularly if sunlight is cut out by snow, water movements are slight because there is no wind action. Small algae, notably motile ones, often are dominant. An example is seen in fig. 8. Here a pool 3m deep had been covered by ice and snow for over two months. The deep green water seen in the bottle had come from just under the ice. It consisted almost wholly of one species of a free-swimming alga called *Chlamydomonas* (p. 80). The whole population was in the top few centimetres, gasping, as it were, for the faint light filtering through the snow-covered ice.

In giving a small, sketchy introduction into the ecology of phytoplankton, it should again be emphasized that the type of lake, its geographical position and peculiarities in its seasonal cycles produce great diversity from place to place. Our imaginary lake has a well-marked seasonal cycle of stratification and algal growth, other lakes do not. Stratification in tropical lakes is likely to have a different pattern to that in temperature ones; as one approaches the poles so the seasonal cycle is dominated more and more by winter darkness; thermal stratification in a pool or even a shallow lake is likely to be irregular – and so on. However, the basic factors determining algal growth are the same, only their intensities in space and time differ.

Chapter Two

Chlorophytes (green algae)

Chlorophytes contain a larger number of genera and species than any other algal group and embrace a wide variety of structural forms.

Most Chlorophytes are green because of the predominance of green coloured photosynthetic pigments in their cells. Sometimes these green pigments are masked by others, usually giving the cells a yellow-orange or reddish colour. In some species such changes are typical of certain stages in their life cycles.

By life cycle is meant the cyclical phases an organism can go through, for example youth, age, asexual or sexual reproduction. Life history is almost synonymous with life cycle, meaning all the states into which a given alga may enter, for example those mentioned as examples of life cycles and others such as special states in which an alga can remain alive under conditions unfavourable for growth or which would kill the alga if it did not take on a protective phase.

An example of a life cycle familiar to all is the germination of a seed of a flowering plant, followed by the growth of the seedling into an adult plant which produces flowers. The flowers are fertilized and seeds produced. The life history of such a plant may have additional phases, for example, all but a resistant organ or part of a plant may shrivel up and die during drought. The plant may pass unfavourable periods such as drought or winter as seeds, bulbs, corms or leafless shoots (e.g. deciduous trees).

Unicellular and colonial forms

Figures 12 and 13 show cells of two globose unicellular species. The cells of the *Eremosphaera* (fig. 12) commonly are about ten times as broad and so 1000 times as big (i.e. in volume) as those of *Chlorococcum* (fig. 13).

In *Eremosphaera* (fig. 12) there are many, small discoid bodies, the chloroplasts, located within strings of protoplasm radiating from the central region of the cell to its outer surface. The chloroplasts are in sharpest focus at the lefthand side of the cell (at about 6.30 o'clock on the clock-face basis). Chloroplasts are the bodies (cell organelles) which contain the pigments (e.g. chlorophyll *a*) which make photosynthesis possible.

The cells of *Chlorococcum* (fig. 13) have a single chloroplast lining the inner surface of the cell. Within the chloroplast is a prominent globose body, the pyrenoid. Pyrenoids are involved in the production of starch which in this case can be seen as continuous or broken rings around the pyrenoids. Each chloroplast of *Eremosphaera* also contains a small pyrenoid.

When talking about the inner surface of a cell we are referring to the layer just within the thin membrane surrounding the living part of the cell. Beyond this membrane usually there is non-living material, the cell wall, seen as a prominent white ring around the cells of *Chlorococcum*. The cell wall can be penetrated by living matter, so that it is not dead in the ordinary sense of the word. It is simply not one of the "working parts" of the cell (e.g. chloroplasts and pyrenoids) but an external coating.

Eremosphaera is especially common in acid, boggy places and *Chlorococcum* in soil (fig. 3), though it is also widespread in aquatic habitats.

In *Sphaerocystis* (fig. 14), *Chlorococcum*-like cells live in groups within mucilage. The word mucilage covers a variety of watery and usually soft, colourless substances, few of which have, as yet, been chemically analysed. If a mass of mucilaginous algae is big enough to be handled, it will feel slippery or slimy. Little is known for certain about the function mucilage plays in the life of the algae we cover here. In the case of *Sphaerocystis*, one obvious function is to hold the cells together in a group, or, to use the technical term, a colony.

The number of cells present in a *Sphaerocystis* colony basically is a multiple of two. As a result of successive divisions of its contents each mother cell usually produces 4, 8 or 16 daughter cells. The individual daughter cells then surround themselves with a wall and each one becomes a copy of the mother cell. When cell multiplication has ended, the wall of the original mother cell breaks up, leaving the baby daughter cells in a group. Since all the cells in the colony are separate entities held together by mucilage, any one or all of them may be in the process of reproduction at a given time (fig. 14, lower colony). One of the groups of daughter cells present in this colony (at 2 o'clock) is in the process of being released from it. As these daughter cells grow, mucilage is produced and pushes them apart to form the type of colony seen above in fig. 14. This is a common type of reproduction in colonial Chlorophytes. It is also one of the ways in which *Chlorococcum* reproduces. In some algae, instead of this successive method of multiplication, the whole mass of protoplasm divides up simultaneously. The species of *Sphaerocystis* shown is one of the commonest and most widespread of planktonic algae. Recently it has been removed from this genus and placed in one called *Eutetramorus*.

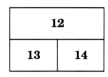

Fig. 12. *Eremosphaera* (190 μm diam.) on each side of which is a filament of *Zygnema*.

Fig. 13. *Chlorococcum* (cells 12-17 μm diam.). Note pyrenoids surrounded by a layer of starch.

Fig. 14. *Sphaerocystis*. Two colonies, the larger one, 81 μm diam. **I**.

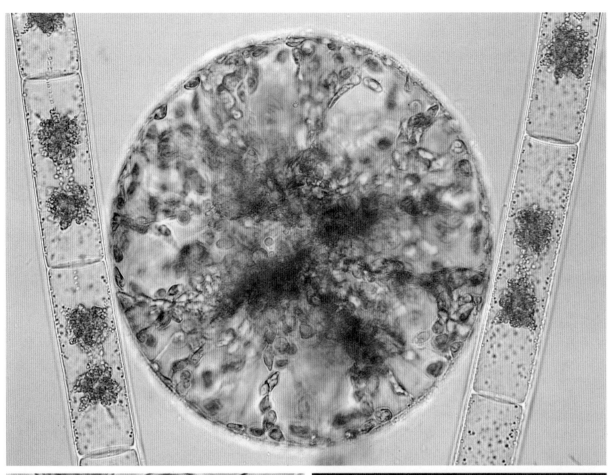
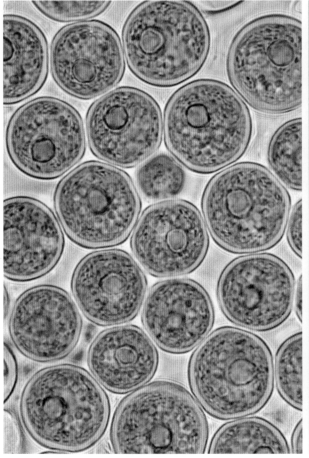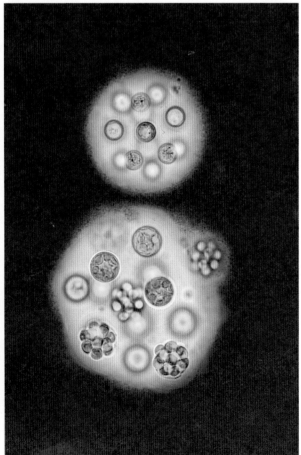

23

Those who are not taxonomists often complain at the frequency with which those who are taxonomists change the names of plants (see also splitters and lumpers, p. 6). Algae are no exception and in the last 20 years or so there have been more changes of names than ever before. Usually there are good reasons for such changes but this is not so with changing the name of the *Sphaerocystis* shown here to *Eutetramorus*. Apart from the fact that the name *Eutetramorus* is based on an imperfectly described alga, this name change could have been avoided. The alga concerned is very common, often abundant and has been known and recorded as *Sphaerocystis* for almost 100 years.

Asterococcus (fig. 15) can exist as single cells but more often the cells are in small groups surrounded by stratified mucilage, that is by mucilage consisting of two or more concentric layers. Like *Sphaerocystis*, the cells multiply by producing two or more (commonly four) copies of the mother cell. Unlike *Sphaerocystis*, the chloroplast is stellate, not peripheral. The pyrenoid is in the centrally located part of the chloroplast, from which lobes radiate out towards the periphery. The *Asterococcus* shown is common in boggy waters.

Apiocystis (fig. 16) lives attached to filamentous algae and other underwater objects. Its colonies are balloon or bladder-shaped. In the picture, two large and several small, young colonies are attached to a filament of the alga *Oedogonium* (p. 62). Several fine white threads can be seen to extend beyond the margins of the two large colonies. These consist of mucilage within which is an extremely delicate thread called a pseudocilium. The latter is invisible at the low magnification of this figure. Each cell of *Apiocystis* bears two such mucilage-covered pseudocilia. Why these structures are called pseudocilia is explained in the glossary.

The cell of *Lagerheimia* shown in fig. 17 is in the process of producing four daughter cells. In life, *Lagerheimia* is a unicellular alga and each individual cell of the species depicted bears four, firm, non-mucilaginous spinous structures. Following division, each daughter cell develops its own four spines. Both the cell and spines are surrounded by mucilage.

Micractinium (fig. 18) unlike *Lagerheimia* is colonial. The number of cells in the colony is variable and often more than the four seen in our specimen. Its cells too are spinous. The spines are thinner than those of the *Lagerheimia* shown and there are several per cell.

Apiocystis, *Lagerheimia* and *Micractinium* are common in lakes and pools, the latter two in the plankton.

Dictyosphaerium (figs 19, 20) also is a very common planktonic genus. The cells are embedded in mucilage and joined to one another by threads. In reproduction, each cell divides into four (sometimes into two) daughter cells which are released from the mother-cell after its wall splits into

15	16
17	18

Fig. 15. *Asterococcus*. Four-celled colony (69 μm diam., cells 25 × 19 μm) surrounded by peat and filamentous algae.

Fig. 16. *Apiocystis*. Balloon-shaped colonies attached to an algal filament. Largest colony 325 × 225 μm. **I**.

Fig. 17. *Lagerheimia*. Cell (16 × 10 μm; spines 20 μm l.) in process of forming four daughter cells. **I**.

Fig. 18. *Micractinium*. Four-celled colony (cells circa 5 μm diam.). **P**.

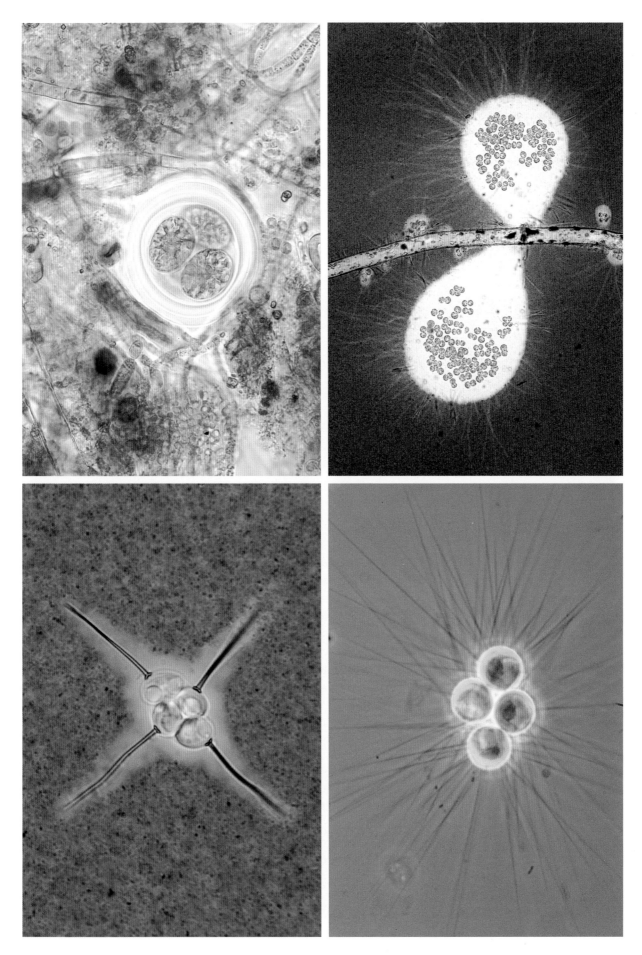

four parts. Each daughter cell retains one of these four portions of wall which becomes the branched stalk holding them together. The numerous black rodlets in the mucilage external to the cells in fig. 20 are bacteria. It is not known what kind of bacteria they are; why they occupy a ring midway between the algal cells and the outer margin of the mucilage external to the algal cells, why they are arranged parallel to one another or what, if any, are the interactions between them and the alga, apart from the fact that they are not parasites. Other examples of such bacterial rings are shown in figs 63 and 395.

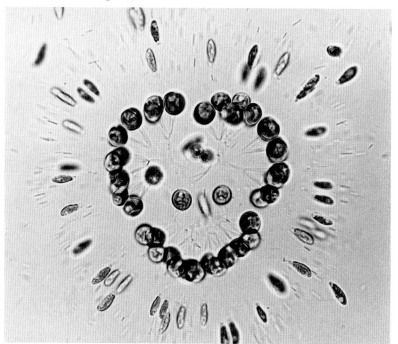

Fig. 19. *Dictyosphaerium* (× 640). Elongate cells of a species of *Chlamydomonas* are attached to or lie within the mucilage surrounding the ring of spherical *Dictyosphaerium* cells.

In fig. 19 nearly all the rod-shaped bacteria are out of focus. In this colony, they are accompanied by a second inhabitant of the mucilage, a Chlorophyte called *Chlamydomonas* (p. 80), the cells of which are located at various depths within or are attached to the outside of the mucilage. The *Chlamydomonas* cells all have their narrower and largely colourless (anterior) ends pointing towards the centre of the *Dictyosphaerium* colony. This is because the anterior end bears the flagella with the aid of which they attach themselves to or penetrate into the *Dictyosphaerium* mucilage. This *Chlamydomonas* may be found on the surface of or within the mucilage of various planktonic algae. It is simply occupying a space, hence it and other organisms with similar behaviour are sometimes called space parasites, an unfortunate appelation because they are not parasites in the true sense of the word (discussed on p. 279).

Only a few of the many species of *Chlamydomonas* live in or on the mucilage of other algae.

In *Coelastrum* (figs 21, 23) adjacent cells are joined together by processes projecting from their walls or simply

Fig. 20. *Dictyosphaerium* (colony 137 × 112 µm). Note threads joining the cells together and ring of rod-shaped bacteria in the mucilage surrounding the *Dictyosphaerium* cells. **I, P.**

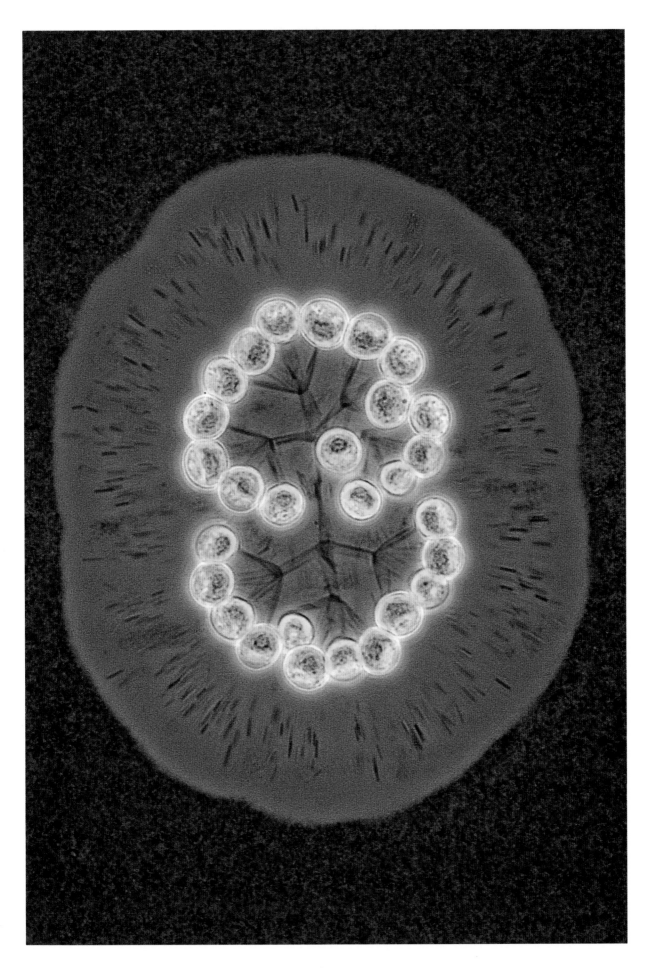

adhere to one another to form hollow spherical or polyhedral colonies.

In *Tetrastrum* (fig. 22) and *Crucigenia* (figs 24, 25), four cells are united together to form a flat or slightly curved plate. The walls of some species of *Tetrastrum* are spinous. In *Crucigenia* four, four-celled colonies are usually joined together (fig. 24).

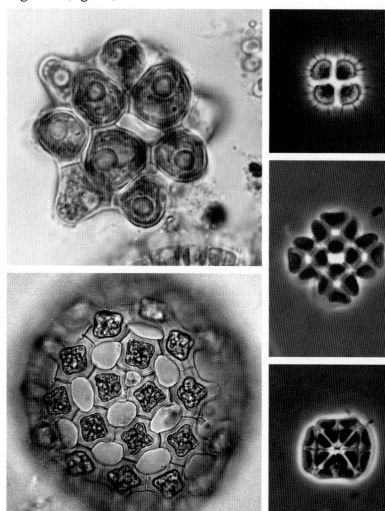

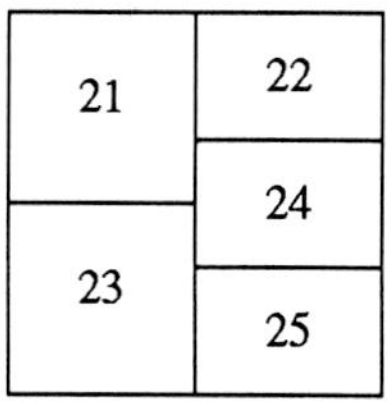

Fig. 21. *Coelastrum* (× 1600).

Fig. 22. *Tetrastrum* (× 1600). **P**.

Fig. 23. *Coelastrum* (× 640).

Figs 24, 25. *Crucigenia* (× 1600). **P**.

The cells of *Scenedesmus* (figs 26-28) usually are arranged in a row to form 4- or 8-celled colonies. Two to 16-celled colonies can occur and very rarely there are more than 16 cells per colony. Under certain conditions, the cells are not attached to one another. The terminal cells of some species are spinous (figs 26, 28). In other species spines are present on several or all of the cells. The number and types of spines and their distribution over the cell surface are characteristic of the species concerned but can also vary in relation to the environmental conditions. The cells of some species (e.g. fig. 27) never have spines.

There are hundreds of named species and varieties of *Scenedesmus* which is one of the commonest genera of freshwater algae. It is very doubtful if some of the

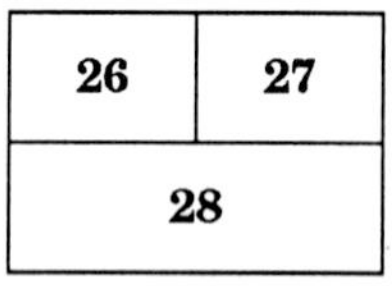

Figs 26-28. *Scenedesmus*.

Fig. 26. Four-celled colony (30 μm l.), spines up to 22 μm l. **I**.

Fig. 27. Colonies (up to 19 μm l.) of a species without spines. **P**.

Fig. 28. Colonies composed of four or eight cells (to 62 μm l.). **D**.

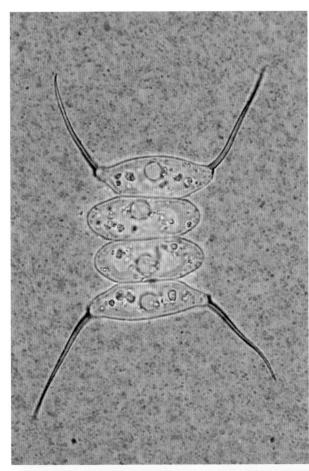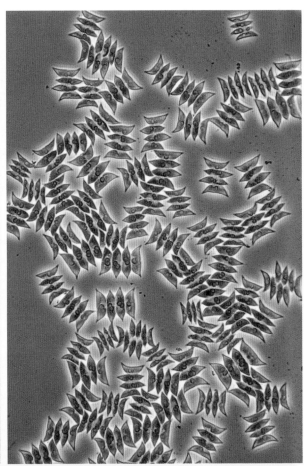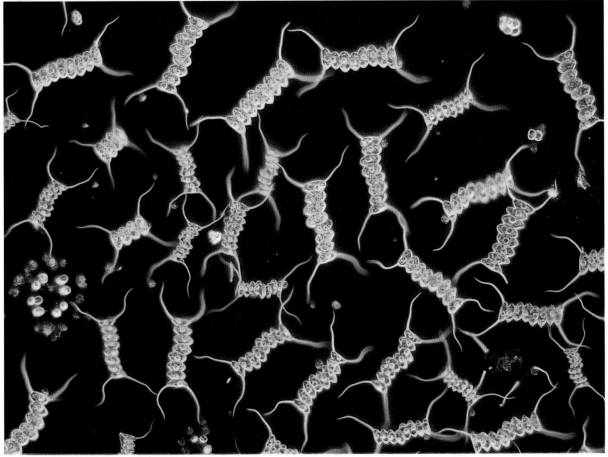

described species are separate entities because of the plasticity of shape and form which is now known to be possible in the more thoroughly studied species. Certain structural features are only clearly revealed by electron microscopy, which is now being used to identify species.

Waters rich in plant nutrients, particularly shallow ones, often are coloured green by a mixed plankton containing species of *Lagerheimia, Micractinium, Dictyosphaerium, Scenedesmus* and other related genera such as those seen in fig. 31.

Kirchneriella (fig. 31, bottom right) has several or many curved lunate cells embedded in mucilage (not shown). In this colony there are two large mother cells and several groups of smaller cells. Each packet of four small cells has been produced by the division of a large cell. The content of the lower of the two large cells is just starting to divide. The two faint white lines across this cell show where subdivision of its content has taken place. The daughter cells so produced will take the same shape as the mother-cell before they are liberated from it.

The four scattered groups of long, pointed and slightly curved cells in fig. 31 are colonies of *Ankistrodesmus* and the cogwheel-shaped colony is a species of *Pediastrum* (see later). In the colony of *Dictyosphaerium* (top right) two cells have divided recently into four pear-shaped daughter cells. Eventually they will grow into the same shape as the large (adult) cells present. When occurring in large numbers, cells of *Ankistrodesmus* often live separately and then are easily mistaken for those of another genus, *Monoraphidium*, which is always unicellular.

Fig. 31. Plankton of colonial Chlorophytes. *Kirchneriella*, lunate cells; *Ankistrodesmus*, elongate curved cells; *Pediastrum*, discoid cogwheel colony (44 μm diam.); *Dictyosphaerium*, colony with larger, globose (fully grown) and smaller, ovoid daughter cells. **P**.

Fig. 29. *Selenastrum* (× 800). **P**.

Fig. 30. The "so-called" *Selenastrum capricornutum* (really *Kirchneriella*, see text) in the unicellular state, as usually used for bioassay (× 640). **P**.

Selenastrum (fig. 29) only differs from *Ankistrodesmus* in the strong curvature of the cells and the way in which they are arranged "back-to-back". However, there are all grades in the degree to which these arise between species included in the two genera so that, like many taxonomists, we think the separation of *Selenastrum* from *Ankistrodesmus* is unjustified. The most famous species of *Selenastrum* is a so-called *S. capricornutum* (fig. 30) which has been used worldwide as a bioassay organism. However,

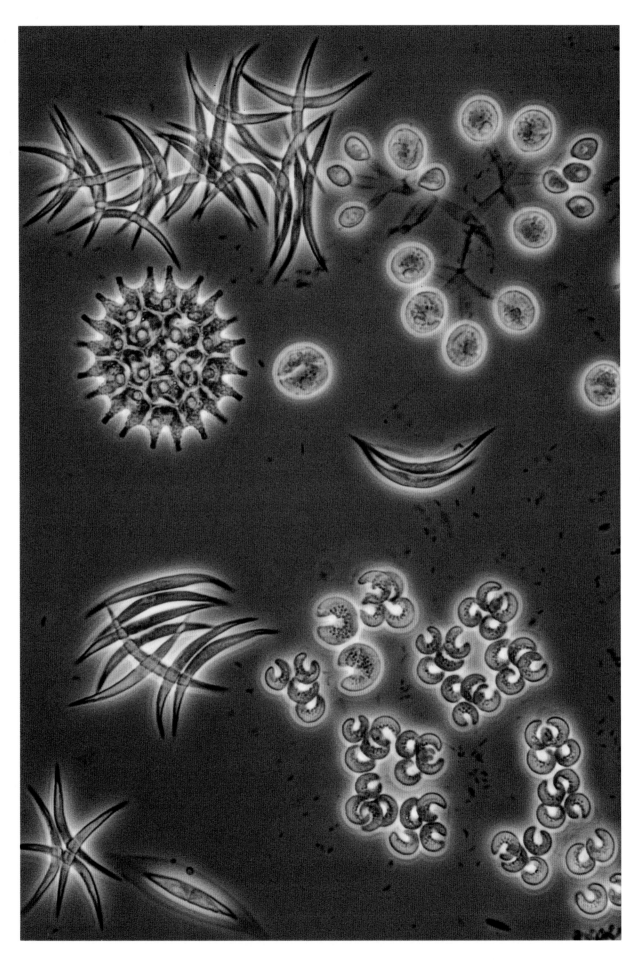

this bioassay organism was wrongly identified; it is not the true *S. capricornutum* but a species of *Kirchneriella*. Algal bioassay is the use of a specific population, originally derived from a single cell or asexual individual (a clone, see p. 304) to test the effect of a given substance (e.g. a chemical element or compound), mixtures of substances (e.g. lake water, sewage, industrial or agricultural effluents) on that test alga. From the results obtained, inferences may be made about the effects of these substances on other algae and on the potential fertility of a waterbody.

Pediastrum (figs 35-37) has flat or somewhat curved discoid cog-like colonies. Each cell contains a single pyrenoid surrounded by a ring of starch. The outer cells of a colony are usually notched and in some species deeply so (figs 35, 37). The resultant lobes commonly are ornamented. The species in fig. 36 possesses pin-headed, stalk-like appendages. In fig. 37 one of the fine rod-like processes with which the cells of this species are furnished can be seen projecting from the upper lobe of the cell at 1 o'clock. Every cell in a *Pediastrum* colony is capable of producing a new colony.

Hydrodictyon reticulatum the water net (figs 32-34, 38) forms a cylindrical sac-like net that can be over 20 cm in length, but usually is somewhat shorter. Nets over 30 cm long have been found in nature and yet bigger ones in laboratory cultures. The greatest recorded size is 114 cm long and 4 – 6 cm broad. Incidentally, this fascinating alga is very easy to culture.

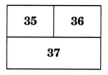

Figs 35-37. *Pediastrum* colonies, 53, 93 and 95 μm diam. respectively. In figs 35, 36 note central pyrenoids. Figs 36, 37, N.

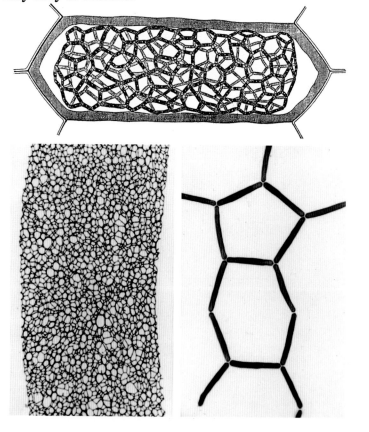

Figs 32-34. *Hydrodictyon.*

Fig. 32. One of the first drawings of a daughter colony still within its mother cell.

Figs 33, 34. Micrographs of cells from young and old nets Both × 10.

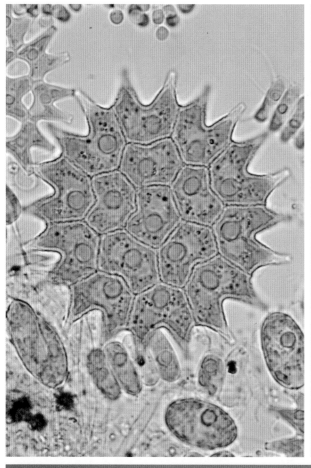
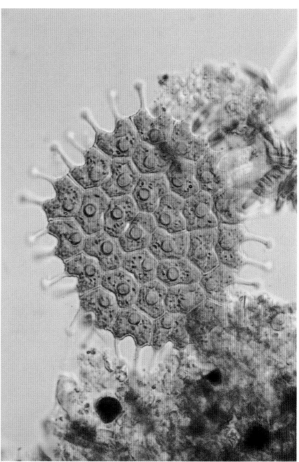
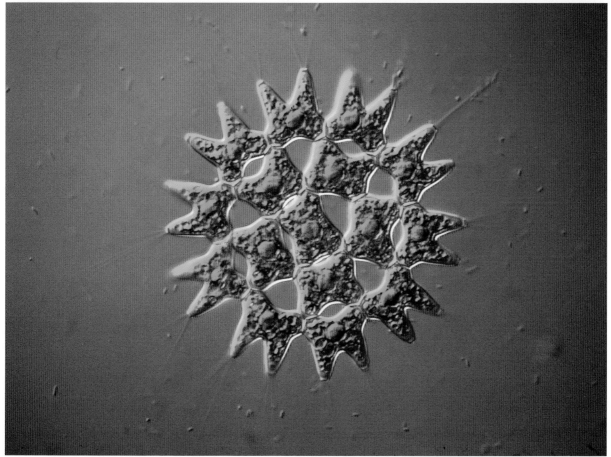

33

The meshes of the net commonly have five or six sides. Each side consists of a single cylindrical cell which at the time of reproduction can form a minute daughter net within itself (fig. 32). In this asexual method of reproduction very large numbers of minute spores are formed in each cell (usually several thousand). Each of these minute globose daughter cells enlarges into a cylindrical cell and where the ends of the separate cells meet they stick together. This is how the meshwork of cells is formed. It can be shown mathematically that the enlargement of these tightly packed globose cells within the mother cell must produce this type of net. As the baby net enlarges, the wall of the mother-cell disintegrates, so releasing the daughter net.

A water net contains thousands of cells and most or even all may be producing new nets at much the same time so that the whole net breaks up. If, as can happen under conditions favouring rapid growth and reproduction, most or all the nets in a pool are reproducing at the same time, the multiplication in number of the alga can seem to be explosive. In a short time, a few hundred nets become hundreds of thousands of nets. This gives the impression that the water net grows very fast. In fact, it does not grow any faster than many other algae. They increase just as fast, in some cases faster or even much faster than the water net. However, the increase in their numbers is going on all the time and not, as in the situation mentioned for the water net, in sudden, dramatic outbursts of multiplication.

The likelihood that this beautiful plant could be disliked may seem bizarre but that is just what has happened in New Zealand. This is a country which has suffered from the entry and subsequent explosive multiplication of a variety of alien plants and animals. The water net is the latest example. Actually, it is surprising that it has so recently arrived because New Zealand's freshwater algal flora consists mainly of species present all over the world. It is one of the few regions from which the water net had not been recorded.

The water net favours waters rich in plant nutrients (eutrophic, see pp. 12, 220) and the highly developed agriculture of New Zealand offers many such habitats even in areas not enriched by human and industrial sewage effluent. The water net is of sporadic occurence but when it is present, it usually is very abundant. Among well known places for such abundance are rice paddys, waters receiving effluents from fish farms and irrigation ditches. In New Zealand it has invaded popular tourist lakes and their marinas, clogging the intakes of boat's engines, forming green tangles over other water plants and on the shores of lakes. When these growths die they are unsightly, stink and use up essential oxygen in the water. The way in

Fig. 38. *Hydrodictyon*. Part of a juvenile colony. Cells 16-25 μm l. N.

which it smothers and ousts other plants makes it a serious threat to the conservation of wetlands.

At the time of writing much research is being carried out on how to eradicate or control it in New Zealand but no solution to the problem has been found. It may well be that one of two things will happen; organisms which parasitize it or eat it in preference to other plants will be found or it will in time become less abundant for some unknown reason. There are well-known examples of biological control – generally better than chemical control – of pests and of alien pests eventually becoming a common but not harmful part of the flora or fauna of the country concerned.

Botryococcus (figs 39-41) is present world-wide but not often in large numbers. When it is very abundant it produces spectacular superficial scums (waterblooms; fig. 11). It is found both in fresh and somewhat saline water. The colonies vary in colour from green (fig. 39) to yellow or orange (figs 11, 40, 41) or even reddish-orange. Old colonies, or those which are not actively growing usually are not green but yellowish in colour. Each cell is embedded in a cup of tough mucilage, clearly a very different kind of mucilage from the mucilages of previously mentioned algae. Several colonies can be joined together by strands of this mucilage (fig. 39).

Botryococcus is rich in oil which can be squeezed out of it by moderate pressure without destroying the cells (fig. 41). Because the mucilage is so resistant to decay, *Botryococcus* is a common fossil in certain geological formations. It has a long geological history but can be found also in recent deposits (e.g. peat). Certain types of coal and bitumen consist almost wholly of fossilised *Botryococcus*. Hence, it seems it must have been much more abundant than now in certain periods of geological time. Since *Botryococcus* contains so much oil, these deposits are likely to be rich in oil, as for example some oil shales. It is the main constituent of boghead coal and torbanite (named after Torbanhill in South Scotland) and the black, rubbery and highly inflammable coorongite (Coorong, a district in South Australia). The paraffin industry in Scotland depended on torbanite. It flourished during the second half of the last century. The industry was destroyed by a combination of falling reserves of high quality torbanite and the import of cheaper Russian and American oil. In Australia, the discovery of coorongite in the 1850's led to the belief that it was derived from seepages of oil. Much time and money was spent in boring for this non-existent oil field. It was many years before it was generally accepted that coorongite is composed of vegetable matter.

The oily matter from *Botryococcus* contains unique hydrocarbons, derivatives of which have been discovered in commercial oil. This is an unusual, possibly unique case of

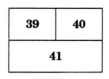

Figs 39-41. *Botryococcus*.

Fig. 39. Green colonies (bottom left colony, 125 × 156 μm.)

Fig. 40. Part of a yellow colony.

Fig. 41. Edge of colony from which oil has been expressed; the cells are unharmed. Cups enclosing cells 17-19 μm. l.

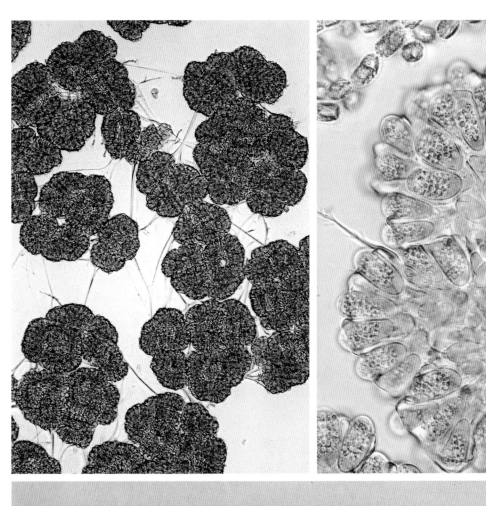
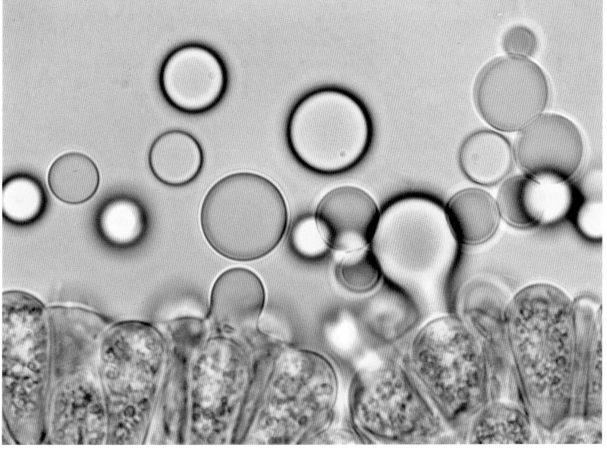

37

identifying a component of oil to the level of genus. In a Sumatran oil up to 1% consists of these compounds.

The Australian coorongite was formed from vast masses of *Botryococcus* which accumulated on the shores of lakes. A sample of *Botryococcus* from an Australian lake was cracked, that is refined like crude oil. The material obtained from it consisted of 67% petrol fraction, 15% aviation turbine fuel and 15% diesel oil. The present abundant supplies of oil, like those of any fossil fuel, must decrease and eventually become scarce, though it may be a long time before scarcity is reached. It might be that then *Botryococcus* would become an important industrial alga. Consequently, attempts already are being made to grow it in large quantities. However, up to now nobody has been able to make it grow fast enough for possible industrial use. In addition, the "greenhouse effect" has to be borne in mind. It may well be that by the time the technology for using *Botryococcus* has been mastered, dependence on fuels producing gaseous wastes causing the warming of the earth's atmosphere will no longer be necessary because alternative sources of "safe" energy will be sufficient for all but a fraction of human needs.

Conjugate algae

Desmids

Cell-form

Looking at figs 42-80 it will not be surprising to learn that desmids are much loved by microscopists because of their variety and beauty of shape and structure. Over 3500 different species have been described from freshwaters all over the world. They do not occur in the sea and very few have been recorded from brackish water. They are most abundant in numbers of cells and of species in waters which are "soft", that is, poor in salts of calcium and magnesium as well as in total content of salts. Such waters are also usually neutral or acid (e.g. pH 7.0 – 5.0). In very acid waters the number of species present decreases but the number of cells of those species which are present still may be large. A similar situation can exist in alkaline waters (e.g. pH 7.5 – 10.0) but desmids are poorly represented in the algal flora of highly alkaline waters where the total salt content also is likely to be high.

Two examples of desmids frequently present in collections from boggy places are shown in fig. 42. The elongate cells are species of *Closterium*, the flat discoid cells belong to *Micrasterias*. The oblong cells are diatoms (see Chapter 5).

The cells of *Micrasterias*, like those of *Closterium*, are composed of two halves or semicells but in *Micrasterias* the two half cells are separated by a narrow waist or

42

Fig. 42. Desmids. *Micrasterias* (discoid cells, largest 250 × 212 µm) and *Closterium* (curved cells, largest 525 and smallest 150 µm l.). Brown rectangular cell, a diatom (*Pinnularia*).

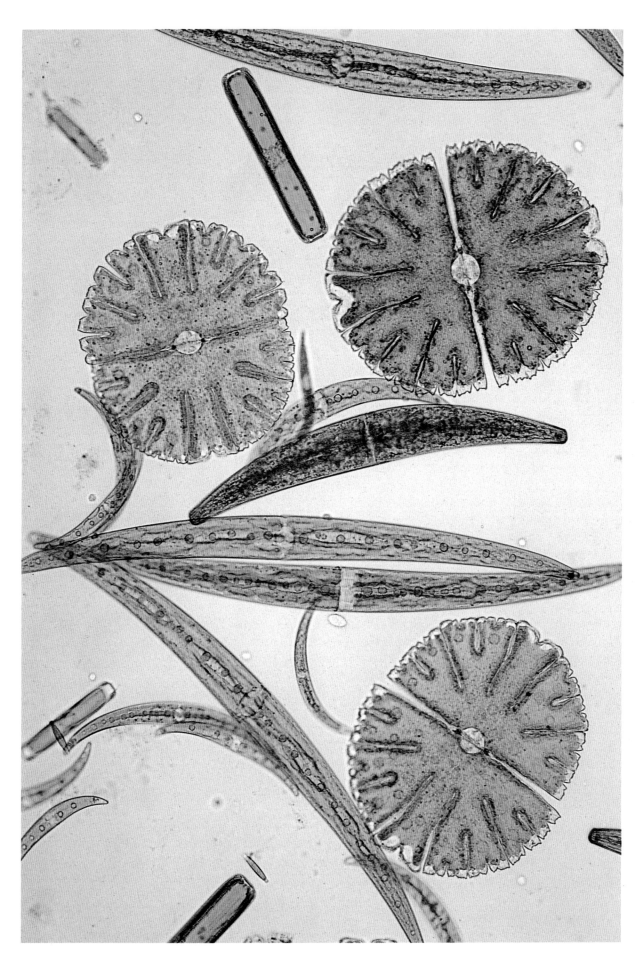

isthmus. The majority of desmid genera have this same type of clear distinction into two halves (e.g. figs 52-70.

The largest genus of desmids and of Chlorophytes as a whole is *Cosmarium* (figs 44, 45). Over 2000 species have been described. *Euastrum* (figs 46, 49) typically has cells with wavy margins and terminal, often notch-like indentations.

The cells of nearly all the species of *Closterium* (figs 42, 43, 47) are curved and narrow towards their extremities. They are composed of two halves which are difficult to see except when it divides. The chloroplasts nearly always terminate just before either end of the cell is reached. In these terminal regions is a round body or vacuole containing one to many granules (fig. 47). The granules are crystals which for over 100 years were believed to be composed of calcium sulphate (gypsum). Recently, it has been shown that, with few exceptions, they consist of barium sulphate (barytes, barite). The amount of barium in water nearly always is low and far less than that of calcium. Nevertheless, even in waters rich in calcium, almost always it is barium and not calcium sulphate crystals which are deposited in these vacuoles. Why barium sulphate is amassed and crystallised in this way and in this position is unclear. The possible function of the crystals also is unclear. Some other desmids also have such crystals in their cells though not necessarily in terminal vacuoles. Scattered crystals of barium sulphate can be found in a variety of organisms.

Halfway along the length of the large *Closterium* cell in fig. 47 is a colourless area free of chloroplasts. Within this is a round colourless body and within that a smaller round body. These two bodies form the cell nucleus which contains the main mass of the cell's hereditary matter (e.g. chromosomes, genes, DNA) and is the controller of cell growth and reproduction. Chloroplasts also contain their own genetic material but are under the overall control of the nucleus. The smaller body, the nucleolus, is a part of the nucleus with certain special functions. Within the chloroplasts, ten pyrenoids are disposed at intervals in a line along the length of the cell. The number of pyrenoids can be a little more or less than 10 in this species. In the small species on the right the number is about 5; here too the region of the nucleus is clearly visible.

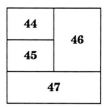

Fig. 44. *Cosmarium* (cell, including mucilage envelope, 87 × 91 μm), a smooth walled species.

Fig. 45. *Cosmarium* (cell 65 × 94 μm), a species with a tuberculate wall.

Fig. 46. *Euastrum* (175 μm l.).

Fig. 47. *Closterium* (larger cell 200 μm l), note nucleus and in it the nucleolus, a row of pyrenoids and terminal vacuoles containing crystals.

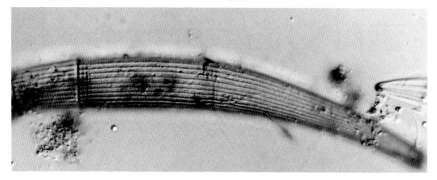

Fig. 43. *Closterium*. Only one complete semicell. Longitudinally and transversely (see lefthand end of semicell) striated cell wall (× 640). N.

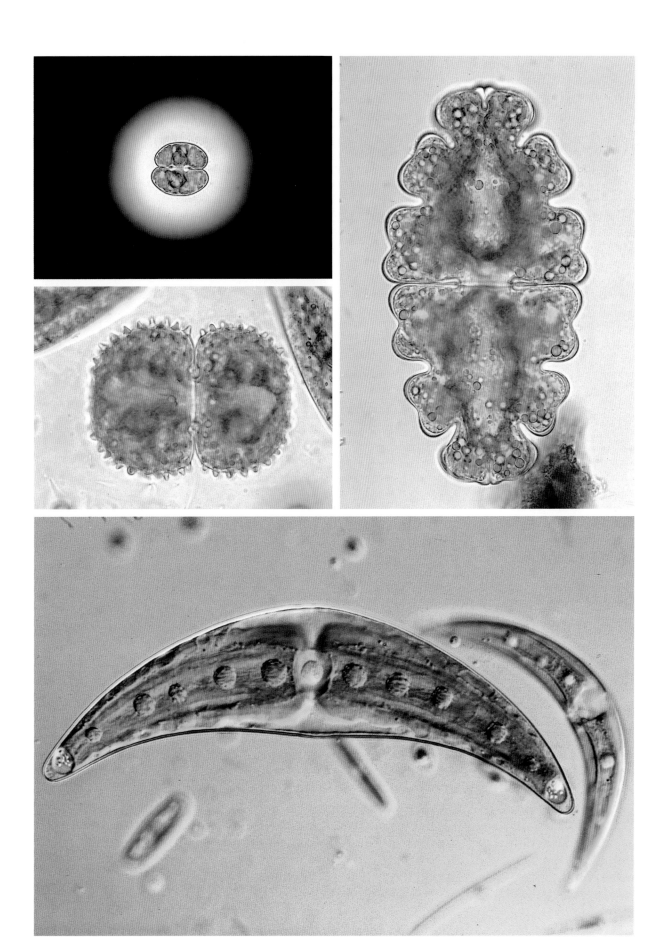

The cell wall of many species of *Closterium* is longitudinally striate; a few transverse striations may also be present (fig. 43). In the species of *Penium* shown (fig. 48) regular longitudinal markings are joined by many finer irregular transverse ones to form a network. The cell walls of many desmids are strikingly ornamented or bear spines (e.g. figs 48-51, 53).

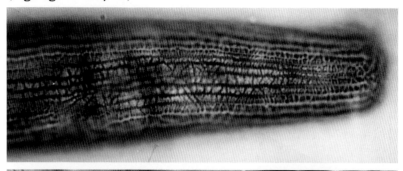

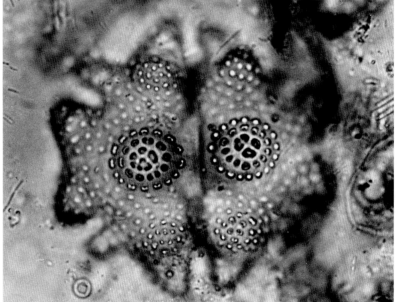

Fig. 48. *Penium*. Coarse longitudinal striations with fine connecting ones (× 1600).

Fig. 49. *Euastrum*. Cell wall ornamentation (× 845).

Fig. 50. *Pleurotaenium*. Cell wall ornamentation (× 640).

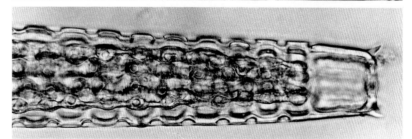

Triploceras (fig. 51) has cells looking like a fish bone. *Pleurotaenium* (figs 50, 53) has cylindrical cells which usually are much longer than broad. Most *Pleurotaenium* species have less elaborate ornamentation of their side walls than the two shown here. The outlines of the discoid *Micrasterias* cells (figs. 52, 54) are so deeply dissected that the cells appear spinous.

The photographs of the four desmids in figures 51-54 are taken from material preserved in formalin to which a little iodine has been added.

Fig. 51. *Triploceras* (328 μm l.). **D**.

Fig. 52. *Micrasterias* (188 × 172 μm. diam.). **D**.

Fig. 53. *Pleurotaenium* (367 μm l.). **D**.

Fig. 54. *Micrasterias* (188 μm diam.). **D**.

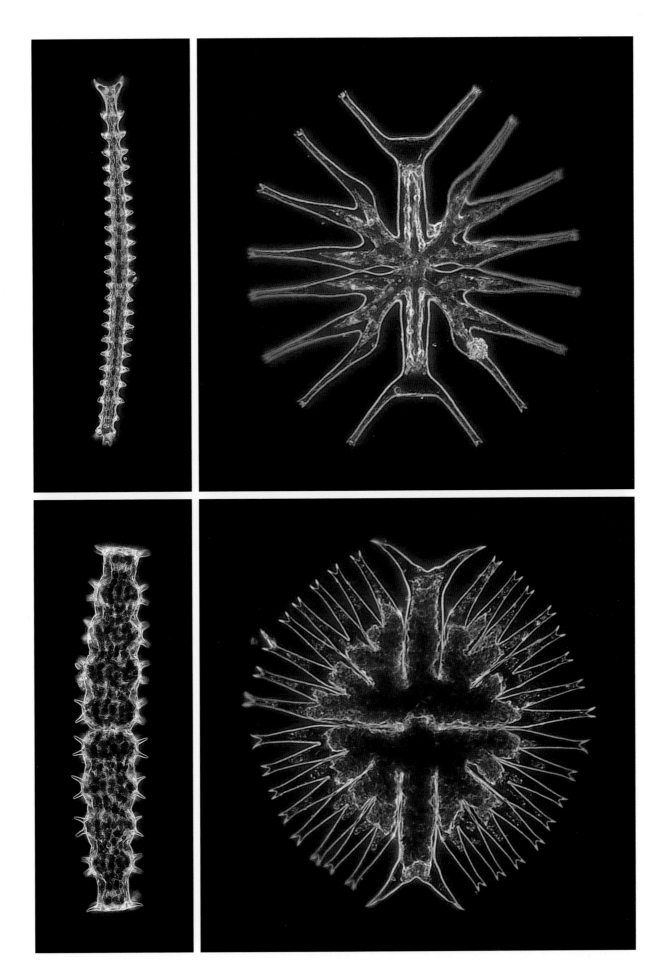

The cells of those desmids which are cylindrical in shape look the same seen from two sides, the so-called front and side views, but very different if seen from their ends. In other desmids, the cells look different in front, side and end-views. In the sample of desmid plankton in fig. 56, all but one of the cells of *Staurastrum* are seen from the broadest (front) face. At the top of the figure, almost at 12 o'clock, is a single cell, orientated at right angles to the others (i.e. top or end-view). By its triangular shape it shows that the cells are three-sided. This shape is also apparent in the empty half-cells of this desmid (fig. 62) which have been left behind when their cells have reproduced sexually (see p. 52).

Several triangular end-views of a different species of *Staurastrum* are seen in another desmid dominated sample of plankton in fig. 57. The corners of the semicells are prolonged into "arms". Some species of *Staurastrum* (e.g. fig. 58) have numerous "arms". *Staurastrum* is the second largest genus of desmids with over 1,000 species. An allied genus *Staurodesmus* (figs 61, 64) used to be included in it.

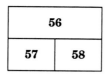

Fig. 56. *Staurastrum*, from sample of plankton. Cell in end view (triangular shape) plus mucilage around it, 78 μm diam. I.

Fig. 57. *Staurastrum* from sample of plankton. Length of central cell, including mucilage but excluding arms, 25 μm. I.

Fig. 58. *Staurastrum*, species with many "arms". Cell plus mucilage around it, 131 × 155 μm. I.

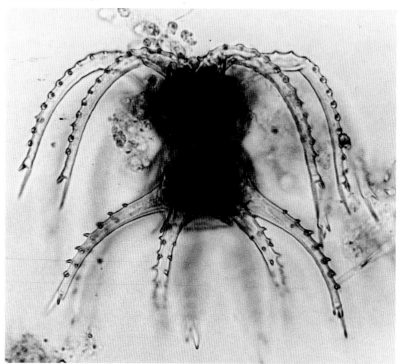

Fig. 55. *Amscottia* (× 942).

The genus *Amscottia* (fig. 55) was created for a species of *Staurastrum* which does not conform to the morphological rule that one semicell is the mirror image of the other. In both semicells the spinous arms are oriented in the same direction. If the mirror image rule prevailed, then the arms on one semicell would be pointing upwards and those on the other semicell downwards (compare fig. 58 with fig. 55). *Amscottia* represents a most unusual type of desmid. It occurs in the tropics.

The cells of many desmids are partially or wholly embedded in colourless mucilage. The outline of this

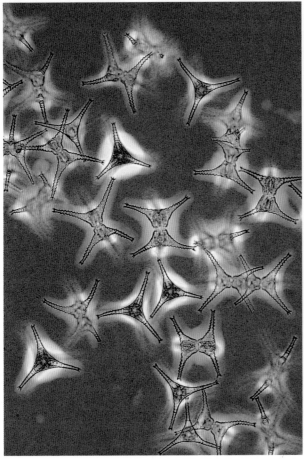
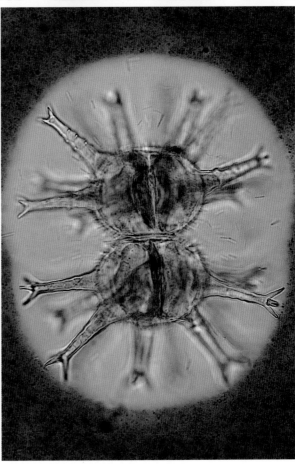

mucilage may be visible using phase-contrast microscopy but the best method for seeing the extent without harming a cell is to mount the material to be examined in Indian ink (e.g. figs 56-58, 63, 66, 79). Alternatively, stains can be used (e.g. fig. 60 of a species of *Staurastrum*) where the mucilage may be seen to be less homogeneous in structure than it appears in Indian ink. Some stains, if used very highly diluted, will stain mucilage without immediately harming a cell.

It has been suggested that an envelope of mucilage may reduce the rate at which a cell would sink without it. The mucilage of desmids is an extremely watery substance and so will have a low density which in turn will lead to a decrease in the overall density of the organism. On the other hand, increased size means an accelerated rate of sinking. Hence it is necessary to have data about the density of the cell and that of the mucilage to be certain that the possession of such an envelope is of advantage in decreasing the rate of sinking. It could have some other function. It may be that the increased size imparted to the desmid saves it from ingestion by the common planktonic animals (e.g. Crustacea p. 274) which filter out particles. On the other hand, if the cell is in the benthos, and many desmids live there, there are many larger animals such as aquatic worms and midge larvae which can ingest relatively large particles (e.g. see fig. 538). Even if mucilage does have some protective function against animals it is no barrier to fungal parasites (Chapter 13) which so often devastate desmid populations. Mucilage also distances the cell from its nutrient supply which would seem to be a serious disadvantage. Yet desmids flourish in nutrient poor (oligotrophic) waters. All the "ifs" and "buts" illustrate the fact that much more needs to be known about mucilage.

Tetmemorus (fig. 59) is a very common benthic desmid, particularly in acid, boggy waters. The boat-shaped cells have a notch at each end. Though the cells are surrounded by little or no mucilage, they can produce broad strings of mucilage of considerable length as is seen in the picture.

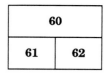

Fig. 60. *Staurastrum* (45 × 45 μm), stained with brilliant cresyl blue to show fibrillar mucilage.

Fig. 61. *Staurodesmus* (cell, including spines, 58 μm. br.).

Fig. 62. *Staurastrum*. Sexual reproduction, note walls of semi-cells nearby (upper zygote 9.5 × 10 μm.).

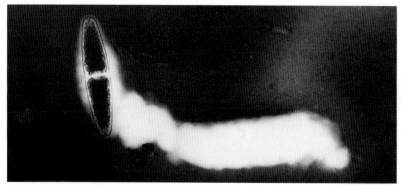

Fig. 59. *Tetmemorus*. The cell (150 μm l.) has produced a broad string of mucilage. **I.**

The strings of mucilage so produced push the cell along. Cells of *Closterium* also produce strings of mucilage. Though the mucilage production neither leads to fast nor

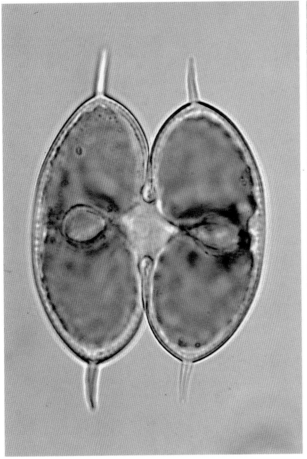
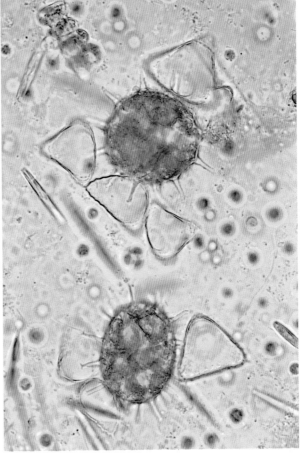

extensive movement, it does enable the cell to change position. Since all the species of *Tetmemorus* and most of those of *Closterium* live on the bottoms of waterbodies, they are liable to be buried in material sedimenting on to them or when the sediment they are on is disturbed by water movements or animals. Such buried cells can push themselves out of the deposit and so back into the light. Mucilage extrusion may also lead to the avoidance of excessive irradiation. The thin discoid cells of certain species of *Micrasterias* may move away from strong light or present their narrow sides to the light. In the latter case, a cell "stands on end" like a coin placed on its edge. The spectral quality of the light and the length of time a cell is exposed to a given degree or quality of light will affect mucilage production. Several other desmids have the same capacity for oriented movement through the localised extrusion of mucilage. In addition, cells can be attached to a suitable substratum by this mucilage. Experiments on cells of *Closterium* attached to glass showed that considerable force was necessary to remove them.

In some desmids, the cells are joined together in chains (figs 63-65). All the species of *Spondylosium* (fig. 63) are filamentous. In the common species shown, the whole filament is embedded in mucilage and in the mucilage of this specimen a ring of bacteria is present. Rings of bacteria can be found in the mucilage around several algae (e.g. figs 20, 395). A curious feature is that such rings so often are situated about halfway between the alga and the margin of the mucilage envelope. Though nothing is known about these bacteria, they are not the same in every case.

Like *Spondylosium*, there are several other wholly filamentous genera. However, in some of the genera in which the great majority of the species are unicellular (e.g. *Cosmarium* and *Euastrum*), a few may be filamentous. Nevertheless, these species usually form very fragile filaments so that the majority of the individuals can also be found as single cells. The *Staurodesmus* in fig. 64 is such a species in a predominantly non-filamentous genus. The cells are connected by a thick strand of mucilage which arises from the base of each spine. The short club-shaped pads projecting from the walls of the *Staurodesmus* (best seen at the apex of the lefthand semicell) are also made up of mucilage which in all cases issues from pores in the cell wall. These pores are frequently difficult to observe in live material and are not visible in the figures. Similar pores, however, can be seen as dark rod-like lines passing through the wall of the *Staurodesmus* in fig. 61 and the *Cosmarium* in fig. 637. In empty, dead cells of desmids the pores are usually more readily seen as small, occasionally large dots on the surface of the cell wall.

The *Micrasterias* shown in fig. 65 forms firmer filaments by the interlocking of parts of adjacent cells. This is the only filamentous *Micrasterias*.

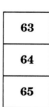

Fig. 63. *Spondylosium* (width of filament, including surrounding mucilage, 94 μm.). Note ring of rod-shaped bacteria. **P, I.**

Fig. 64. *Staurodesmus* (cells, excluding spines, 26 × 31 μm.). **P.**

Fig. 65. *Micrasterias* (86 μm. br.). **P.**

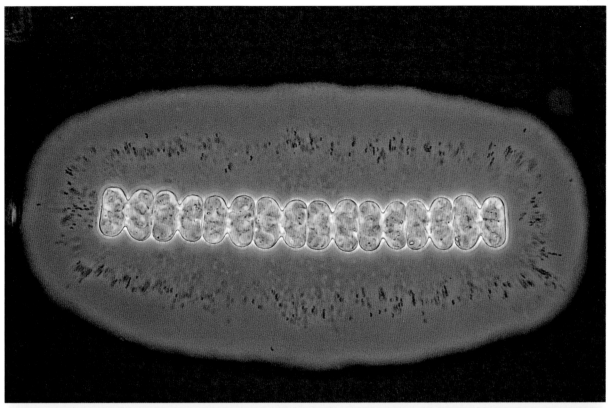
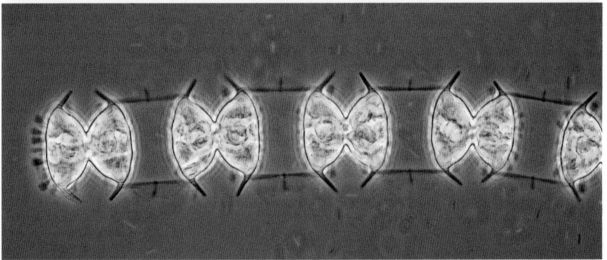
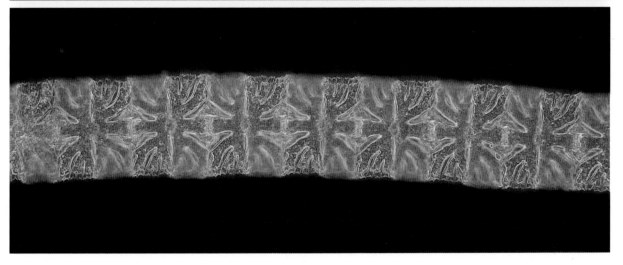

49

Cell multiplication

Cells of desmids which consist of two halves reproduce themselves by a process of "budding", stages of which are seen in figs 68 and 70. One of the desmids concerned (figs 67, 68) belongs to the genus *Micrasterias*, different species of which are seen in the earlier figures 42, 52, 54 and 65 and the other (fig. 70) to the genus *Xanthidium*. Fig. 69 although a different species of *Xanthidium* nevertheless indicates how the desmid in fig. 70 would have looked at its isthmus before cell multiplication commenced.

As in *Closterium* (fig. 47), the nucleus lies in the central region of the cells of *Xanthidium* and *Micrasterias* which in these latter two genera is at the isthmus.

The division of the cell nucleus marks the start of cell multiplication. At the same time as nuclear division and separation of the two daughter nuclei is taking place, the cell isthmus begins to bulge or elongate because of outgrowths produced from each half of the mother-cell. These outgrowths, which at first are colourless, soon become separated from one another by a cross-wall, as can be seen in fig. 68. At the stage shown in fig. 68, each new half-cell is "budding out" new lobes which eventually will become the long, serrate lobes of a new, adult half-cell.

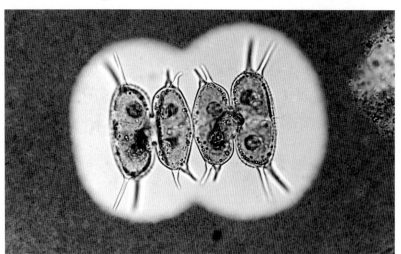

The *Xanthidium* cell has a simpler outline, so its young new half-cells (fig. 70) also have a simpler outline (for later development of spines in the same species, see fig. 66). The chloroplast of each mother half-cell is extending out into the new, young half-cell and this too will happen in the developing lobes of the *Micrasterias* cell. Just the start of this process is seen in fig. 68.

As a result of this kind of multiplication the two halves of an adult cell belong to different generations. The new daughter half-cell also is the mirror image of the mother half-cell out of which it grew. Large unicellular desmids such as those of *Micrasterias* offer scope for studying how cell development is carried out and controlled.

Fig. 67. *Micrasterias* (200 × 230 μm). N.

Fig. 68. *Micrasterias*. Early stage in cell division. N.

Fig. 69. *Xanthidium* (100 × 112 μm, including surrounding mucilage). I.

Fig. 70. *Xanthidium*. Early stage in cell division. I.

Fig. 66. *Xanthidium*. Late stage of cell division; note that spines of daughter semicells not yet fully formed (× 490) I.

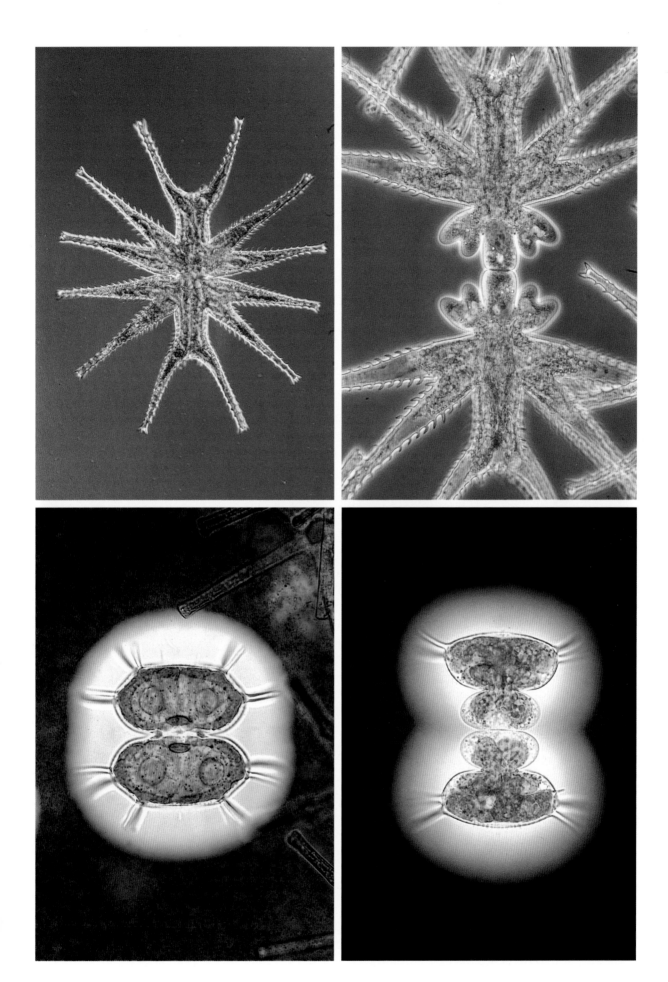

Sexual reproduction

Desmids can reproduce sexually (figs 62, 71-75), but it is not known whether this is true of all of them. It should be mentioned that some of the algae described earlier also have sexual stages, though different in kind from those of desmids. The method of sexual reproduction in desmids, called conjugation, is peculiar to them and members of the *Spirogyra* group. The living matter (protoplast) of each of two cells lying close to one another and usually enveloped in a common mass of mucilage, issues from the cell, leaving behind the cell wall in which it was enclosed. The two protoplasts fuse as do their nuclei. The two sexual partners are called gametes and the product of their fusion and that of their nuclei is called a zygote. The zygote at first is "naked" that is, like the gametes, it is not covered by a cell wall. Though the word naked is used, the cells are not devoid of a membranous covering. If they had none they would collapse or break up. By analogy with human beings one may say that they have no clothes but their bodies are covered by a skin.

Once sexual fusion is complete, the zygote forms a wall around itself. In the species of *Staurastrum* shown (fig. 62) the wall is spiny. The two halves of the cells producing the gametes usually separate as the gametes are released, the walls of these now empty half-cells can be seen in the picture. This desmid is the same one as that shown in fig. 56. Other zygotes are seen in figs 72-74, in each case together with the four empty semicells of the gametes. *Closterium* does not have semicells demarcated by an isthmus but many species break up into two halves when they conjugate. In species like the one illustrated (fig. 71), conjugation takes place shortly after cell division and so before the new semicell has reached its full length. The gametes issue through these immature semicells; that is why the zygotes lie opposite the shorter ends of the *Closterium* cells (in fig. 71).

The walled zygote generally becomes a "resting" zygote or cell. The reason for the use of the word "resting" is that zygotes usually do not germinate for some time after their formation; it may be months or over a year before they do so. This can be of biological advantage to the desmids, since the environmental conditions after conjugation may be unfavourable for growth or even kill ordinary cells, that is vegetative cells. Zygotes may be able to withstand the cold of winter under ice and snow and also drought when the water dries up. The latter ability is not confined to the zygotic stage alone as some desmids can live for long periods in the damp mud below the surface of waterbodies which have no standing water left in them.

Conjugating individuals or zygotes are not very often encountered in nature. In some common planktonic

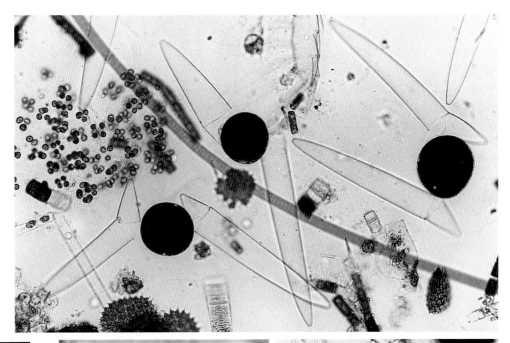

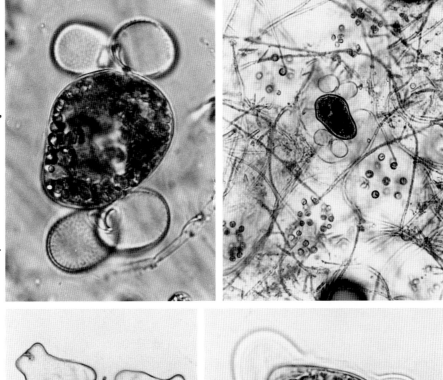

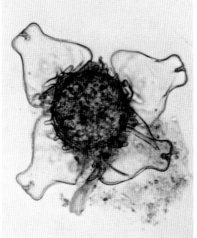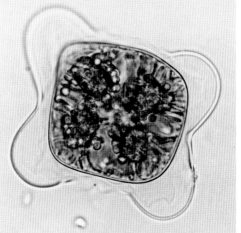

Figs 71-75. Sexual reproduction, zygotes.

Fig. 71. *Closterium* (× 160).

Figs 72 (× 640), 73 (× 256) *Cosmarium*. In fig. 73 among detritus and other algae.

Fig. 74. *Euastrum* (× 504).

Fig. 75. *Cylindrocystis* (× 1600).

species, conjugation has never been observed despite samples being collected at short intervals (e.g. weekly or fortnightly) over long periods. In a lake from which such frequent collections have been made for 49 years, the species in fig. 57 is a good example. Many thousands of specimens have been seen but sexual reproduction has not been observed. This suggests that, if sex occurs at all, it is a very rare event or only occurs at long intervals. However, for conjugation to take place in a planktonic species, a substratum (e.g. the bottom of the lake) may be necessary to ensure that cells remain together long enough for the gametes to make contact and fuse. Sometimes if collections of desmids are left in a cool, little illuminated place over a period of time, zygotes may be formed in the material. That is how the zygotes of the planktonic desmids in figs 62 and 72 were obtained.

The algae shown in figs 76-80 are also called desmids but have certain features suggesting that they are more closely related to the next group of *Spirogyra* and allied genera, than to the desmids shown previously.

Mesotaenium (fig. 78) is common in wet and damp places. The cell contains a single chloroplast with one or more pyrenoids. *Spirotaenia* (fig. 79) has spirally twisted ribbon-shaped chloroplasts running round the outer part of the cell. The cells of *Netrium* (fig. 80) have one or more chloroplasts which have a central rod-like portion from which arise ridges. These ridges run parallel to the central rod-like portion of the chloroplast. *Netrium* is common in bogs and other places frequented by desmids. Though there are more species of *Spirotaenia* than of *Netrium*, the genus is of less common occurrence.

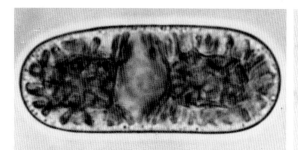

Fig. 78. *Mesotaenium* (cells 19-20 × 11-12 μm).

Fig. 79. *Spirotaenia* (cells 152-158 μm l.). I.

Fig. 80. *Netrium* (290 μm l.).

Figs 76, 77. *Cylindrocystis*. Fig. 76. Side view (× 1600) and fig. 77, end view (× 800).

Cylindrocystis (figs 76, 77) has two chloroplasts per cell, each with a pyrenoid; the nucleus is situated in the area between them. These chloroplasts are basically the same as those of *Netrium* (fig. 80) but lack the crested ridges. Finger-like processes arise from the central rod-shaped portion of the chloroplast. Seen from a cell end-on (fig. 77), that is, in transverse optical section, the chloroplast is stellate. Fig. 75 shows a zygote of *Cylindrocystis*. The cell wall not being composed of two halves, does not break up in sexual reproduction, though the wall of each of these two gametes has been bent in the process. *Cylindrocystis* tends to occur in the same habitats as *Mesotaenium*.

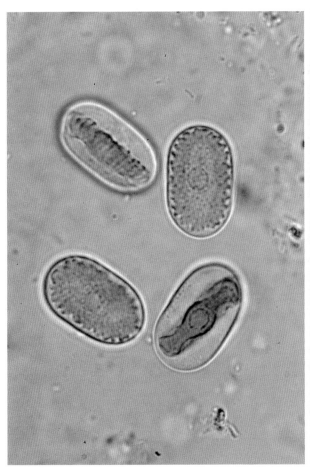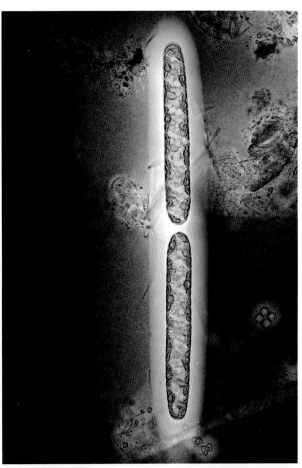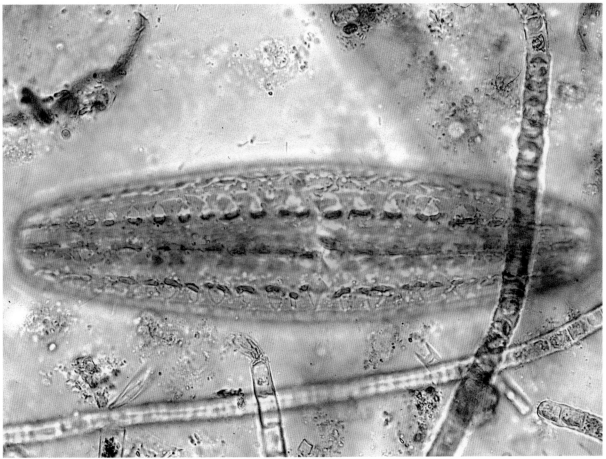

The Spirogyra *group*

Spirogyra, Mougeotia and *Zygnema* (figs 81-92) singly or together are among the commonest algae forming green, slippery or slimy masses in ponds (e.g. fig. 4), bog pools, quiet bays in lakes or backwaters of rivers. They, like desmids, are conjugate algae and similarly are absent from the sea. Many species in this group, also like those of desmids, favour an acid environment and have become more prominent than they used to be in waters affected by acid rain. *Spirogyra* is the largest and best known genus with over 400 species.

The cells of those desmids which form chains (filaments) are attached to one another in various ways (e.g. figs 63-65). No matter how close the cells are to each other, they are never, as it were, fused together, that is they remain as single entities which can separate from one another relatively easily and even without harm. The situation is different in the *Spirogyra* and succeeding groups of filamentous Chlorophytes. In this group, the cylindrical cells are joined to form a row, as is the case in the filamentous desmids, but each cell is united to the cells on either side by a partition, the transverse or cross-wall. This wall is common to both cells and continuous with the longitudinal walls of the filament.

The difference between cells arranged end to end and cells united by a common wall can by analogy be compared to that of two children standing back-to-back and Siamese twins united back-to-back. In the first case separation is easy, in the second case a difficult, dangerous proceeding.

If a cross-wall is broken mechanically the contents of the two cells united together by it will flow out and the cells will die. However, sometimes, the chemical structure of the inner part of a cross-wall alters in such a way that the cross-wall separates into its two outer parts. These outer parts of the cross-wall then remain intact and in contact with the longitudinal walls, hence in this case, the cells are unbroken. This separation of the common cell cross-wall can take place in some or all of the cells forming a filament. Then, the filament can break up into groups of cells or even single cells, all of which potentially are able to form new filaments.

Filaments of species of the three commonest genera in this group, namely *Spirogyra*, *Zygnema* and *Mougeotia*, are seen occurring side by side in fig. 81. They each have a different kind of chloroplast. *Spirogyra*, (fig. 82) has, as its name suggests, chloroplasts which twist spirally around the cell like those of *Spirotaenia* (fig. 79). *Zygnema* (fig. 83), like *Cylindrocystis* (figs 76, 77), has two chloroplasts per cell, each one consisting of a central part from which strands or lobes extend towards the cell wall so giving the chloroplast a more or less stellate appearance. Several species of *Spirogyra* and *Zygnema* are present in figs 81-83.

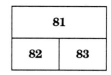

Fig. 81. From above downwards, *Zygnema, Spirogyra, Mougeotia, Spirogyra*. All filaments circa 20 μm br.

Fig. 82. *Spirogyra*, several species. Widest filaments, 31 μm; narrowest, 17 μm.

Fig. 83. *Zygnema*, breadth of widest filament, 31 μm, of narrowest one, 17 μm.

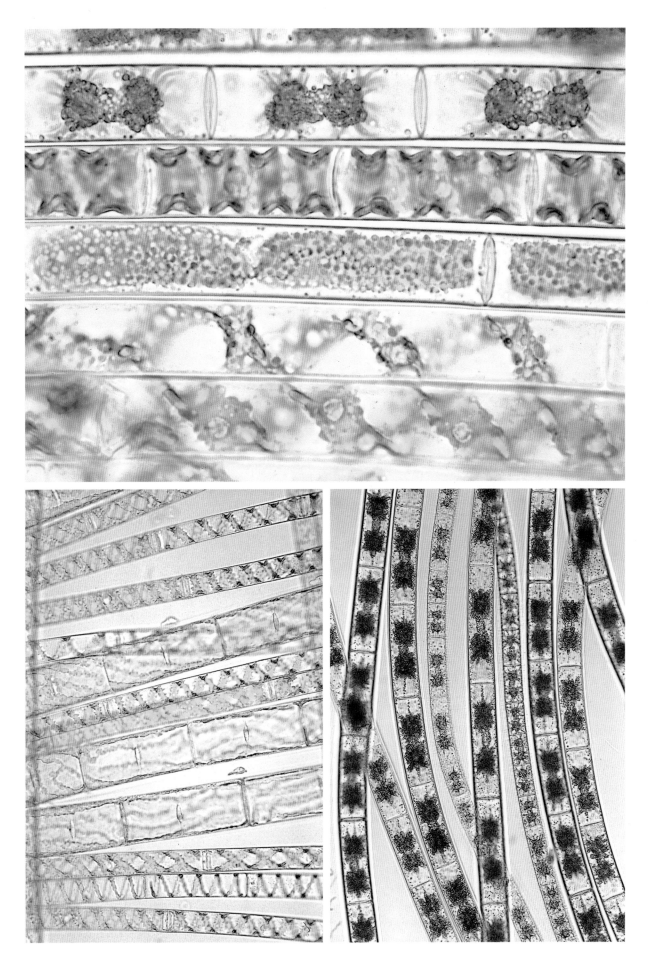

Mougeotia (figs 81, 89) has band-like chloroplasts, like elongate versions of those of *Mesotaenium* (fig. 78). In fig. 89 the chloroplasts of *Mougeotia* can be seen from both the broad and narrow sides. This was not necessarily so before the material was collected and a piece picked out for examination under the microscope. When a filament is undisturbed, the side of the chloroplast facing the light source will be determined by the nature of the light falling on the cell. The chloroplast can move in relation to alterations in the light reaching it. In general, in low light the broad side of the chloroplast faces the light source and in high light, the narrow side of the chloroplast. A great deal of work has been done on how the intensity and spectral quality of the light determines the orientation of the chloroplast of *Mougeotia*.

Large species of *Spirogyra* are good objects for seeing nuclei and cell division (figs 84-86). This is one reason why *Spirogyra* has so often been used for school biology lessons. The discoid nucleus with its nucleolus is situated in the centre of the cell and joined by protoplasmic threads to the thin layer of protoplasm lining the outer part of the cell (fig. 86). In fig. 84 the nucleus has just ended its division. The resultant two daughter nuclei are still close together and no cross-wall has been formed between them. Nuclear division has ended but cell division, at the most, has only just started. In fig. 85 later stages in cell division are seen in the lower filament. Between the recently divided nuclei of the lefthand cell, a cross-wall is developing but it is not fully formed, hence its unclear outline. The wall which joins the righthand cell to the lefthand one is a fully formed cross-wall. Note that in the wide filaments in figs 84 and 85 the discoid *Spirogyra* nucleus is seen from its narrow side. In the upper, narrower filament (fig. 85), a small nucleus is faintly visible left of centre. It is seen from the broad side as is the one in fig. 86.

The main part of the cell is filled with a watery sap in which there are various small bodies. If those bodies near the outer margin of the cell are viewed carefully, it can be seen that they are moving around the cell, being carried along with a stream of protoplasm. Protoplasmic streaming is a very common phenomenon but not always easy to see.

Spirogyra can be found in almost any kind of freshwater. It is eaten in Thailand, though it may be doubted if those who collect it can distinguish it from other green slimy masses of algae, so that what is eaten is probably a mixture of genera.

Sexual reproduction

As mentioned on p. 52, sexual reproduction in the *Spirogyra* group is similar to that in desmids (conjugation).

In one method, the process starts when cells of two filaments are close together. Bulges appear from their

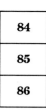

Fig. 84. *Spirogyra*. End of nuclear division and beginning of cell division. Filament, 27 μm br.

Fig. 85. *Spirogyra*. Lower filament (27 μm br.). Lefthand cell, new nuclei still close together but cross-wall developing which eventually will locate them in separate cells. In the lower filament the discoid nuclei are seen from the narrow side.

Fig. 86. *Spirogyra*. Nucleus seen here from the broad side, and strands linking it with the peripheral protoplasm. Cell 97 μm br. N.

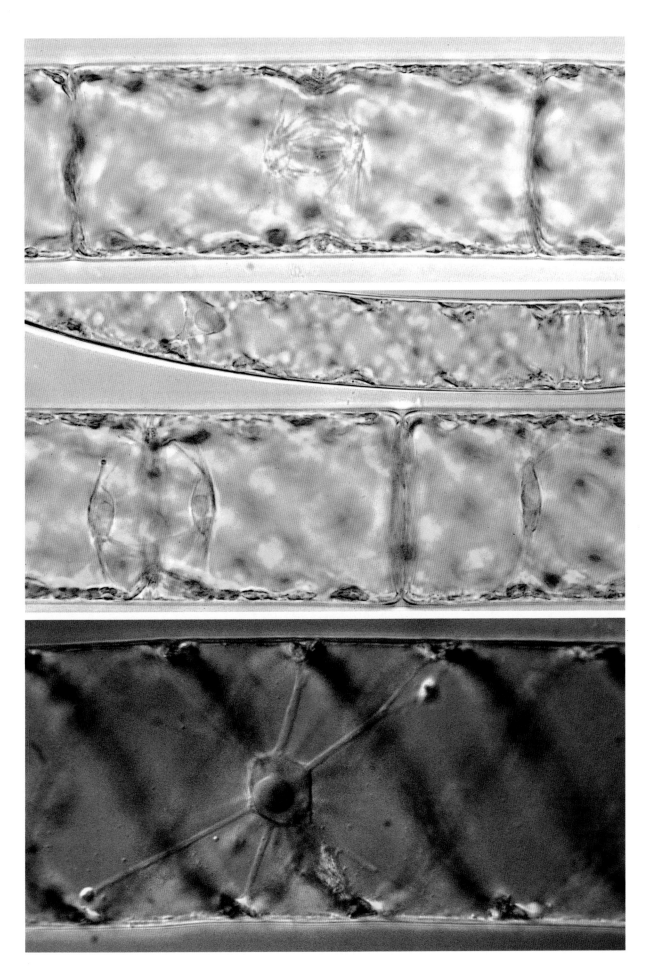

adjacent sides (fig. 89 *Mougeotia*, see bottom right and centre of picture). These cell outgrowths elongate and eventually the parts of their cell walls in contact disintegrate so that there is an open canal between the cells. One of two things then happens; the contents of both cells (i.e. the gametes) pass into the canal and fuse there or one gamete passes through the canal to fuse with the other gamete in that gamete's cell.

In fig. 90 there are three filaments of *Spirogyra* lying parallel to one another but zygotes are only present in the middle filament. Cells of the filaments on each side have conjugated with the middle one. However, in each of the side filaments there is one cell (note presence of unaltered chloroplast) which has not conjugated with a cell in the middle filament. Nevertheless, in each case the presence of a tube issuing from the cell towards an opposing cell in the central filament shows that it has tried to do so. Since gametes from both the side filaments have fused with those of the central filament, it is possible to think of the cells of the two former as male and those of the latter as female. This type of reproduction is often referred to as ladder-like or scalariform from its Latin derivation.

An alternative type of conjugation (lateral) is seen in the *Spirogyra* in fig. 91. The contents of two adjacent cells in the same filament fuse. The first stage in this form of conjugation is for bulges to appear on one end of the common crosswall between the cells. The part of this crosswall in the bulge then breaks down, thus permitting the contents of one cell to pass into the other and so to form a zygote.

As in desmids, the zygotes can also act as resting cells. Their walls often are dark coloured (figs 91, 92). On germination, the protoplasm of the zygote issues as a tube (fig. 92) which on elongation divides into two cells by the formation of a crosswall (fig. 87). Then, further elongation and cell divisions eventually lead to the formation of a typical filament (fig. 88).

89	90
91	92

Fig. 89. *Mougeotia*. Early stages in conjugation. Filaments 23 μm br.

Fig. 90. *Spirogyra*. Scalariform conjugation, zygotes immature. Filament, 11 μm br.

Fig. 91. *Spirogyra*. Lateral conjugation, zygotes (44-52 × 30 μm), mature. I.

Fig. 92. *Spirogyra*. Germinating zygotes, central germling (94 μm l.).

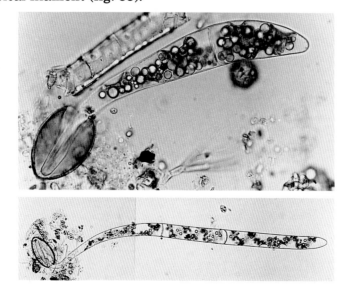

Figs 87 (× 415), 88 (× 165). *Spirogyra*. Stages in the germination of zygotes and growth of the resulting filament.

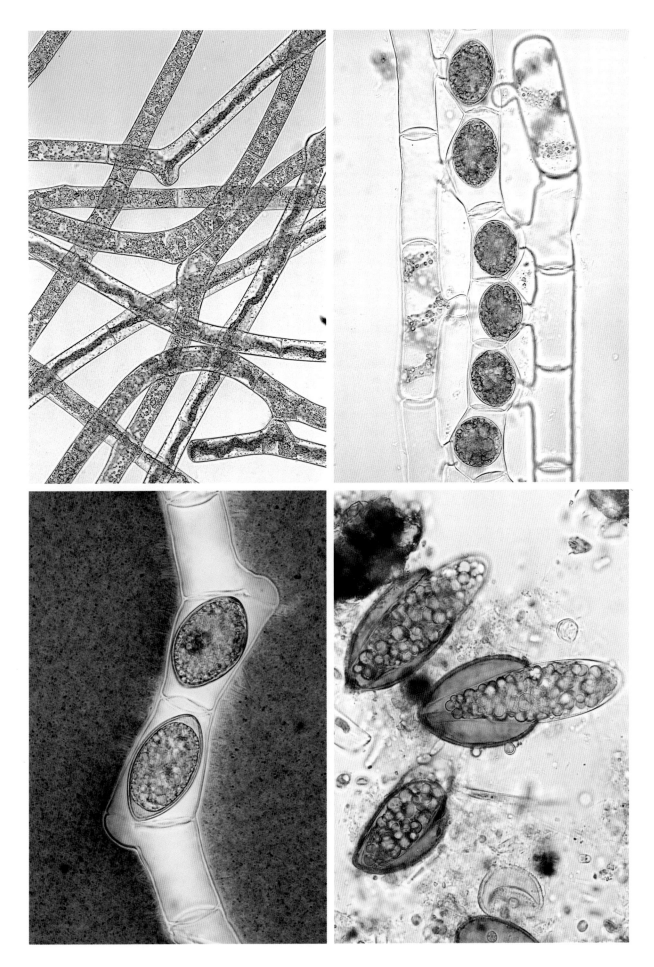

61

The Oedogonium *group*

The filaments of most species of the *Spirogyra* group are unattached to underwater objects, whereas those of the *Oedogonium* group are attached. Nevertheless, parts of filaments may break off and continue to grow, an ability seen in many other filamentous algae which, at least in their young stages, are attached to a surface. Bundles of filaments of *Oedogonium* attached to a stone are seen in fig. 93. The filaments are unbranched and usually rough to the touch.

In fig. 94, a single whole cell of *Oedogonium* and parts of the cells to each side of it are seen. The cells divide by a characteristic method. A description of the process is beyond the scope of this book but a consequence of it is seen in the end of the bottom cell. There, rib-like bands cross the cell. In fact, they go all round the cell. They are parts of the wall of a cell which has undergone division. Once formed, a cell may never divide again or does so only once in its lifetime. In this case there is no such "rib" or only one such "rib" passing round the cell and close to one end of it. On the other hand, a cell may divide several times and the number of times it has done so is indicated by the number of "ribs" present. The cell in fig. 94 has five such "ribs", so we know it has divided five times. This kind of cell division is peculiar to the *Oedogonium* group. Hence, if such transverse ribs are seen in cells of a filamentous Chlorophyte, then it belongs to the *Oedogonium* group.

Oedogonium is a very common freshwater alga; *Bulbochaete* (figs 97, 99-101), though less common and with fewer species is by no means rare. It differs from *Oedogonium* in having branched filaments and bulbous hairs on its cells.

Earlier, it was mentioned that many Chlorophytes reproduce by motile spores. In the *Oedogonium* group there are both asexual and sexual (sperms, see later) free-swimming spores of unique appearance. The spore is propelled through the water by the motion of a ring of fine threads (flagella or cilia, see p. 80). This ring is seen in optical section in fig. 96 and lies below the colourless anterior end of the spore. As a result, the spore looks like a monk's head with its shaven crown. The old scientific name for the group, Stephanokontae, was based on this resemblance (Greek: *stephanos* – a crown).

Another green coloured alga (*Vaucheria*, p. 108), though not a Chlorophyte, has spores propelled by many such threads. However, there the threads are not arranged in a ring but distributed in pairs all over the cell's surface.

Eventually the motile spores of *Oedogonium* and *Bulbochaete* settle down and germinate to form new plants (figs 95, 97, 98). The basal cell of a filament usually differs from the others to some degree in order to assure a firm grip on the substratum. The *Oedogonium* germlings in fig. 95 have a star-shaped attaching disc or holdfast.

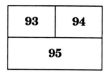

Fig. 93. *Oedogonium* growing on a stone.

Fig. 94. *Oedogonium* cell (33 μm br.) with rings (see text).

Fig. 95. *Oedogonium* germlings with colourless adhesion discs called holdfasts (25-28 μm diam.). N.

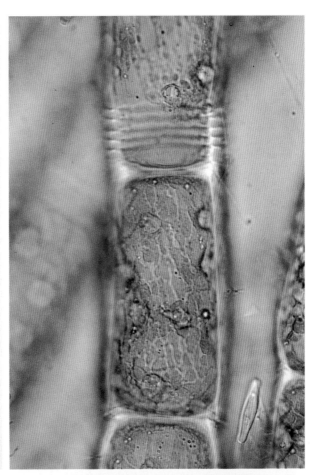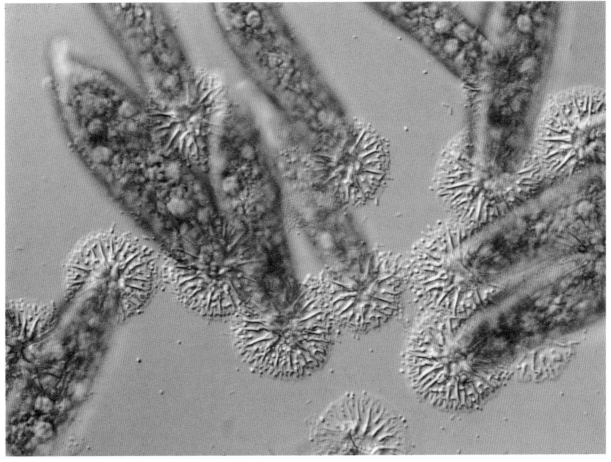

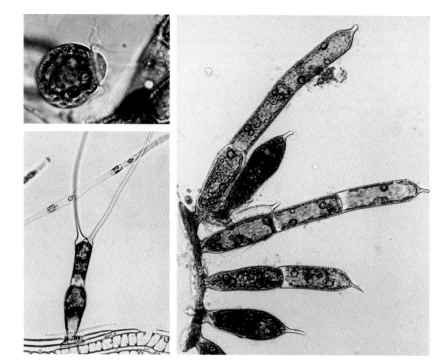

Fig. 96. *Oedogonium*. Motile spore with some of the flagella in the subapical ring in focus (× 640).

Fig. 97. *Bulbochaete*. Young filament (× 250).

Fig. 98. *Oedogonium*. Young filaments (× 320).

Sex in the *Oedogonium* group involves the union of a small male cell (sperm), of similar structure to the above described asexual spore, with a much larger female cell, the ovum or egg enclosed in an egg-cell. Fig. 101 shows part of a filament of *Bulbochaete* with two egg-cells each containing a fertilized ovum or egg, the contents of which have changed in colour from green to brown. The sperms, usually 2-4 per cell, are produced in short cells elsewhere on the same filament or on another filament. Like human and other animal sperms they can swim. Just beyond the upper egg-cell in the *Bulbochaete* filament (fig. 101) and immediately below the base of a bulbous hair can be seen two empty cells which are separated from the egg-cell by a cell with contents. These short, quadrate, now empty cells are male cells. In the *Bulbochaete* in fig. 99 there is a young, still green globose egg-cell. In fig. 102 there is a brown-coloured, fertilized egg-cell of *Oedogonium*. These brown-coloured thick-walled cells also function as resting spores and are called oospores ("egg spores").

In some species of both *Oedogonium* and *Bulbochaete* the sperms are produced from cells formed in special structures called "dwarf males" which have originated from a special type of motile spore. In *Bulbochaete* (fig. 100) such a "dwarf male" is visible attached to the righthand side of the egg-cell. It consists of a long basal attaching cell and above a short apically open cell which is the empty male cell. At the top right (at 1 o'clock) of the female cell near to the male cell, is a white papilla-like excrescence. This outgrowth marks the spot where the sperm entered the egg-cell. The hole through which the sperm went now is closed by this white plug. Sexual stages are often encountered especially in late summer or autumn.

Fig. 99. *Bulbochaete*. Young egg cell (19 μm br.).

Fig. 100. *Bulbochaete*. Fertilised egg cell (34 × 42 μm) with an attached dwarf male cell on its righthand side.

Fig. 101. *Bulbochaete*. Two fertilised egg cells have matured into orange coloured spores (50 × 25 μm) called oospores. Note male cells above upper oospore and below terminal bulbous hair.

Fig. 102. *Oedogonium*. Mature spore (oospore; 44 μm diam.) with a dark thick wall outside which is the thin, colourless wall of the egg cell.

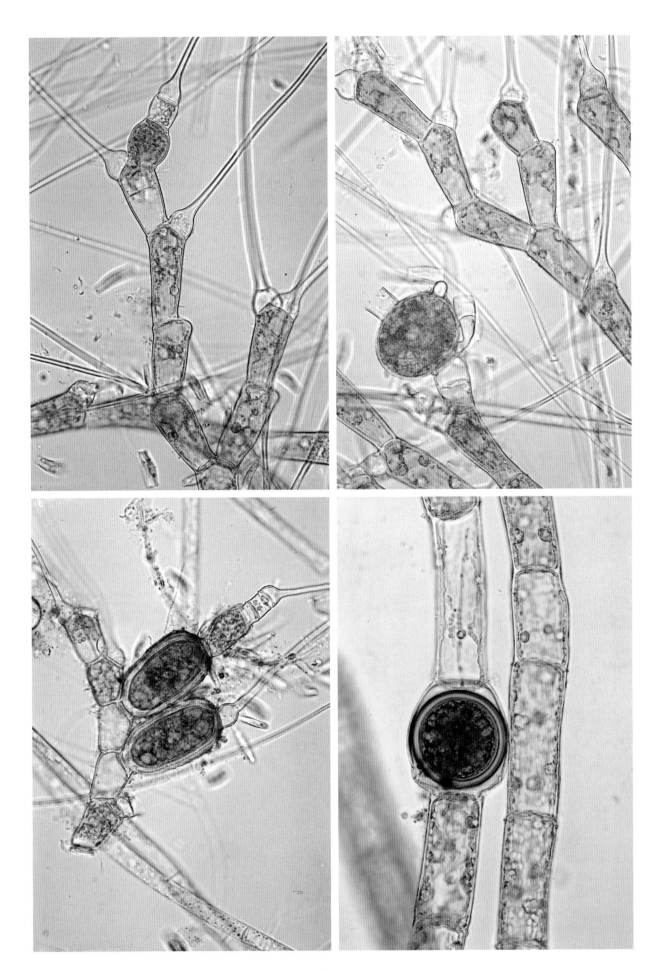

The Ulothrix *group*

Ulothrix (figs 103-105) belongs to a different group of Chlorophytes and has unbranched filaments. Species of *Ulothrix* too are very common in lakes and rivers. The one depicted is particularly common in spring and early summer on the wave-washed shores of lakes in temperate and cold regions.

The cells have a ring-shaped chloroplast. The ring may be closed or split. This is the typical state of a vegetative, that is non-reproducing cell and can be seen in three of the six filaments in fig. 104. The cells of the other three filaments are in the process of reproduction, indeed the holes through which the reproductive bodies have issued can be seen in some of their empty cells. At higher magnification (fig. 105) many of the daughter cells into which the cell contents have divided are visible. Each daughter-cell or spore contains a red-coloured body which is in focus in several of them. This body is called a stigma (*pl.* stigmas or stigmata) or eye-spot, about which more is said later (p. 88). These reproductive bodies can be seen issuing from a cell in fig. 103. They are still within a vesicle but later will become free and swim away. Like the motile spores and sperms of *Oedogonium*, the cells are propelled by flagella but only two or four in number.

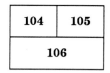

Fig. 104. *Ulothrix*. In the third, fourth and sixth filaments from the lefthand side the cell contents are rounding off, preparatory to producing spores. Filaments, 20-37 µm br.

Fig. 105. *Ulothrix*. Developing spores, note red eye-spots (1.5 µm l.)

Fig. 106. *Geminella*. Cells of lowest filament, 5-6 µm br. I.

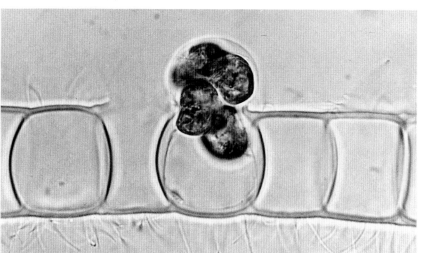

Fig. 103. *Ulothrix*. Spores issuing from a cell but still enclosed in a delicate vesicle (× 1008).

There are two kinds of these motile spores. One, which is asexual (non-sexual) will settle on some object and grow directly into a new filament. The other type is sexual and these spores unite in pairs to form a zygote. Pictures of sexual union of this kind are shown later (figs 134-137). The zygote germinates to release spores, usually four or more in number, which grow into new filaments.

Geminella (fig. 106) is a genus of somewhat uncertain taxonomic status. It is very like *Ulothrix* but does not reproduce by motile spores. The planktonic species shown is surrounded by a wide envelope of mucilage. The cells often have rounded ends and may separate from one another.

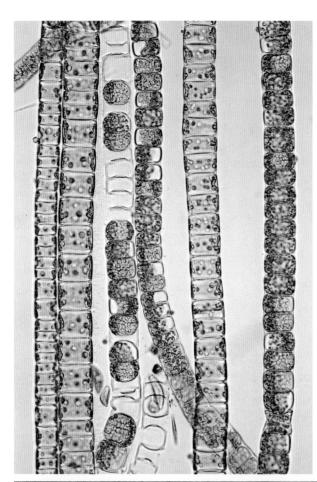
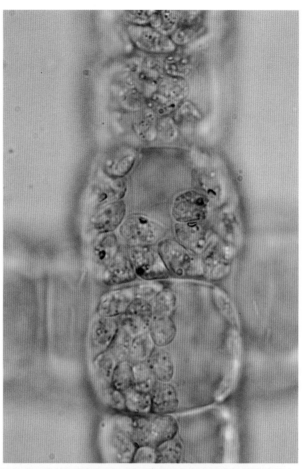
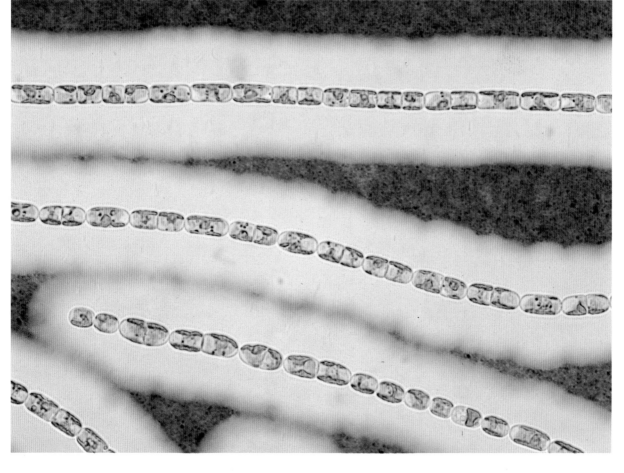

The Chaetophora *group*

The filaments of *Chaetophora* (figs 108, 110) are similar to those of *Ulothrix* but are branched. The species shown forms cushions which can be large enough to be just visible to the naked eye as green, jelly-like blobs on stones, underwater plants, twigs, etc. The whole alga is embedded in mucilage.

Draparnaldia (figs 107, 109, 111) also has *Ulothrix*-like cells (note the ring-shaped chloroplasts of the large cells in fig. 111) and like *Chaetophora* is mucilaginous. However, it does not form jelly-like blobs but feathery structures. It also has branches of two kinds. The cells of the main branches are relatively large. At intervals along these main branches are tufted branches with narrower cells which are much shorter than the main branches and reminiscent of a *Chaetophora*.

Chaetophora and *Draparnaldia* commonly reproduce by motile spores like those of *Ulothrix* but they can also produce non-motile spores which in *Draparnaldia* (fig. 107) turn dark-brown. These latter spores are formed from the whole content of the cell.

Fig. 108. *Chaetophora* tuft (552 μm br.). **D**.

Fig. 109. *Draparnaldia*. Main axes and short, bushy lateral shoots (up to 450 μm l.). **D**.

Fig. 110. *Chaetophora*. Part of plant seen in fig. 108 at higher magnification (cells in centre circa 6 μm br.).

Fig. 111. *Draparnaldia*. Part of plant seen in fig. 109 at a higher magnification (main axis, 30 μm br.).

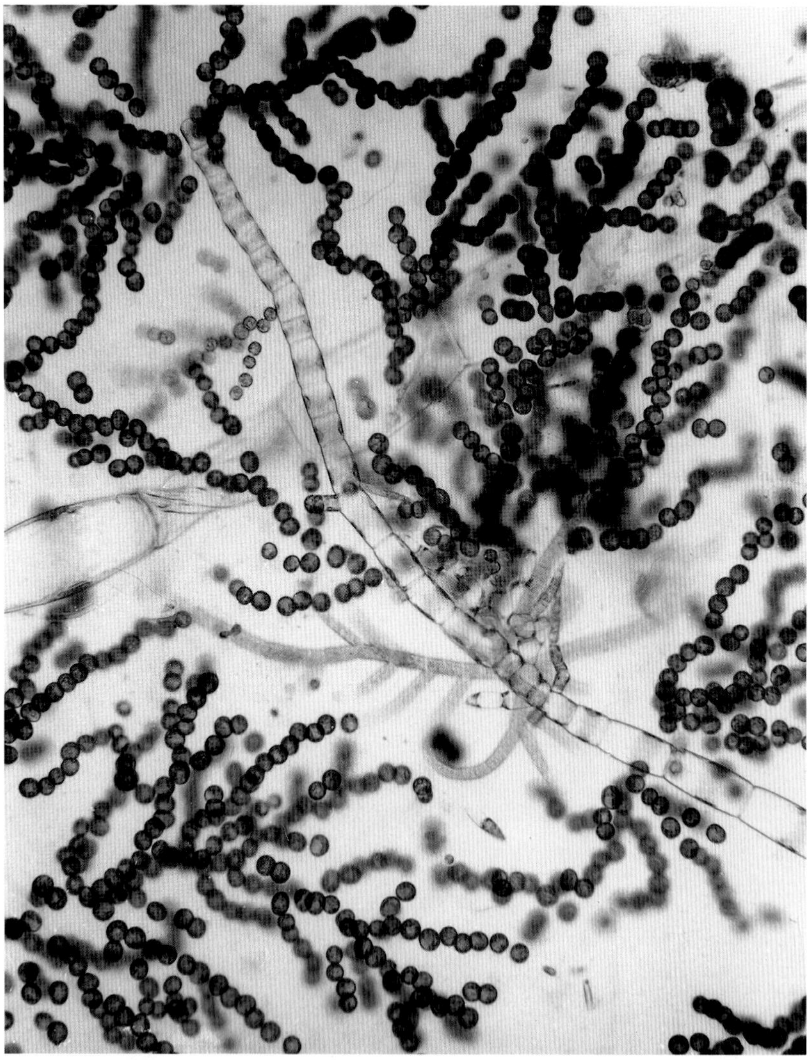

Fig. 107. *Draparnaldia*. In all but the main branch, the contents of the cell have been transformed into dark coloured spores (× 180).

69

Some members of the *Chaetophora* group consist wholly of filaments projecting from the substratum (so-called erect filaments), apart from the cell or cells which attach them to the substratum. Others have both a prostrate system growing over the substratum and an erect one growing out of it (e.g. *Stigeoclonium*, p. 310). The degree to which either system predominates depends on the nature of the species concerned (i.e. its genes), the age of the plant or the environmental conditions. In yet other genera there is no erect system.

Aphanochaete (figs 112-114) is a common genus. It consists solely of prostrate filaments growing over a variety of substrata (e.g. as here on *Oedogonium*). The filaments are unbranched or irregularly and often sparsely branched. The filaments are relatively short, often tapering at each end (fig. 114). Some of the cells bear hairs with a bulbous base (fig. 112).

Chaetosphaeridium (figs 115-117) is unicellular, as in the species shown, or a few cells are connected together by tubular structures, so that even in the latter case it is doubtful whether it is a true, filamentous member of this group. In the widespread species shown here, the cells usually are in small groups (figs 116, 117) attached to various surfaces (here, like *Aphanochaete*, growing on *Oedogonium*). Each cell bears a long hair (fig. 117), up to about 1 mm long, the base of which is encased by a cylindrical sheath (figs 115, 116).

Chaetosphaeridium and certain other unicellular genera have long been considered to be "reduced" members of this group. By "reduced" is meant – arising from a clearly filamentous chaetophoralean ancestor. It is not known whether this hypothesis is justified or whether *Chaetosphaeridium* and other genera grouped with it should be excluded from this group. Molecular biology, notably DNA studies, should establish the affinities of these algae.

Hairs are a feature of many chaetophoralean algae, indeed they represent the only feature common to all the unicellular forms. They are present in some species of *Chaetophora*, *Draparnaldia* and *Stigeoclonium*. Hence, presence or absence of hairs has been used as a distinguishing character. However, relatively recent studies have shown that the presence or absence of hairs can depend on the chemistry of the water in which the alga grows. For example, if a species of *Stigeoclonium* has hairs when collected and then is kept in media rich in phosphate, it will become hairless. The presence or absence of hairs can be altered at will – with a low concentration hairs will be present, with a high concentration they will be absent.

All the members of this group so far mentioned are common in lakes and streams; *Trentepohlia*, described on p. 246 is terrestrial.

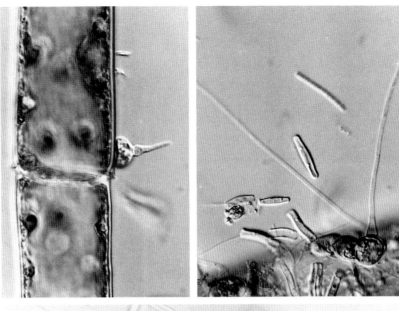

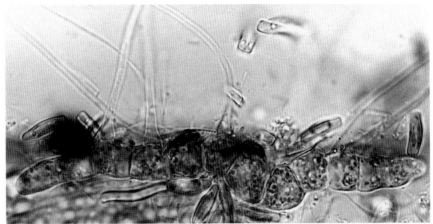

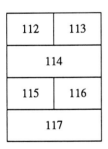

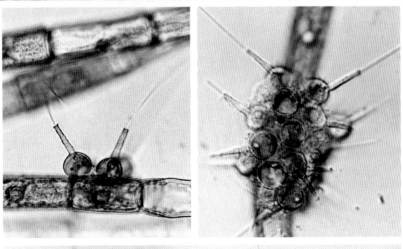

Figs 112-114. *Aphanochaete*. Fig. 112. Cell with developing hair (× 640) N; fig. 113, a three-celled filament (× 640) N; fig. 114, part of a larger filament, the arcuate cells are diatoms (*Achnanthes*) (× 1008).

Figs 115-117. *Chaetosphaeridium*. In fig. 117 the great length of the hairs is indicated. Figs 115, 116 (× 640); fig. 117 stained with brilliant cresyl blue (× 400). In figs 112-117 the algae are growing on *Oedogonium*.

The Cladophora *Group*

These Chlorophytes have branched or unbranched filaments which are not slimy but usually rough to the touch because they are not covered by mucilage. The very roughness of the cell walls makes it easier for other algae to attach themselves to them rather than to those algae with slippery mucilaginous walls. As a result, occasionally these basically rough filaments may feel slippery because of the large numbers of other algae which themselves are slippery and have attached themselves to a cladophoran alga.

Unlike those of most freshwater algae (e.g. *Spirogyra*, fig. 86) the cells of *Cladophora* contain several nuclei. The cells almost always are longer than broad (fig. 120), sometimes much longer (fig. 121 lefthand filament). The shortest and broadest cells generally are found near the base of a plant. These cells can be barrel-shaped (fig. 121 righthand filament) and have thick walls. Such cells can act as overwintering parts of the plant. In winter, much or most of the upper parts of the plant may be lost.

Cladophora (figs 118-121) is common and often abundant in rivers, lakes and smaller bodies of water. It is also common in estuaries and on the seashore. The freshwater species can withstand considerable amounts of salt and some penetrate into the brackish water of estuaries. The commonest freshwater species, *Cladophora glomerata*, has been adapted under laboratory conditions to a concentration of salts equal to half the strength of that in seawater. This was done by growing populations (cultures) for sometime in each of a series of nutrient media of increasing salt content.

Cladophora is uncommon or absent in freshwaters poor in such major plant nutrients as calcium, phosphorus and nitrogen and in acid water, chemical characteristics which often occur together with low concentrations of essential nutrients. It forms short or long tufts or tresses attached to rocks etc. The tresses can be very long, even several metres in length. The filaments usually are richly branched (fig. 120) but the degree of branching is affected by the environmental conditions. In addition, some species are less richly branched than others under all conditions. On the wave-washed shores of lakes and in fast flowing rivers richly branched plants normally predominate. Little branched or almost unbranched and often unattached forms are commonest in the still waters of pools, ditches, dykes and quiet bays of lakes.

Cladophora *balls and blanket weed*

In certain lakes, *Cladophora* grows as mossy balls (figs 122, 123) sometimes admixed with fibrous or other matter. This is a different species from that shown in figs 118-121

118	119
120	121

Figs 118-121. *Cladophora.*

Fig. 118. On wave-washed rocks.

Fig. 119. A tuft on a stone in still water.

Fig. 120. Branched filaments (circa 20 μm br.).

Fig. 121. Filaments with long cylindrical cells (22 μm br.) and short barrel-shaped (27-31 μm br.) ones.

though free-living irregular tangles of that species often are present in lakes. Some botanists consider the ball-forming species to be so different from other species of *Cladophora* that it should be put into a separate genus called *Aegagropila*. The filaments grow out radially from the centre of the ball. In the larger balls the central part usually is dead.

Cladophora balls typically live on the bottom of a lake, sometimes many metres below the surface. In the latter case, their presence may only be detected when they are dislodged from the bottom and washed ashore. The balls vary in diameter from a few mm to 10 or more cm. Ball formation depends on the plants being rolled over the bottom by water movements. If the plants become fixed to the bottom but still washed by the water, then they grow into mossy carpets or cushions. If they are washed into still water so that they are no longer rolled about, then they lose their regular ball shape and can disintegrate. The biology and life history of this *Cladophora* still is very imperfectly understood, despite the fact that the occurrence of balls in a lake almost always is recorded or commented upon.

In the Hokkaido district of Japan, there is a lake with especially fine *Cladophora* balls which form part of a summer festival connected with the folklore of the local Ainic people. Judging by the issue of a special stamp (fig. 122) and the picture postcard (fig. 123) "*Cladophora* worship" seems to have become a tourist attraction. Moreover, there is (or was) a bar in Tokyo called Marimba, the Japanese name for these balls, where plastic *Cladophora* balls are on sale. It seems that the mythology surrounding these balls involves a young man and girl who drowned in the lake, their hearts turning into *Cladophora* balls. So popular have *Cladophora* balls become in Japan that they are now protected plants. It is said that plants of other non-ball forming species, such as the one in figs 118-121, are rolled by hand into balls and sold as true Marimba.

Cladophora has a common name, blanket weed. The name has arisen from the fact that large masses of dead or dying *Cladophora* washed on to or left behind on the banks of rivers or shores of lakes can turn grey or brown and resemble a piece of woollen blanket. However, similar accumulations of some other algae can be called blanket weed. The commonest but much less common algal component of blanket weed is *Oedogonium*. So far as *Cladophora* is concerned the name blanket weed commonly is applied nowadays also to healthy, bright green living populations. In certain parts of Russia, blanket weed is so abundant that it has been made into paper and cartons. However, it has not come into general industrial use.

Rich growths of *Cladophora* can be a nuisance, for example to fishermen who get their hooks entangled in them instead of in a fish's mouth. Naturally, it is the fly-

Fig. 122. Japanese First Day Cover with *Cladophora*-ball stamp.

Fig. 123. Japanese picture postcard, *Cladophora*-ball festival.

fisherman who suffers most. The appearance of abundant *Cladophora* in the form of long tresses often is a sign of eutrophication (see p. 12) but fishermen and others are wrong to object to or be worried by a moderate growth of *Cladophora*; it is a natural component of waters of high quality.

Cladophora can also become a pest in ornamental pools, dykes and on the shores of lakes.

Vast masses of *Cladophora* and certain other algae are weeds in the sense that farmers and gardeners use the word, that is too many plants of certain kinds in the wrong places. Some algae, as we shall see, kill certain animals. Large amounts of *Cladophora* can look unsightly and become offensive when stranded on the shores of lakes or banks of rivers, so destroying amenity. In the Great Lakes of America, *Cladophora* has become a serious nuisance. The water as a whole may not have become greatly enriched by eutrophication but near urban centres it can be. Rich growths of *Cladophora* are torn off by the waves or come off as they die. They can then be transported over long distances in such huge lakes, to accumulate, die and decompose in some bay or sheltered region which is a popular area for picnicking, fishing, swimming or boating. In many lakes, eutrophication is correlated with deterioration of reed beds. One of the causes of damage to the outer parts of reed beds seems to be the great masses of algae washed on to them. *Cladophora* usually is a major component of such mats; other algae are *Hydrodictyon* (p. 34), *Oedogonium* (p. 62), *Enteromorpha* (p. 78) and *Spirogyra* (p. 56).

It has been suggested but not proved that *Cladophora* is a cause of swimmer's itch, a skin infection with reddish spots and much irritation arising after swimming. The commonest cause of swimmer's itch is microscopic animals which spend one part of their lives in snails and another part in birds or fishes. It is the reproductive bodies of the snail stage in their life histories which are released into the water and make the mistake of burrowing into the skin of a swimmer rather than into that of the animal they parasitize. Fortunately, they get no further. A related tropical parasite of snails does have man as its alternative host and is a serious pest causing the disease called schistosomiasis or bilharzia. *Cladophora* is especially common in moderately or strongly calcareous water. Snails need a large amount of calcium in order to produce their shells so that abundance of *Cladophora* often is accompanied by an abundance of snails, many of which may be parasitized by the animal causing swimmer's itch.

Pithophora (figs 124, 125) looks like *Cladophora* when purely vegetative but is strikingly different when reproducing. *Cladophora* reproduces by motile spores in much the same way as *Ulothrix* (figs 103, 105). *Pithophora* reproduces solely by unicellular, non-motile spores produced at intervals, sometimes in series along a filament. The contents of a cell pass into one end, almost always the upper one, which swells out and is cut off by a transverse wall from the empty or almost empty part below.

Unlike flowering plants, most freshwater algae are cosmopolitan, but some are tropical or subtropical. *Pithophora* is most common in such regions (e.g. India) but can also be rare or absent in them (e.g. Africa). Further, it is common in non-tropical parts of the U.S.A. (e.g. Indiana). In what was the USSR it is uncommon everywhere. In Europe it is recorded from tanks in hot-houses, warm-water aquaria and near hot industrial effluents. There also are records from European rice fields but there too the evidence points to it being introduced. It would be interesting to know what has produced this somewhat curious geographical distribution. Clearly, temperature is involved, since it does not flourish in cold regions but it is tempting to believe that sea and land barriers (e.g. mountain ranges, deserts) also are involved in its distribution. Yet most other algae have overcome all such barriers. *Pithophora* can be a pest in small lakes, fishponds (especially nursery ponds) and rice fields.

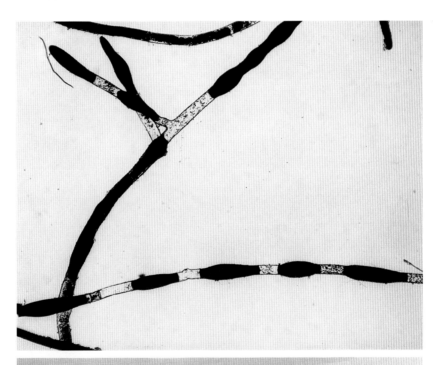

Fig. 124. *Pithophora*. Filaments and spores (× 40).

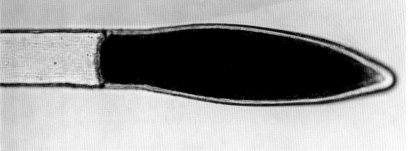

Fig. 125. *Pithophora*. Terminal spore (× 256).

The Enteromorpha *Group*

This group is less common in freshwater than in the sea, where on the seashore a well-known species is *Ulva lactuca* which looks like a small lettuce and is called the sea lettuce. Most of the members of this group have their cells united into a type of plant construction similar to that of *Enteromorpha*.

Enteromorpha (figs 126-130) is easily seen by eye. In freshwater it is most frequently found in rivers and dykes. It is also common on the seashore. Indeed, most of its species are marine and are especially abundant near sources of organic pollution (e.g. sewage effluent). The freshwater species, like some of those of *Cladophora* also can grow in both fresh and salt water. *Enteromorpha* often is found together with *Cladophora* in rivers the waters of which have been enriched by sewage effluent and agricultural wastes so that, like *Cladophora*, large masses of it (fig. 127) usually are an indication of eutrophication. The plants resemble the shape of intestines (fig. 129), hence the name of one of the common species, *Enteromorpha intestinalis*. Indeed, the generic name *Enteromorpha* is derived from two Greek words meaning "intestine" and "form" respectively. The plant is cylindrical, hollow and branched. In freshwater, the young plants can be attached to the substratum but the main mass of a population usually is unattached, often entangled among other plants etc. and liable to be washed downstream or out to sea in floods. Any part of the plant which is broken off can continue to grow. The structure of *Enteromorpha* is different from that of the preceding filamentous algae. The small cells are firmly united together to form a sheet or tissue which is one cell thick (figs 126, 130). This tissue of cells forms the outer "skin" of a hollow cylinder.

Figs 127-130. *Enteromorpha*.

Fig. 127. Mass growth in a river.

Fig. 128. Surface view of part of such a floating mass.

Fig. 129. Intestinal appearance of filaments, circa 1/5 natural size.

Fig. 130. Tissue-like arrangement of the cells.

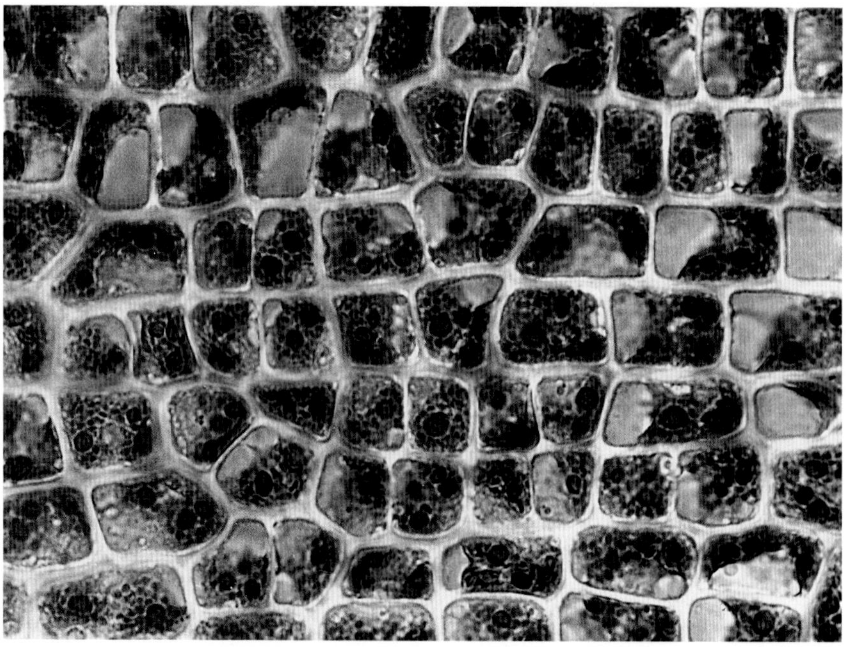

Fig. 126. *Enteromorpha*. Arrangement and varied shapes of cells (× 1008).

Free-swimming forms

Introduction

The previously depicted Chlorophytes do not swim in the vegetative (non-reproductive) stages of their life-cycles but many of them produce spores which do swim.

Members of the present group of Chlorophytes swim in the vegetative state and produce spores which also swim. They are propelled through the water by the beating or undulations of fine threads (e.g. figs 131-137) called flagella or cilia. There is no fundamental difference between a flagellum and a cilium in any plant or animal possessing them. It is past history which has produced the present illogical custom of using different words for the same thing. In general but not universally, the word flagellum is used when the organisms concerned have few such propelling threads, and cilia for those whose cells have many such threads. Thus, a variety of unicellular or colonial plants and animals with one or a few such threads per cell are called flagellates, whereas other unicellular animals with many such threads per cell form a major group called ciliates, examples of which are shown later.

This use of two words for the same thing is so unreasonable and confusing that one may be surprised that it did not end shortly after electron microscopy revealed that flagella and cilia were the same. Further, some bacteria swim by means of fine threads called flagella which have a quite different structure from the flagella (or cilia) of plants and animals, thus adding to the confusion.

In a recent major multi-authored work on microscopic animals and plants a strong case was made for replacing the words flagellum and cilium by the single word undulipodium so far as non-bacterial organisms are concerned. The author of this proposal somewhat ruefully accepts the fact that many or even most specialists will continue to use the words flagellum or cilium for some time to come; indeed some do so in that very book. We too use both words.

Unicellular forms

Chlamydomonas (figs 131, 134-137) is the largest genus, with over 400 species, one or other of which can be found in almost any kind of waterbody. It is also a common alga on soil or in its upper layers.

In the *Chlamydomonas* in fig. 131 can be seen two flagella of equal length and a large black body the pyrenoid surrounded by starch which turns black when stained in iodine. An apical white round area indicates the position of a contractile vacuole. There are in fact two contractile vacuoles per cell but often, as here, only one is visible at a

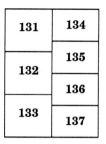

Fig. 131. *Chlamydomonas*, lightly stained in iodine solution, hence black colour of the starch surrounding the pyrenoid (cells, 6.5 × 6 μm). **P**.

Fig. 132. *Dunaliella* (17 × 5 μm). **P**.

Fig. 133. *Brachiomonas* (20 × 17 μm) stained in iodine solution. **Pn**.

Figs 134-137. *Chlamydomonas* and stages in its sexual reproduction; note quadriflagellate zygote. Stained in iodine solution. **P**.

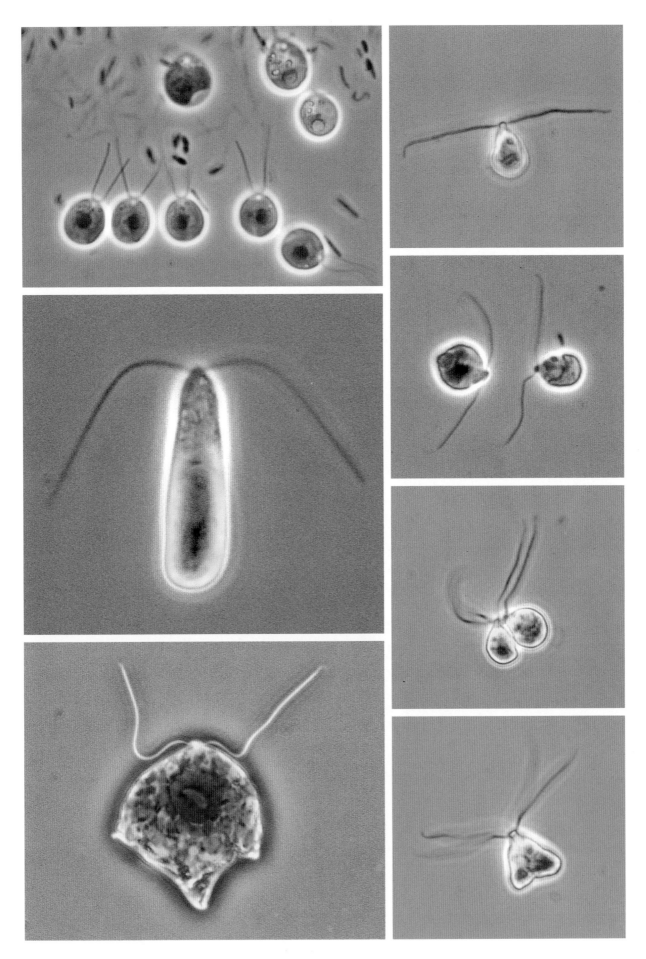

time. These bodies (organelles) fill with liquid, contract, so expelling the liquid and then reform again in a continuous process. Frequently, the two vacuoles fill and empty at different times. Hence when one is fully distended (open) as in fig. 131 the other has contracted (shut) and disappeared. They occur in the cells of all freshwater free-swimming Chlorophytes but usually are absent in those living in the sea or inland saline waters. The function of contractile vacuoles is related to the water and salt content in the cell and the salt content of the water it lives in. There is evidence that they can serve other functions too.

In general, contractile vacuoles are present in flagellate forms and flagellate reproductive bodies in all freshwater algae. They also often are found in algae which, though themselves normally non-motile, are very similar and so presumably closely related to motile species. For other pictures of contractile vacuoles, see figs 241, 300, 552.

Asexual reproduction commonly takes place by the cell content dividing up into several parts which become copies of the mother cell. Such cells are called zoospores. The same name is given to the free-swimming spores of *Oedogonium* (p. 64, fig. 96) and other Chlorophytes. The word zoospore means "animal" spore and reflects an old view that it is animals which swim, not plants.

The sexual cells (gametes) also are flagellate (figs 134-136) and look like zoospores. The gametes may be of identical size, of somewhat different sizes (fig. 135) or always of clearly different size. In a few species, the differences in size between the gametes is so great and the motility of the larger of the sexual partners so slight that the terms female and male can be applied to them. Figs 135-137 show stages in the union of two gametes of somewhat different size. After the two cells have fused together, the new quadriflagellate cell (fig. 137) may continue to swim for a time. Eventually, the flagella are lost and the cell nuclei fuse. As in other Chlorophytes the zygote may function both as part of a sexual process and also as a resting cell.

Chlamydomonas is an individual consisting of one cell. This and its basically plant-like nature make it a very powerful experimental tool for understanding the biochemistry, molecular biology and genetics of plant cells. The following are some of the reasons why *Chlamydomonas* is under investigation in laboratories all over the world. It can multiply at a great rate; only relatively cheap chemicals are needed as nutrients; cultures are easily maintained and occupy little space; sex can be prevented or induced at will and zygotes made to germinate quickly. The whole life history can be encompassed in a fortnight or even less time. Huge numbers of individuals (cells) means great opportunities for finding mutants and isolating, modifying or inserting genes.

The vast majority of the species of *Chlamydomonas* live in freshwater whereas most species of *Dunaliella* (fig. 132) live in salt water. *Dunaliella* looks like *Chlamydomonas* but differs in there being no outer coating or wall to the cell. Because of this, asexual reproduction does not take place within the non-living wall to form two or more daughter cells which eventually escape from the mother-cell. The cell of *Dunaliella* simply splits longitudinally into two parts. This method of reproduction is seen later in a different group of green algae (Euglenophytes) whose cell content also is not enclosed in a wall. Such a splitting of a cell into two parts is called binary fission.

The species of *Dunaliella* shown is common in saline inland waters, including those which are more salty than the sea. A problem of life in very salty water is maintaining a balance between internal and external water. If one adds salt to a freshwater alga it will lose water and the contents are likely to shrivel up. *Dunaliella* keeps a balanced water and salt system by producing glycerol in its cells. Within the salt limits tolerated, the more salty the water it is in, the more glycerol will be formed within the cell. Since salty water and plenty of unused space is characteristic of arid regions, the possibility of growing *Dunaliella* in large outdoor waterbodies and of producing glycerol commercially from it is being investigated. However, a more promising industrial product is the more highly priced substance β-carotene. Certain strains of *Dunaliella* can have 10% of the cell's dry weight composed of β-carotene. Since β-carotene has long been used to colour foodstuffs and as a source of vitamin A in animal diets, it is not surprising that commercial exploitation of *Dunaliella* is a possibility. In addition, its β-carotene may have medicinal properties. One species of *Dunaliella* is on the market as a health food.

Brachiomonas (fig. 133) also is commoner in salty than in fresh water but not in the very salty water in which some species of *Dunaliella* flourish. It is common in rock pools near the sea, especially those above high tide mark. The salt content of such pools can fluctuate widely and rapidly. In stormy weather such a rock pool may be very salty because of the spray blown off the sea. In quiet weather or with the wind blowing off the land and heavy rain falling, the pool can contain fresh water. Hence, *Brachiomonas* has to be able to withstand rapid and massive changes in salinity.

Haematococcus (figs 138, 139) can also be found in such rock pools but is much less tolerant of high salt content. Its characteristic natural habitat is a rock hollow intermittently filled with rainwater. This natural habitat is mirrored by bird baths, (fig. 9), other garden ornaments and receptacles containing rainwater. Like *Brachiomonas*, any alga in such natural or artificial habitats has to withstand rapidly varying and often extreme environmental conditions. Though inland rainwater pools may never contain the amount of salt that is common in seaside rock hollows, the salt content of the rainwater and anything blown in as dust, left by visiting birds (e.g. faeces) or from the container itself (e.g. rock) will become more and more concentrated as the water evaporates. When dry, the alga will be exposed to the sun's rays and heat and in winter to sub-zero temperatures.

The motile cells of *Haematococcus*, like those of the previous algae, have two flagella (fig. 138). The cell contents are concentrated in the central regions from which fine protoplasmic threads pass out towards the cell wall. Parts of some of these threads can just be seen in fig. 138. The cell content is green with a small or large addition of orange or brick red pigment. If conditions are favourable for growth, the cells will not be very red in colour. Under the specially favourable conditions of culture which can be produced and maintained in the laboratory the cells can be wholly green. In nature, this is not a normal situation and the waters in a recently rain filled rock hollow or bird bath are unlikely to be rich in nutrients, though a great deal depends on where the rock hollow or bird bath is located and how often birds splash about in it.

In industrial regions, rain can be rich in sulphuric and nitric acid. Though both sulphur and nitrogen are essential nutrients and the latter an especially important one needed in considerable amounts, the acidity of industrial rain may be harmful. Before the British Clean Air Act reduced the industrial and domestic production of smoke, rock hollows containing *Haematococcus* often were black with soot.

As conditions become more and more unfavourable, either because of lack of nutrients or the drying up of hollows or bird baths, so the cells become redder and more and more of them lose their flagella, round off and produce thicker walls. The cell content now fills the whole cell cavity. These are resting cells and commonly are called cysts (figs 9, 139). There now is evidence to support the long held belief that the red pigment protects these cysts from bright sunlight and, in particular, ultraviolet light. This protection may become more and more important as industrial products (e.g. CS gases) and chance volcanic eruptions destroy the ozone layer protecting us all. Parts of the deposit and cysts in such dry hollows may flake off and be blown away, so dispersing the cysts. Earlier, when water is present, birds may carry them away. When conditions are favourable once more, new swimming cells arise from the cysts.

Recently, the red pigment, astaxanthin, in the cells has become a potentially valuable product. The red colour of fish such as salmon and trout can only be assured in fish farms if astaxanthin or a suitable alternative is added to their food. In these "green" days, natural products are preferred to those made industrially. In the same way, the

Fig. 138. *Haematococcus*. Free-swimming cells and two cysts (23 and 21 μm). **P.**

Fig. 139. *Haematococcus*. Cysts, largest: 36 μm diam.

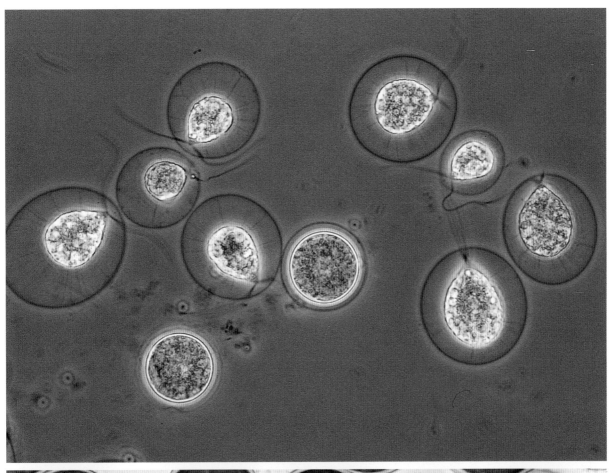
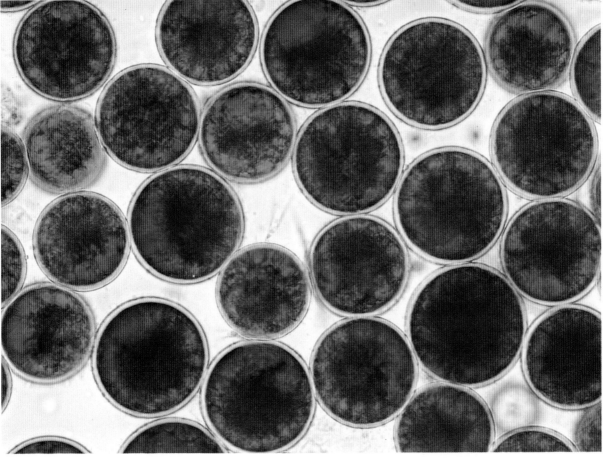

"natural" β-carotene of *Dunaliella* can be preferred to "artificial" β-carotene or other industrial colouring agents. The extraction and use of astaxanthin from *Haematococcus* is being studied and the methodology of at least one production process has been patented. It is claimed that the cost of natural astaxanthin from *Haematococcus* can be comparable to that of synthetic astaxanthin.

Haematococcus is very common in the habitats mentioned and often the dominant organism in bird baths. Occasionally, it appears in quantity in rivers or the margins of lakes when drought exposes large areas of rock rich in hollows. In other kinds of temporary pools or puddles it usually is absent.

Colonial forms

In *Gonium* (figs 140, 142, 143) four to sixteen cells are embedded in mucilage to form a flat or slightly curved plate. Each cell is like that of a *Chlamydomonas* and has two flagella. These two flagella arise close together and pass out through a papilla (fig. 140). In this too *Gonium* resembles *Chlamydomonas*. The shape of the papilla varies from species to species in both genera.

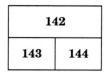

Fig. 142. *Gonium* (41 × 39 μm), note flagella. **P.**

Fig. 143. *Gonium*. Asexual reproduction. Colony plus surrounding mucilage, (137 × 137 μm l. and br.). **I.**

Fig. 144. *Eudorina* colonies, up to 156 μm diam. **I.**

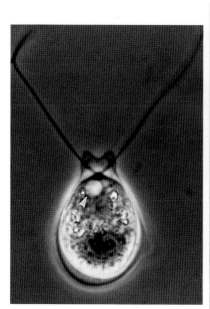
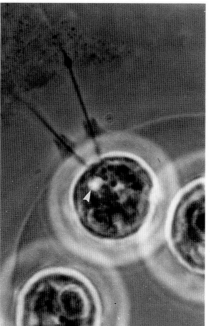

Fig. 140. *Gonium*. Exit of flagella through an apical thickening of the cell wall (× 1600). **P.**

Fig. 141. *Eudorina*. Exit of flagella through the cell sheath and tubular structures (× 1800). **P, I.**

Contractile vacuoles arrowed.

As the plate of cells swims it also rotates. Every cell can produce a new colony. In fig. 143 one cell in the colony has not started reproduction; another cell (below it, left) has divided into 4 and the rest into 8 or the full complement of 16 cells typical of this species. Since the typical number of cells is 16 and every cell can produce a new colony, there should be 16 cells or daughter colonies in the specimen in fig. 143. Instead, there is one undivided cell and eleven developing daughter colonies. The missing four daughter colonies presumably have freed themselves from the mother colony and swum away.

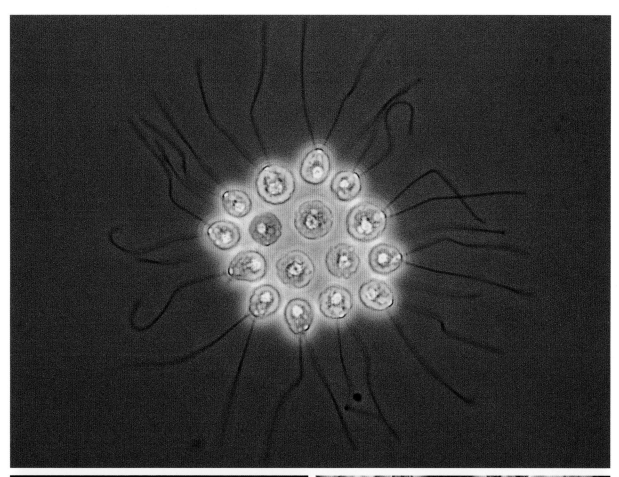
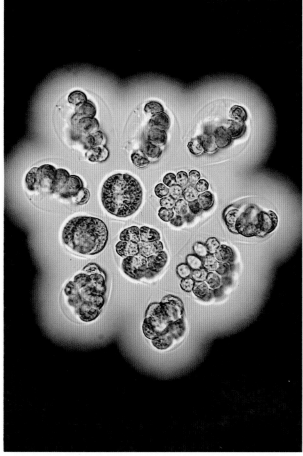
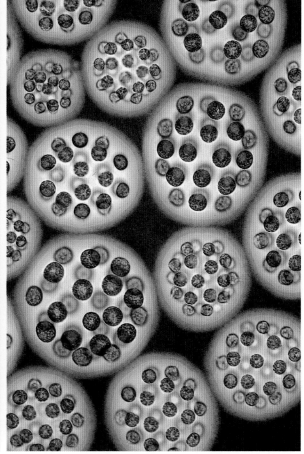

Eudorina (fig. 144) and *Pandorina* (fig. 146) have globose or ellipsoid colonies of biflagellate cells. In the *Eudorina* (fig. 141), the two flagella arise wider apart than in *Gonium* and pass through two short tubes at the surface of the colony. The *Pandorina* shown has cells which are so closely appressed to one another that they are rounded conical in shape. The colony looks like a miniature mulberry, hence its specific name *P. morum* (Greek: *moron*).

The cells of almost all free-swimming Chlorophytes possess a bright red body, usually located near their apices (fig. 147, *Eudorina*; fig. 148 *Volvox*). These organelles are called eye-spots or stigmas. They are involved with the perception of and reaction to light. The actual "light organelle" is beneath the stigma which can be considered to be a light filter. It is remarkable how colonial forms such as species of *Gonium, Pandorina, Eudorina* and *Volvox* (see later) with their separate cells react as a unitary organism to a directional stimulus such as light. Frequently the stigma present in the cells at the "front end" of a colony is larger than in those at the rear.

Asexual reproduction in *Pandorina* and *Eudorina* is similar to that in *Gonium* apart from the fact that in some species of *Eudorina* a few cells, usually anterior ones, always divide later than the others or do not divide at all. Some people put species of the last kind in a separate genus called *Pleodorina*.

In *Volvox* (fig. 145), the largest free-swimming, colonial Chlorophyte, only a minority of the cells in a colony can produce new colonies. These cells are detectable at an early stage in colony growth because they are larger than the cells which cannot reproduce. The small dark spheres in fig. 145 are daughter colonies.

Fig. 146. *Pandorina*. Largest colonies 44 × 36 μm.

Fig. 147. *Eudorina* cells (17 μm diam.), note large, red eye-spots and striate chloroplast.

Fig. 148. *Volvox* cells (10 × 7.5 μm), also with large eye-spots.

Fig. 145. *Volvox*. Colonies (× 40) containing daughter colonies.

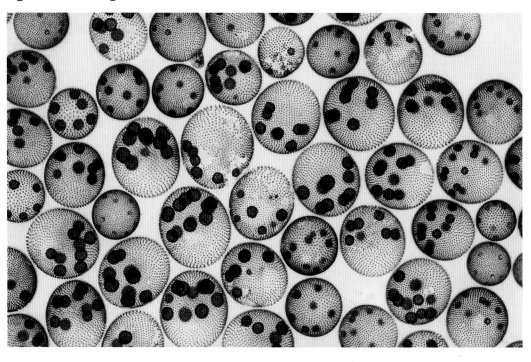

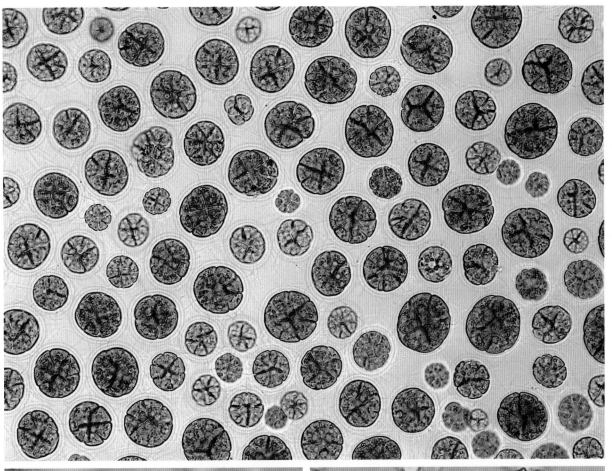
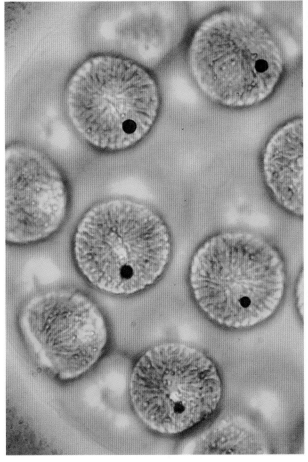
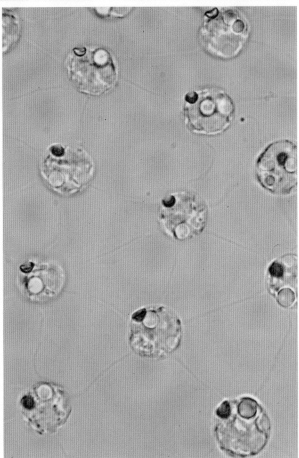

Whereas the number of cells in *Eudorina*, *Pandorina* and *Gonium* is 32 or less, the number in a colony of *Volvox* is several hundred or more than a thousand. Hence, full-grown colonies are visible to the naked eye. The cells are located at the periphery of the sphere. Each cell has two flagella (fig. 149), eye-spot (figs 148, 151), pyrenoid, contractile vacuoles (fig. 152) etc. just like a cell of *Chlamydomonas*. The paired black dots in fig. 150 represent the holes or pores in the colony membrane through which the flagella pass to the outside. In fig. 152 the contractile vacuoles appear as small pale blue circles best seen in the cell at the uppermost righthand corner, its adjoining cell left and the cell below this one. In young colonies the cells are connected to one another by protoplasmic strands (figs 151, 152) but in a few species these strands are lost before the colony is full grown.

At first sight, the colony appears to have no polarity and that the random beating of the flagella of hundreds or thousands of cells around a sphere could not produce organised, oriented movement. However, the cells do not

Fig. 151. *Volvox aureus*. Cells (circa 9 μm diam.) with connecting threads. **P**.

Fig. 152. *Volvox globator*. Cells with connecting threads (circa 5-10 μm l.). **P**.

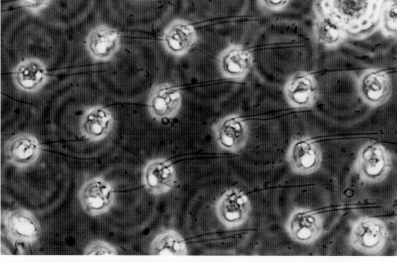

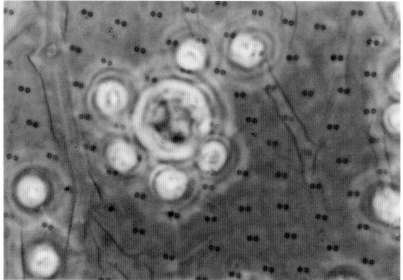

Fig. 149. *Volvox*, showing the paired flagella (× 800). **P**.

Fig. 150. *Volvox*. The pairs of black dots mark the holes in the cell through which the flagella pass (× 1600). **P**.

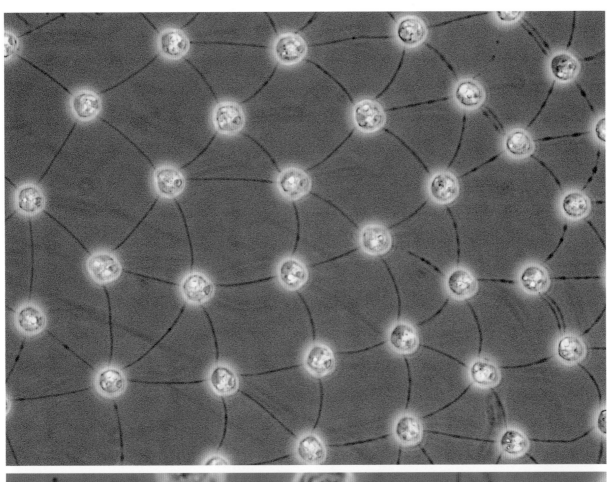

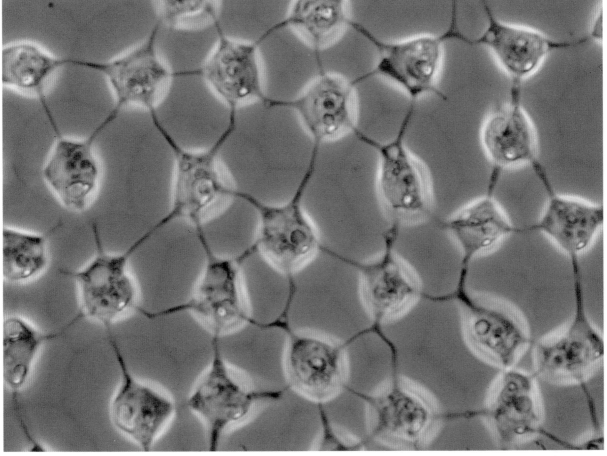

behave in a random manner, and the eye-spots in one part of the colony are larger than in another, so delimiting anterior and posterior regions. Most of the daughter colonies arise in the posterior half of the colony.

Because *Volvox* colonies are a millimetre or more in diameter their movements can be followed when large populations are present. In a dish it can be seen that they move towards the light, provided it is not too bright.

All the free-swimming, colonial Chlorophytes mentioned can reproduce sexually as well as asexually. In *Gonium* and *Pandorina* the gametes are like the vegetative cells and are similar to one another in size, as in the species of *Chlamydomonas* shown (fig. 135). In *Eudorina* and *Volvox* there is a clear distinction between male and female cells. In some species, egg and sperm cells are produced in one and the same colony. In others they are produced in separate colonies.

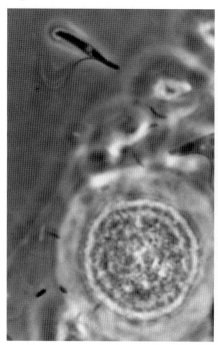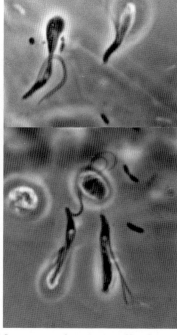

| 153 | 154 |

Figs 153, 154. *Volvox*, the spherical cell is an egg cell, the narrow, elongate biflagellate cells are sperms (× 1600). **P.**

The male cells of *Volvox* divide up to form small plate-like colonies (fig. 155) which are clearly distinct from the globose asexual daughter colonies. They are reminiscent of the young *Gonium* colonies shown in fig. 143. The male colonies in the species pictured break up into separate biflagellate cells or sperms (figs 153, 154) on reaching a female colony. The egg cells are produced singly and at first resemble the large, as yet, undivided cells which initiate asexual reproduction. After fertilisation, the zygote usually becomes yellow or orange in colour (figs 156, 157) and develops a thick wall which, depending on the species, can be smooth (e.g. *Volvox aureus* fig. 156) or spiny (e.g. *V. globator* fig. 157).

Volvox is common in pools and lakes. During winter it usually is absent from the water, overwintering in the zygote stage on the bottom of the waterbody.

155	
156	157

Fig. 155. *Volvox*. Male colony with sperm packets (up to 13 μm). **N.**

Fig. 156. *Volvox aureus*. Fertilized eggs now transformed into spores (oospores; 47 μm diam.). **D.**

Fig. 157. *Volvox globator*. A single oospore (50 μm diam.). **I.**

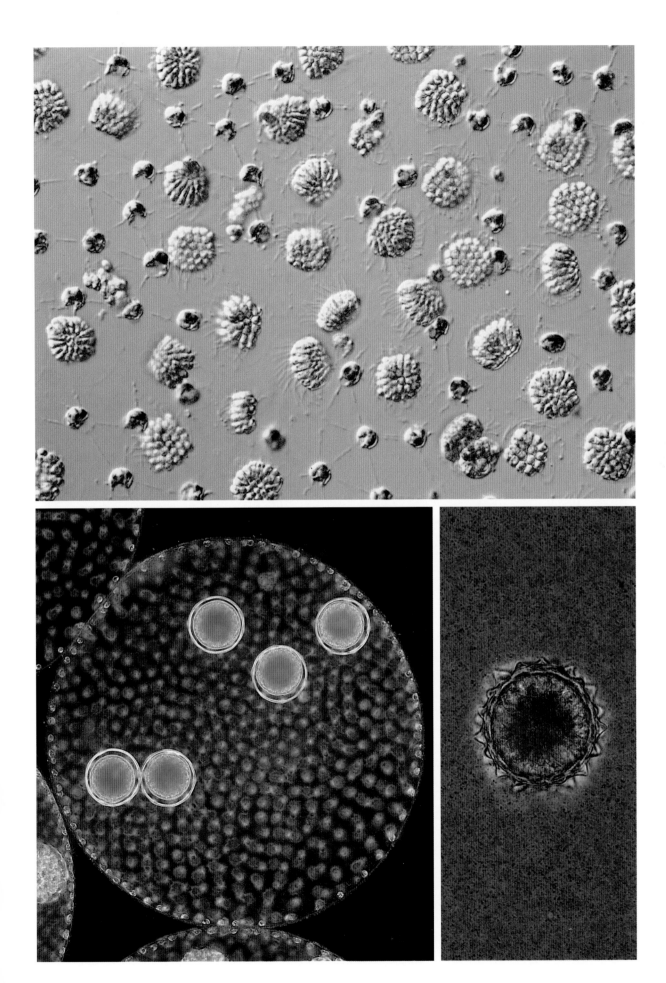

Chapter Three

The Euglenophytes

The photosynthetic members of this group are as green as any Chlorophyte but in other respects the Euglenophyte is a very different kind of organism. The scientific names used for this group differ somewhat according to whether a plant or animal classification is being followed but all start with "euglen" based on the best known genus *Euglena*.

Nearly all Euglenophytes live as solitary, motile cells for much or most of the time. There are no filamentous species and when non-motile they do not form the kinds of colonies seen in Chlorophytes (e.g. like those in figs 21-29; 31-38). Their method of asexual reproduction is unlike that of all but a few Chlorophytes and sexual reproduction seems to be absent, the few alleged records of it being unconfirmed. Their cells have a different kind of outer coating to that of Chlorophytes. The cells of many of them are flexuose and can contract and expand.

Euglena (figs 158-161, 163, 165, 167). The cells of the first species shown (fig. 158) illustrate the ability of most species to change their shape. When swimming, like all euglenas with the aid of a single flagellum, the cells are more or less straight. They can also move when on a solid surface, irrespective of whether they have retained or lost this flagellum. Then, the cells move and change shape in a worm or snake-like manner. Apart from the flexuose changes in shape, the cells can become shorter through either end contracting (see cell at bottom of fig. 158). Reexpansion usually soon takes place. In some Euglenophytes these contractions and expansions produce "bulges" which run rapidly from one end of the cell to the other (figs 177-180). These changes in shape and the associated movement seen in *Euglena* and certain other Euglenophytes are difficult to describe in words but are so characteristic that the term euglenoid movement is used. Most species of *Euglena* can carry out euglenoid movement and also swim but a few rarely, if ever, have functional flagella and so do not swim.

Apart from the many discoid chloroplasts, a red body is present near the round apex of the cell (figs 158-160). This is the eye-spot or stigma which, as mentioned (p. 88) is involved with perception of and reaction to light. The shapes of the chloroplasts and the numbers of them vary markedly from species to species. All but one cell in fig. 158 belong to the same species, the brown cell belongs to a second species.

Fig. 158. *Euglena*. The brown cell (130 µm l.) belongs to a different species from the green ones.

95

A *Euglena* cell is not covered by a wall, that is a structure external to the membrane enclosing all the living matter. Just within this membrane, the cell surface consists of strips running helically around the organism. Where these strips join, striations are produced. In many species the striations are so faint that they are difficult to see. However, in other species they are prominent and ornamented with warts (fig. 159) which may become coloured brown by oxides of iron or manganese (fig. 158 solitary specimen, see also cases of *Trachelomonas*, fig. 171). The strips of material coating the cell do not stop it from changing its shape or dividing into two.

Fig. 159 shows the anterior end of a cell (note the eye-spot). At the very tip is a light coloured area. This marks an invagination (seen later, fig. 185). It is from this invagination that the functional flagellum arises which propels the *Euglena* when in a swimming phase. Besides the functional flagellum there is also a small non-functional one which is extremely difficult to see. The long emergent flagellum can be seen attached to some of the swimming *Euglena* cells in fig. 165. A large number of oval to ellipsoid white bodies is contained within these cells. They are grains composed of a substance called paramylon which in the Euglenophytes takes the place of the starch of Chlorophytes. Paramylon grains often are large and of characteristic shapes. Their presence is also visible in the Euglenophytes shown in figs 174-176.

The cells in fig. 160 not only are capable of changing their shape but in addition, they can attach themselves temporarily to debris or other substrata. They attach themselves by their pointed basal ends. Relatively few species of *Euglena* do this. One cell in fig. 160 is in the process of asexual reproduction. In this, the cell divides from the apex downwards into two. This is the same method as that of *Dunaliella* (p. 83) and characteristic of algal cells which are not encased in a firm outer coat or wall. Division stages are not seen as often as might be expected, even in large populations for two main reasons. First, division commonly starts at night and has finished by early morning. Second, as the population grows, so does the demand for nutrients. Sooner or later the rate of supply of nutrients acts more and more strongly on the rate of growth of the population and so the number of dividing cells.

The *Euglena* cells in fig. 161 have both lost their flagella and euglenoid movement and rounded off. Many have divided and most of the two new cells are still so close together that they are flattened along the plane of cell cleavage. The species of *Euglena* that often go through such a non-motile phase in their life-cycle sometimes produce large aggregates of cells and much mucilage. This is the so-called palmelloid stage (see Glossary). Such large bright green masses are not uncommon near heaps of dung.

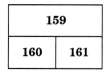

Figs 159. *Euglena*. Cell (10 µm br.); anterior end with red eye-spot.

Fig. 160. *Euglena*. Cells (circa 110 µm l.) attached to debris. One cell in the process of division.

Fig. 161. *Euglena*. Non-motile, rounded off cells (to 19 × 22 µm) embedded in mucilage (palmella stage).

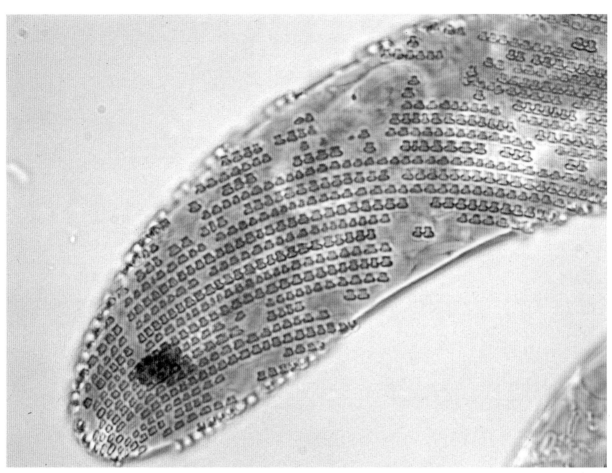
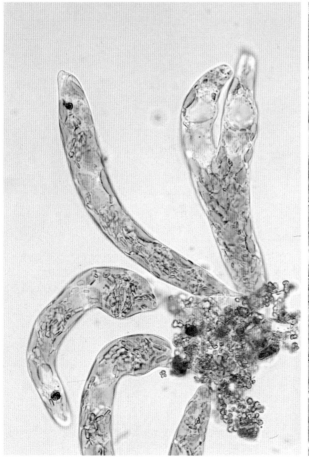
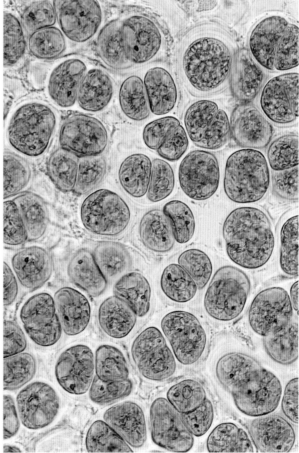

A few species of *Euglena* can be red in colour, like *Haematococcus* (fig. 139). The commonest of these species is the aptly named *Euglena sanguinea* (fig. 167). When large numbers are present they can form a scum on the surface of the water (figs 7, 166). Dry summer weather favours the development of the scum. If something is put over part of a scum so as to exclude direct irradiation of the cells beneath, their colour will change to green or greenish in half an hour or so. The red pigment moves from the outer into the inner part of the cell, so exposing the green chloroplasts previously lying below it. Similar changes can arise at night. The cells in this surface scum can be so closely packed that their shape is changed by mutual pressure. Such dense scums are not easily broken by rain but most scums are broken up by a heavy downpour (e.g. in thunderstorms).

As in *Haematococcus*, the red pigment may protect the cells from the harmful effects of ultraviolet light to which they can be exposed in surface scums. However, the production of the red pigment is not solely or perhaps even primarily a matter of irradiation but one of availability of nutrients. This red pigment is astaxanthin, so like *Haematococcus*, *E. sanguinea* potentially is of commercial value.

Fig. 165. *Euglena*. Cells (circa 44 μm l.) with flagella and containing many paramylon grains (the white bodies). P.

Fig. 166. *Euglena sanguinea*. Close-up of a waterbloom (compare fig. 5).

Fig. 167. *Euglena sanguinea*. Cells (circa 100 μm l.) exhibiting a variety of temporary shapes.

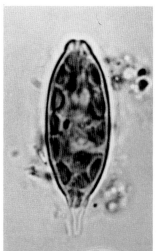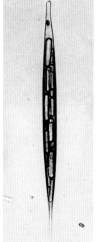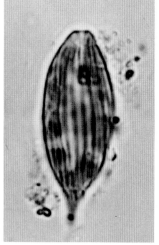

A few species of *Euglena* do not show euglenoid movement and are little flexible (e.g. fig. 163). The cells of *Lepocinclis* (figs 162, 164) are rigid. The species shown is unlike most Euglenophytes in having striations which appear to run up and down the cell (fig. 164), not spiralling round it.

Euglena is much used in physiological, biochemical and genetic research and in teaching.

Colacium (figs 168, 169), like the *Euglena* in fig. 160, can attach itself to a surface. Species of *Euglena* do not show selectivity in what they attach themselves to, *Colacium* does. The *Euglena* in fig. 160 is attached directly to a piece of debris by its posterior end. *Colacium* is attached by a mucilaginous stalk arising from its anterior end. The *Euglena* is only attached for short periods, but

Figs 162, 164. *Lepocinclis* (× 2000).

Fig. 162. Optical, longitudinal section.
Fig. 164. Surface view.

Fig. 163. *Euglena*. A species with more or less rigid cells (× 320).

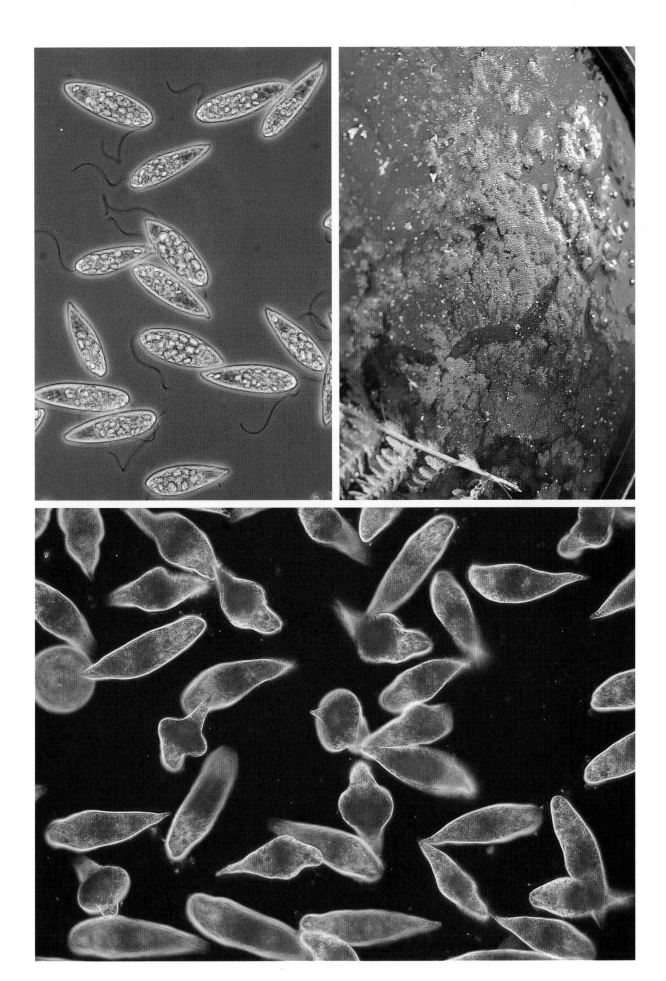

like *Colacium* can divide when attached. *Colacium* is attached for longer periods and as one cell division follows another and new cell stalks are formed, so a dendroid or bushy colony can develop (fig. 168). The external flagellum is lost when a cell is sedentary but can regrow quickly. Hence, cells readily detach themselves from their stalks and swim away; they are then indistinguishable from those of *Euglena*.

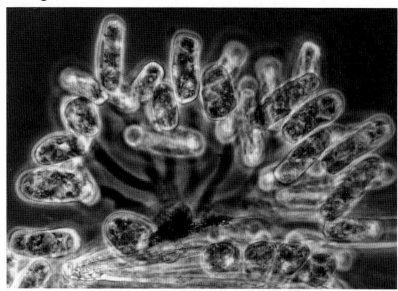

Fig. 168. *Colacium*. A branched colony (× 800). **P.**

Colacium is often found attached to the small planktonic crustaceans *Daphnia* and *Cyclops*. In fig. 169, individual *Colacium* cells are attached to the rotifer *Keratella*.

The cell of *Trachelomonas* (figs 170-173) is like that of *Euglena* but lives inside a case. This case is not a wall surrounding and attached to the cell but a structure separated from it. In species (e.g. fig. 170) in which the cell fills the case more or less completely it usually is not attached to it, but when the cell only fills part of the case (e.g. fig. 172) it may be attached to the base by its posterior end, rather like the *Euglena* in fig. 160. The cell can undergo a limited amount of euglenoid movement within its case or squeeze out through the apical pore.

When a cell divides, one of the two daughter cells leaves the mother-cell through the apical pore and forms a new case around itself. Sometimes both daughter cells leave the case. The opening in the case varies in size and structure and the whole case may be perforated by small pores (fig. 172). In fig. 173, the case, ornamented with small tooth-like spines, has been broken by gentle pressure (note the piece containing the round apical pore) and the *Euglena*-like cell within now released. The case nearly always is brown in colour, sometimes deeply so (fig. 171). What is the biological advantage of a photosynthetic cell living as it were behind a brown filter? It is not clear whether there is any advantage. It may be that it is a price the alga has to pay for living in waters favourable to it in other ways.

Fig. 169. *Colacium*. Cells (13-18 µm l.) attached to the rotifer *Keratella* (178 × 63 µm).

Fig. 170. *Trachelomonas*. Cells of smaller species, circa 13 µm and of larger species, circa 22 µm.

Fig. 171. *Trachelomonas*. Cells (11 × 8 µm).

Fig. 172. *Trachelomonas*. Total length of case, 16.5 µm; breadth to 7.8 µm.

Fig. 173. *Trachelomonas*. Case broken to reveal the euglenoid cell (13 × 9.8 µm) within. Note, at 12 o'clock, fragment containing the pore through which the flagellum passes.

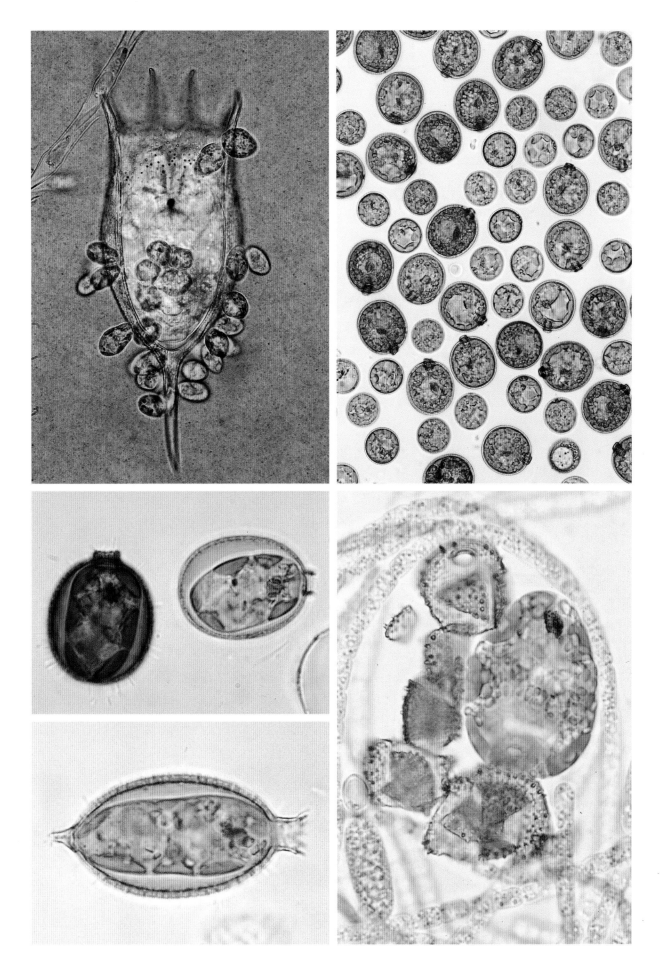

There are many species of *Trachelomonas*; they are of widespread occurrence and at times abundant in habitats rich in organic matter. Such waters often have low, even very low, concentrations of dissolved oxygen. In such poorly oxygenated waters, the concentrations of soluble salts of iron and manganese are much higher than in well oxygenated waters where different and highly insoluble salts of these elements predominate. The soluble salts of iron and manganese are transformed in the case enclosing a cell of *Trachelomonas* into the highly insoluble ones typical of well oxygenated water, possibly by the oxygen released during the alga's photosynthesis. It is these compounds, iron and manganese hydroxides, which give the case its brown colour.

As has been mentioned earlier (p. 98), some species of *Euglena* do not change their shape much if at all. The latter situation is typical of *Phacus* (figs 174-176, 186), another large genus. Here the cells are compressed and so often take on a leaf-like shape. In fig. 174 several non-rigid elongate cells of *Euglena* and rigid ones of a *Phacus* are present. The fish-like cell of the *Phacus* has a ridge running down it (note also the red eye-spot). In fact, as a transverse section would show, it has a flattened triangular shape. The cell contains a large, colourless paramylon grain. The dark spot in the paramylon grain of the uppermost central cell marks an indentation or hole in the grain. Central holes are not uncommon (see figs 176, 186) in large euglenoid paramylon grains. The *Phacus* species in fig. 186 has two large grains one on each side of the cell. Here the long flagellum also is visible. Like the preceding genera, cells of *Phacus* swim by means of a single emergent flagellum. Some species have twisted or corkscrew-shaped cells (fig. 176).

Cells of *Phacus* and those of *Euglena* and *Trachelomonas* are commonest in water rich in organic matter. Indeed, this is true of virtually all Euglenophytes. Even those which can photosynthesize need organic nutrients, hence their abundance in places rich in rotting animal or plant matter. It is often difficult or as yet impossible to grow Euglenophytes in the laboratory in the absence of bacteria which, presumably, produce the nutrients they need in the process of decomposing organic matter. Given suitable organic matter, some of the photosynthetic species can be grown in the dark. There is also a large number of Euglenophytes lacking chloroplasts and so unable to photosynthesize. The nutrition of many "colourless" species is akin to fungi such as that of yeasts and molds. Some feed on microorganisms. Hence, Euglenophytes, like several other kinds of algae, contain both plants and animals. Calling the non-photosynthetic forms colourless is convenient but not necessarily strictly true since cells lacking chloroplasts may contain other pigmented bodies.

Figs 174-176. *Phacus*, fish-shaped cells, and *Euglena* (uppermost, straight cell, 103 µm l.).

Fig. 175. *Phacus*, 84 µm. l.

Fig. 176. *Phacus*. Species with helically twisted cells (circa 13.5 µm l.). Note large paramylon grains.

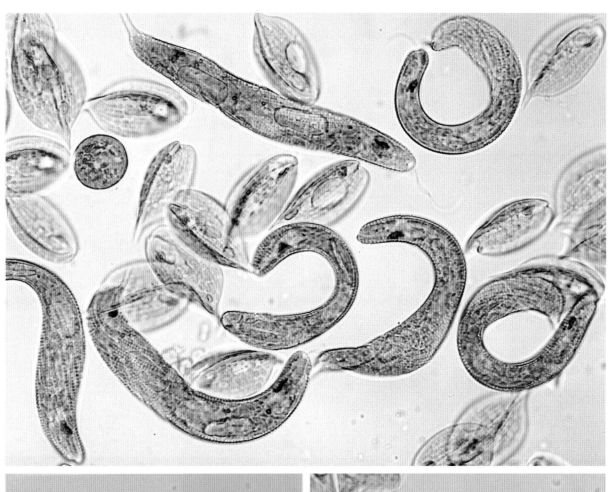
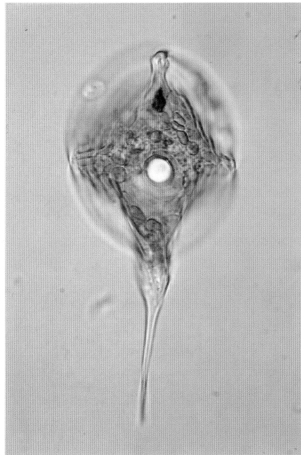
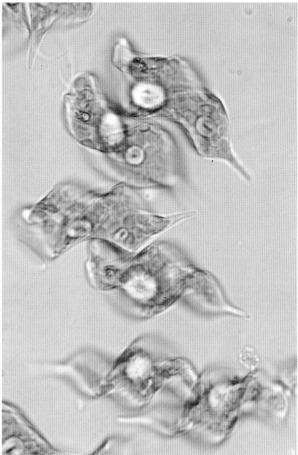

Astasia (figs 177-183, 187, 188) is a common "colourless", saprophytic Euglenophyte. Apart from the lack of chloroplasts, some species are virtually indistinguishable from *Euglena*. Moreover, colourless (chloroplastless) forms of a species of *Euglena* can be produced experimentally and are very similar to a certain species of *Astasia*. It remains to be seen if old claims that cells from natural populations of *Euglena* became colourless is correct. The cells vary greatly in shape due to rapid euglenoid movement (figs 177-180). Cell division (figs 181-183, 188, central cell) is longitudinal as in other Euglenophytes (cf. fig. 160).

Distigma (figs 184, 185) is another "colourless" saprophyte. It differs from all the previous genera in the emergence of two flagella, one longer than the other. In the form shown the shorter emergent flagellum is extremely short. As mentioned earlier (p. 96), cells of Euglenophytes may appear to have only one flagellum, though in fact there is a second, very short, even rudimentary, one within an anterior invagination of the cell. The origin of these two flagella in the *Distigma* can be seen in fig. 185, though they are too close together to be distinguished as separate entities. In *Distigma* and other genera with two emergent

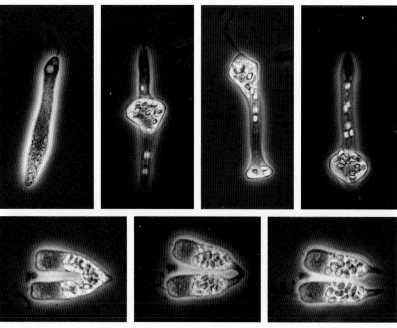

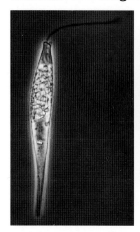

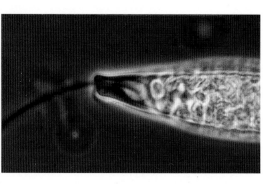

186	
187	**188**

Fig. 186. *Phacus* (16.5 x 13 µm). Note the two large paramylon grains and flagellum. **P.**

Fig. 187. *Astasia.* Cells (longest circa 30 µm) massed around rotting vegetable matter.

Fig. 188. *Astasia*, central cell (25 µm l.) in process of division. **P.**

177	178	179	180

Figs 177-180. *Astasia*, passage of a cell inflation up a cell and down again (x 800). **P.**

181	182	183

Figs 181-183. *Astasia.* Consecutive stages in cell division (x 990). **P.**

184	185

Figs 184 (x 640), 185 (x 1600) *Distigma*; fig. 185 shows apical invagination from which flagella arise. The short flagellum curves backwards. Both **P.**

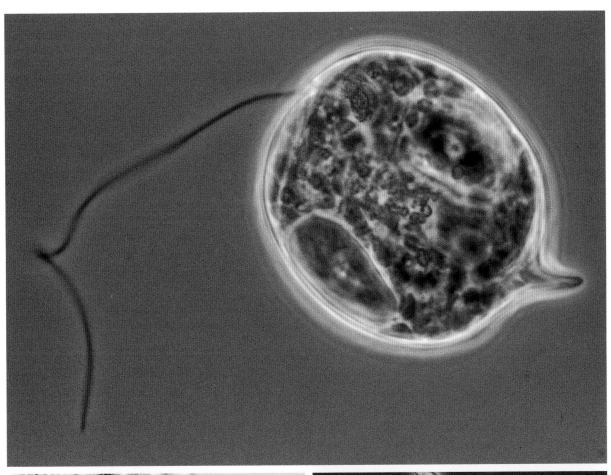
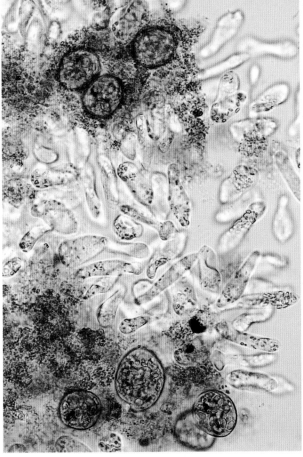 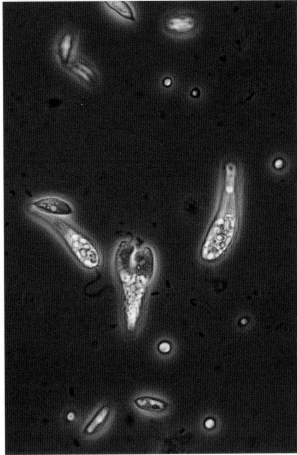

flagella, they pass from the chamber into a narrow canal and through it into the water. In genera with two emergent flagella, their length, thickness, position (e.g. one held out straight and the other curving backwards), fine structure and motion show much variation from genus to genus. A very small minority of Euglenophytes have more than two flagella per cell.

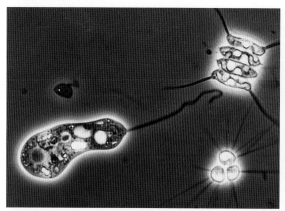
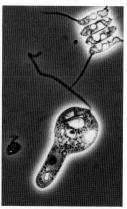
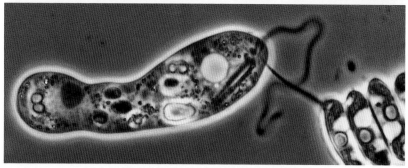

Figs 189-191. *Peranemopsis* (also present: *Scenedesmus, Micractinium*).

Fig. 189. Flagellum held out in motion (× 820).

Fig. 190. Body changing shape (× 640).

Fig. 191. From right to left: apex of cell with flagllum; rod-shaped organ beside which is a contractile vacuole (circular white body); just below each of the latter two bodies is a digestion vacuole with a captured particle (in one case appearing black, in the other white); at base of cell, large black sphere and surrounding greyish area, is the nucleus. (× 1600). Some of these organelles are also visible in figs 189 and 190. All P.

Peranemopsis (figs 189-191) also is non-photosynthetic. It has one emergent flagellum and has been confused with the supposedly well-known *Peranema* (see later). The cell shape varies considerably because of strong euglenoid movement. In free-swimming cells the long, relatively thick external flagellum is held out straight in front (fig. 189); only near the tip does it undergo wavy movements. Nutrition is holozoic, that is, the cell ingests other live or dead organisms or parts thereof. Holozoic nutrition differs from saprophytic nutrition in that the ingested material is digested in the cell. In saprophytism, the organic matter is absorbed, not ingested.

In *Peranemopsis* and certain other genera, the ingestion apparatus includes two rods so close to one another that they look like a single structure (fig. 191). These rods can be protruded from the cell and capture microorganisms such as algae. There is evidence that they can rasp prey open. The captured food is drawn back with the rods into the cell and there enclosed in digestion vacuoles. In fig. 291, see two such bodies nearly halfway down in the cell, one black and the other white, each surrounded by a halo (see also legend to figure). The information about feeding is based on the study of *Peranema* – whatever that genus may be (see below)!

106

Decisions on the correct names for *Peranemopsis and Peranema* arise from a nomenclatural minefield. Since *Peranema* and *Peranemopsis* are both included in the plant and animal kingdoms, their nomenclature is affected by differences in the international codes of botanical and zoological nomenclature. In addition, a recent classification would put them in a kingdom – Protoctista – for which there does not seem to be an internationally agreed code of nomenclature.

Similar problems arise with Cyanophytes (Chapter 10) which, being prokaryotes, could be named according to the bacterial code of nomenclature – some are, but most are named under the rules of the botanical code. As with botany and zoology, the codes of bacteriology and botany are not identical. Moreover, the same words may mean different things in different codes. Clearly there is a need for all scientists of nomenclatural goodwill to come together for the benefit of the rest of us.

The taxonomic problems of *Peranema* and *Peranemopsis* do not end there. The original description of *Peranema* stated that there was one emergent flagellum as in the *Peranemopsis* shown here. Later, what appeared to be the same organism was found to have a second emergent flagellum. This flagellum is difficult to see because it bends backwards on exit from the canal and is closely appressed to the cell surface; it is also thinner than the flagellum held out in front of the cell.

Two nomenclatural "legal" views were then expressed. First, since *Peranema* was described 156 years ago, it is not surprising that the second emergent flagellum was not seen, and so all that is needed is to add this feature to the original categorisation of the genus. The name need not be altered. Second, who can say whether the organism seen originally did not have a single flagellum? If so, a biflagellate *"Peranema"* must have a different name. However, such arguments became irrelevant when it was discovered that the same name had been given to a fern in 1825 and thus had priority over the Euglenophyte name. So the botanical name of *Peranema* was changed to *Pseudoperanema* in 1962. In the meantime, a *Peranema*-like Euglenophyte which did only have one flagellum had been described in 1940 and called *Peranemopsis*.

If, as a zoologist, you consider *Peranema* to be an animal, then the fact that a plant was given this name earlier is irrelevant and the name stands. Since algae clearly include both plants and animals it seems that the name *Peranema* is "legal" or "illegal" according to which nomenclatural code you follow.

Apart from their abundance in organic rich habitats certain Euglenophytes live in the alimentary tracts of a variety of aquatic organisms. They are generally said to be parasites but it may well be that this is not true of a number of them.

Chapter Four

The Xanthophytes (Tribophytes)

The name Xanthophyte should be replaced by Tribophyte to accord with the rules of nomenclature. However, as yet this change has not permeated the whole world of textbooks.

The cells of many Xanthophytes (Greek: *xanthos* – yellow) can be yellowish (see fig. 204) when rich in fat or oil but when little is present are greenish, even as green as those of any Chlorophyte. However, they do contain a different mix of photosynthetic pigments.

As in Chlorophytes, there are unicellular, colonial and filamentous genera. There are very few free-swimming forms but many species reproduce by free-swimming spores (zoospores). The cells contain neither starch nor paramylon and their detailed or fine structure is not the same as that of Chlorophytes. Several Xanthophytes might be mistaken for Chlorophytes at the level of observation possible with a light microscope.

The filaments of *Vaucheria* (figs 192-194) are long tubes without cross-walls (except under certain circumstances). The plant is the equivalent of a many-celled one in that it contains very many chloroplasts and nuclei. Damage to a filament may result in the formation of a cross-wall but normally cross-walls are only produced in relation to reproduction.

In asexual reproduction, a large motile spore is formed from the end of a filament which has been cut off by a cross-wall. Like the filaments, the spores contain many chloroplasts and nuclei. They also have numerous flagella arranged in pairs. The spore germinates by the outgrowth of a typical *Vaucheria* filament (fig. 193).

In sexual reproduction (fig. 194) the male and female structures differ markedly from one another. They may arise separately on a plant or on separate plants. In the species shown, they arise at the top of a short branch growing from the main filament. In fig. 194 the male organ is the empty curved thread above and partially behind the deep green club-shaped body which is a fertilised female cell (oospore), as is the similar cell on the righthand side of the filament. The male cells (sperms) produced from the male filament are minute and motile.

Like those of *Mougeotia* (p. 58) the chloroplasts of *Vaucheria* change postion in relation to variations in the intensity and spectral quality of the light falling on the filament. There is no morphological top, side or bottom in a *Vaucheria* filament which is a simple tube but one can

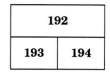

Fig. 192. *Vaucheria*. Filaments 59-70 μm. br. Note small, discoid chloroplasts and absence of cross-walls.

Fig. 193. *Vaucheria*. Germination of spore (filament 650 μm l.). Also present, filaments of *Spirogyra*.

Fig. 194. *Vaucheria*. Sexual reproduction. Two dark green egg cells (oospores; 70-78 μm l.) with a curved empty male organ between them.

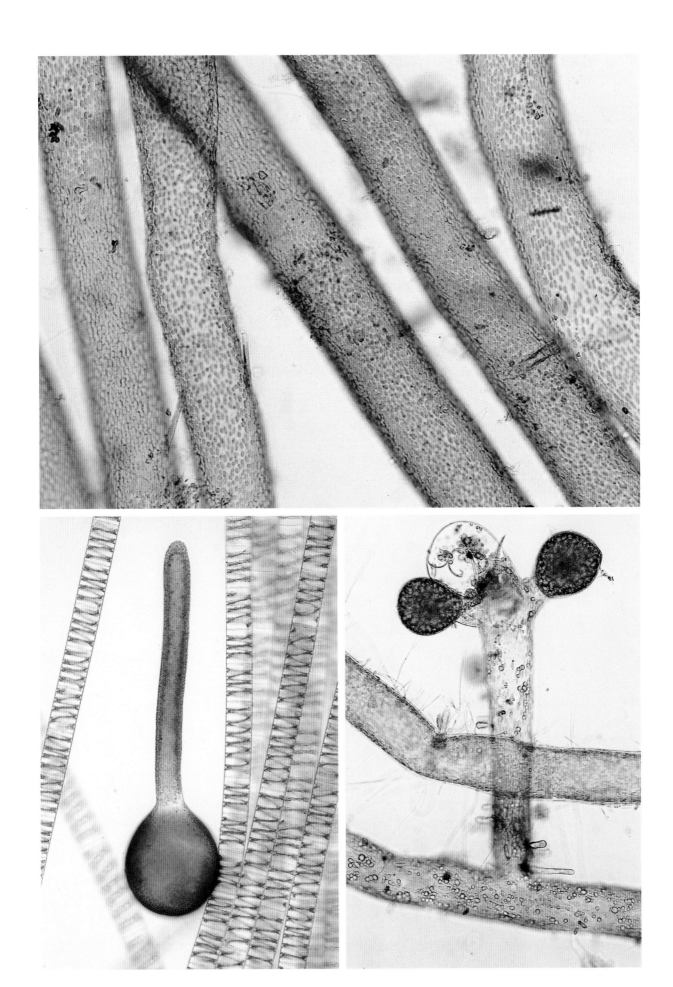

speak of such in a filament's position in relation to a light source. When strong light impinges on the positional top of a filament, the chloroplasts mass along the sides. Thus they form a vertical column in relation to the light source and so receive less light than they would do if they were massed in a layer one chloroplast thick along the upper or lower sides of the thread. The latter kind of layering in relatively weak light, as in the picture (fig. 192), arises when they are illuminated at a moderate intensity by a microscope lamp. In the dark, the chloroplasts are randomly dispersed. In weak light they mass at the side furthest from the light source because the cell acts like a lens focussing the light on the far side. If the filament is illuminated at the same intensity from the side, the chloroplasts will move to the side furthest from the light source. Using the word move is not strictly correct because it has been shown that they are moved in a stream of protoplasm. The chloroplasts of a variety of plants can move in relation to the position or intensity of the light source. In *Micrasterias* (p. 48) the whole cell may move in relation to the intensity or type of the light falling on it.

Vaucheria is a very common alga in ditches, pools and streams where it often forms deep green mats, felts or cushions. Most species favour well aerated water and so habitats where water flows or trickles over them but not so fast that the mats or cushions are washed away. *Vaucheria* is also common on damp ground or soil which is otherwise more or less bare and has not been disturbed recently. Many species live in brackish water or seawater and are especially common in saltmarshes. It can be a nuisance in greenhouses where it may form green coatings on seedbeds or trays and it is said that it can interfere with the germination of seeds and with their early growth. It and *Cladophora* may interfere with water flow in dykes, drainage ditches and similar channels. They can be killed by placing bales of straw, notably barley straw, in the water.

Tribonema (fig. 195) is a widespread filamentous alga. The filaments are unbranched and usually unattached to anything except when young. The outer parts of the cells contain a few or several chloroplasts which are shaped like curved discs or pieces of foil. This type of chloroplast and the presence of more than one per cell is characteristic of most Xanthophytes but relatively uncommon in Chlorophytes. The threads of *Tribonema* are somewhat olive green in colour thus aiding recognition when mixed with other more grass-green filamentous algae.

Most of the species are benthic but a small minority is planktonic, notably in eutrophic waters.

Mischococcus (fig. 196) is less common than *Tribonema* and most often encountered in calcareous waters. It consists of mucilaginous stalks, which may be branched, at or towards the ends of which are a few spherical cells. The

Fig. 195. *Tribonema*. Filaments circa 6.5 μm br.

Fig. 196. *Mischococcus*. Cells 5 μm diam. Stained in brilliant cresyl blue to show mucilaginous stalks. **P.**

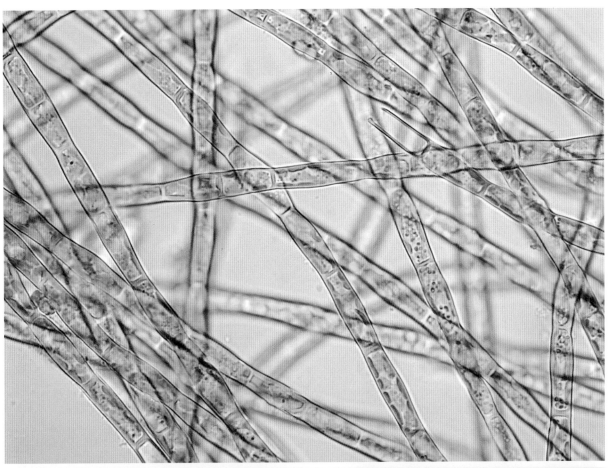

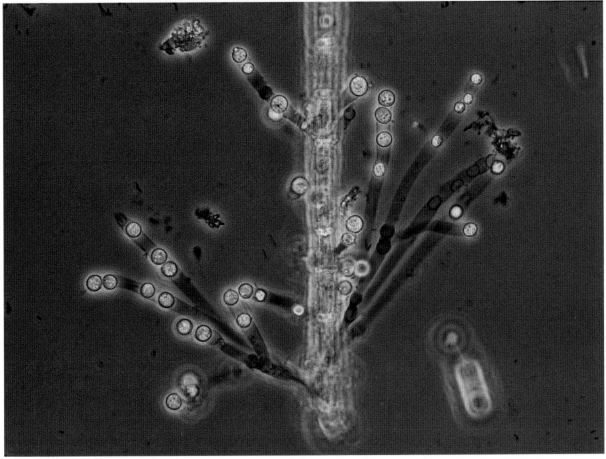

111

base of the stalk is attached to some underwater object, often, as in this case, to a filamentous alga.

Botrydium (figs 197-200) is easily seen by eye as a little green globose vesicle on the surface of wet mud or occasionally soil. These vesicles (about 1-3 mm diameter) generally occur in large clusters. If the mud is rich in lime, the surface of the vesicle may appear spotty (as in fig. 198) because of the presence of grains of calcium carbonate on it.

Careful observation will show that the base of the vesicle consists of a cylindrical stalk disappearing into the mud. Within the mud a network of branched, colourless filaments arise from the base of the stalk (fig. 197). The whole plant consists of a single cell containing numerous

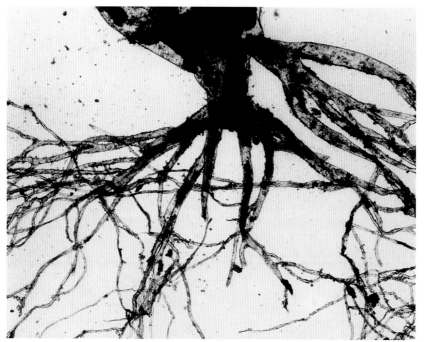

Fig. 197. *Botrydium*. Branched threads which enter the soil (× 35).

small chloroplasts and nuclei. There are two kinds of reproduction, motile cells (zoospores or gametes) and non-motile asexual spores. The latter (fig. 200) are spherical and contain several lens-shaped chloroplasts. In both cases, very large numbers are produced.

Botrydium typically is found on mud which has recently been exposed at the edges of ponds or larger bodies of standing water and rivers. If the mud continues to dry, then eventually the non-motile spores are produced. If the water level rises and submerges the vesicles, then motile spores are produced. Occasionally in severe droughts when large bodies of water dry out, very large areas are covered by *Botrydium*.

The largest population of *Botrydium* ever seen in Britain arose in the great drought of 1984. A reservoir, Hawes Water (area 3.9 km²), in the normally wet English Lake District largely dried up. Remains of the little village drowned 50 years before were uncovered. The BBC and other television and radio companies came. The national

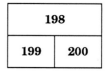

Fig. 198. *Botrydium*. Natural diameter of fully grown plants, 1-2 mm.

Fig. 199. *Botrydium*. Diameter of head 0.9 mm, length of stalk 1.4 mm. Soil partially washed away to reveal the branched filamentous system within it.

Fig. 200. *Botrydium*. Spores (circa 10 µm diam.).

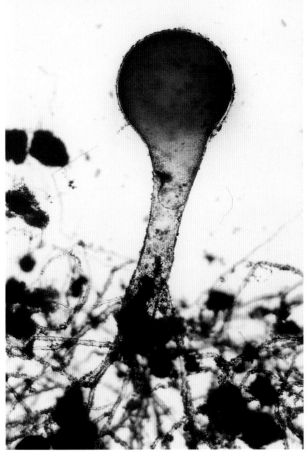
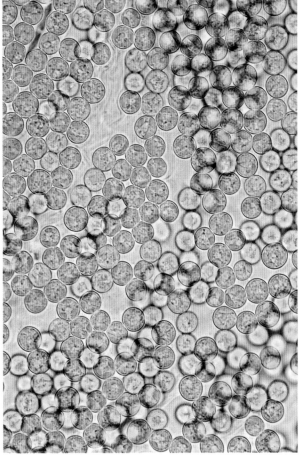

newspapers also took up the story with reports, interviews and pictures. Thousands of people arrived daily in their cars and vans selling ice-cream, soft drinks and other refreshments appeared. Complete traffic chaos reigned until the police closed the road leading to the reservoir. The daily thousands who tramped over the exposed mud little realised that they were also trampling on the untold billions of *Botrydium* vesicles covering it.

The non-motile spores of *Botrydium* (fig. 200) are indistinguishable from the cells of a unicellular Xanthophyte called *Botrydiopsis* which also reproduces by motile or non-motile spores. *Botrydiopsis* is a common soil alga but also can be aquatic. The cells can contain so much oil or fat that in calm weather, or places protected from the wind, they rise to the surface to form a yellowish superficial film or waterbloom (fig. 204).

Fig. 204. *Botrydiopsis* waterbloom.

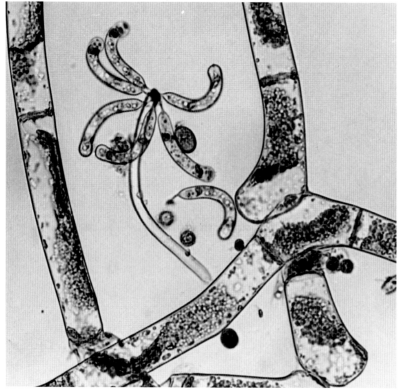

Fig. 201. *Ophiocytium* (× 420), species with curved cells. Also present, filaments of *Mougeotia*.

The cells of most species of *Ophiocytium* (figs 201-203, 205, 206) are cylindrical or club-shaped and straight, curved or twisted. They too reproduce by motile or non-motile spores, released when the cap-like top of the cell comes off. The motile cells often do not swim away but settle around the top of the mother-cell (figs 201, 205, 206).

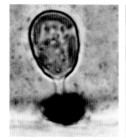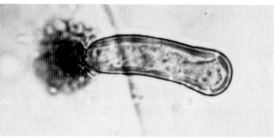

Figs 202 (× 1840) I, 203 (× 1600) *Ophiocytium*. Germlings with holdfasts.

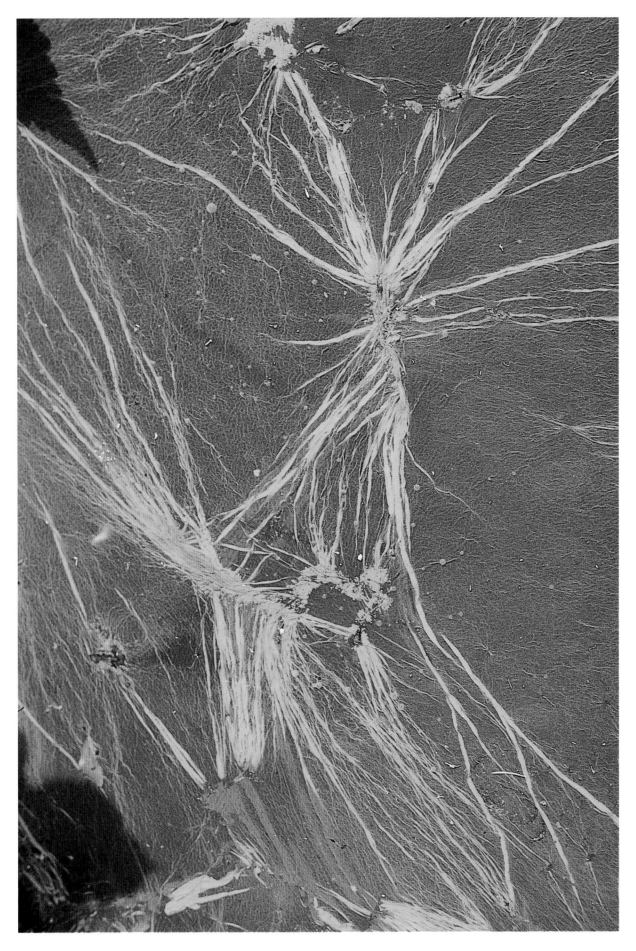

Sometimes a second, and very rarely third or fourth, generations arise on top of one another. However, usually the second generation, like the first, arises from single cells attached to some other substratum. Whatever the substratum, the cells have a short stalk ending in an attachment disc which usually is coloured brown by oxides of iron or manganese. The somewhat swollen ends of the cells in figs 205 and 206 indicate the position of the dehiscence cap.

Ophiocytium is common in small bodies of water and attached to other algae or detritus. Fig. 205 shows a characteristic algal microhabitat containing specimens of *Ophiocytium*.

Fig. 206. *Ophiocytium*. Length of empty mother cell, 90 µm.

Fig. 205. *Ophiocytium* (× 380) among various filamentous algae. In the centre, a cell with its thin, short stalk attached to a filament of *Oedogonium*. At the bottom, two cells, the lefthand one has produced 16 daughter cells.

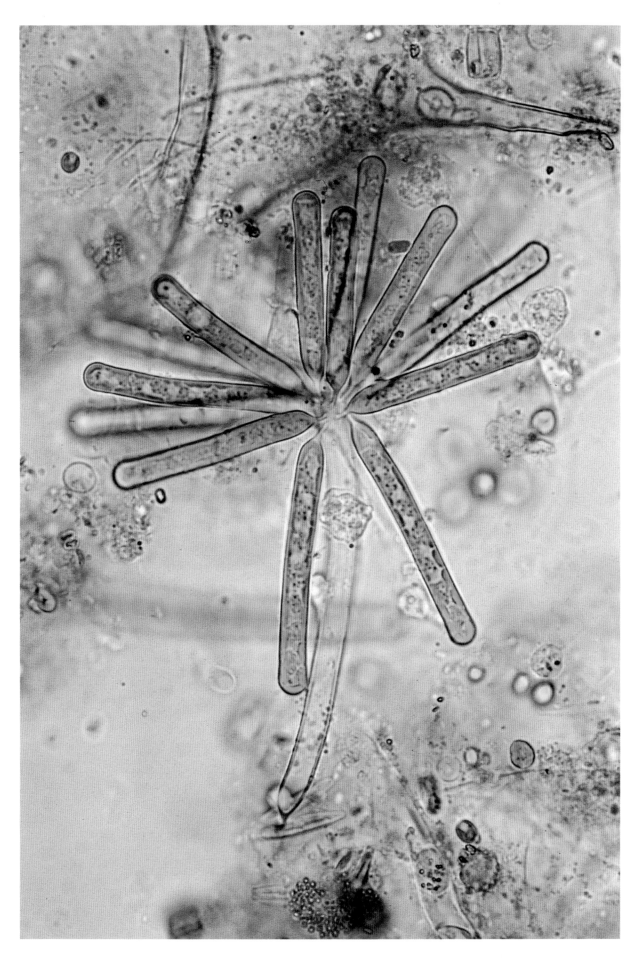

Chapter Five

The Bacillariophytes (Diatoms)

The most striking feature of the diatom cell is the wall enclosing the living contents. The walls of Chlorophytes or Xanthophytes are composed of organic matter (e.g. cellulose and allied substances), those of diatoms are composed of amorphous, opaline silica. Ever since microscopes became generally available, scientists and others have been fascinated by the beauty and intricacy of the architecture and anatomy of these siliceous walls. As microscope lenses improved, so more could be seen, indeed the quality of lenses used to be judged by the degree to which they could reveal the fine markings on the walls of certain diatoms. Though the basic structure of the diatom wall can be seen by light microscopy, the fine structural details are only revealed by electron microscopy. If diatom walls are beautiful when seen by light microscopy, the three-dimensional form revealed by scanning electron microscopy often is stunningly beautiful. Though most genera and species can be identified by light microscopy, some of the species can only be identified by electron microscopy.

The wall consists of two parts which we call the upper and lower valves. The upper half is, like the overlapping lid of a box, wider and longer than the lower. However, the upper half does not fit directly over the lower half. The two halves are joined by a girdle of overlapping connecting pieces, some attached to the upper, some to the lower half of the wall. The overlapping of the two halves of a cell can be seen in fig. 208 though the separate elements of the girdle connecting the two halves cannot be distinguished. This is called the girdle view. The view at right angles to this (fig. 207) is called the valve view. Put another way, when the cell is seen from above or below it is in valve view and when from the side, in girdle view.

The cells of the vast majority of diatoms are yellow or brown in colour. A very few are "colourless", that is they lack chloroplasts and so are non-photosynthetic. At present these colourless species number less than ten amongst the probable 10,000 photosynthetic ones. They are also marine.

We started with the word chloroplast and will continue to use it. The green coloured chlorophylls in the chloroplasts of diatoms are masked by brown pigments. Sometimes, in recently killed cells the underlying green colour can be seen.

The use of the word chloroplast (Greek: *chloros* – the green colour of grass, leaves, meadows etc.) can be

Fig. 207. *Pinnularia*. Valve view (167 µm l.). I.
Fig. 208. *Pinnularia*. Girdle view (131 µm l.). I.

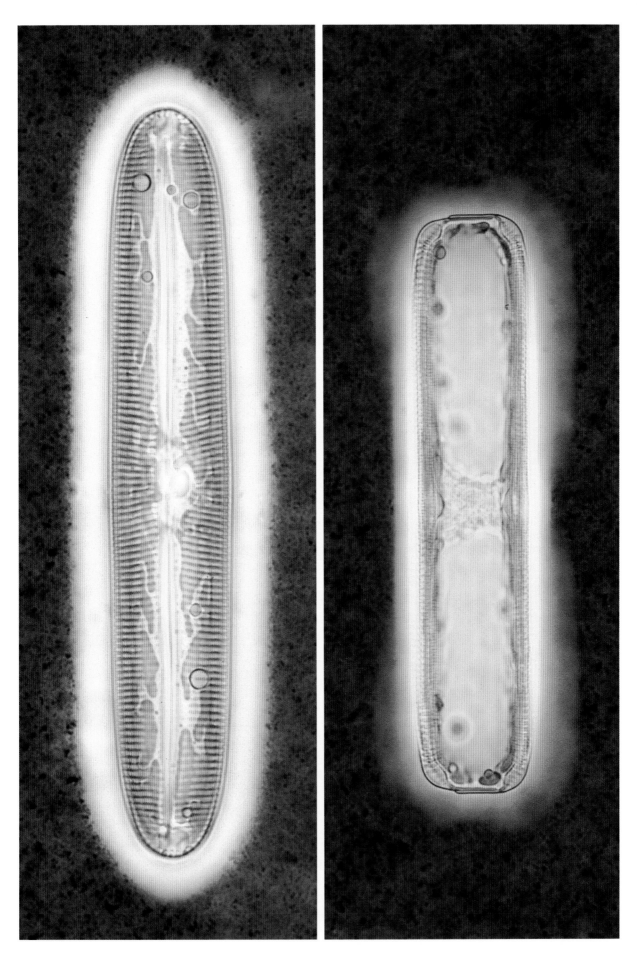

criticised. Commonly, the words chromatophores (Greek: *chroma* – colour) or plastids are used for photosynthetic organelles of any colour or of any other colour than green.

Diatoms live singly or are attached to one another in various ways to form colonies. In some species these colonies are filamentous (figs 210, 211). Like the filamentous colonies of desmids (p. 48), each cell in a chain is an entirely separate entity. Diatom filaments, therefore, are unlike those of *Ulothrix* and certain other Chlorophytes and the Xanthophyte *Tribonema* in which neighbouring cells share a common cross-wall. In other words, diatoms, like desmids, basically are unicellular not multicellular organisms.

There are two major groups of diatoms differing in cell symmetry. In pennate diatoms (e.g. *Pinnularia*, figs 207, 208) the basic cell symmetry is bilateral, whereas in centric diatoms (e.g. *Cyclotella*, figs 209, 213, *Stephanodiscus*, figs 215, 216) it is radial. Based on these bilateral or radial symmetries the cells may have a variety of shapes. In a few cases the cells do not have the usual symmetry of the group to which they belong. This is especially the case in several of the marine genera; the freshwater genera pose fewer problems. However, they can be placed in their correct group on the basis of certain structural details or type of life history.

Most of the taxonomy and so identification of diatoms is based on cell shape and the detailed structure of the siliceous wall to the exclusion of the type of cell content. Long ago, at the turn of the century, it was pointed out by the Russian diatomist Mereschkowsky, that chloroplast shape and structure could be used to distinguish certain diatoms of similar shape and wall structure from one another but, with the vast amount of work done on wall structure, these old observations were forgotten or ignored. However, recently there has been renewed interest in the use of chloroplasts for purposes of taxonomy and so identification. These studies have shown Mereschkowsky to be extremely perceptive and his ideas correct. The chloroplasts of pennate diatoms usually are few in number and elongate (e.g. figs 207, 208), whereas those of centric diatoms are numerous and discoid (e.g. figs 215, 216). Pores and slits occur in the walls of diatoms and are essential for obtaining nutrients and exchanging materials with the surrounding medium. Life would be impossible in a closed opaline box. The siliceous ribs, struts and other parts of the wall can be considered as supporting or strengthening devices.

Cyclotella (figs 209, 213, 214) is a common, mainly planktonic diatom. The species in fig. 209 can be planktonic but often is benthic. In fig. 209, cells can be seen in both valve (circular outline) and girdle (rectangular) views.

Stephanodiscus (figs 215, 216) is another common planktonic genus which differs from *Cyclotella* in the

Fig. 209. *Cyclotella*. Valve (10-17.5 μm diam.) and girdle views.

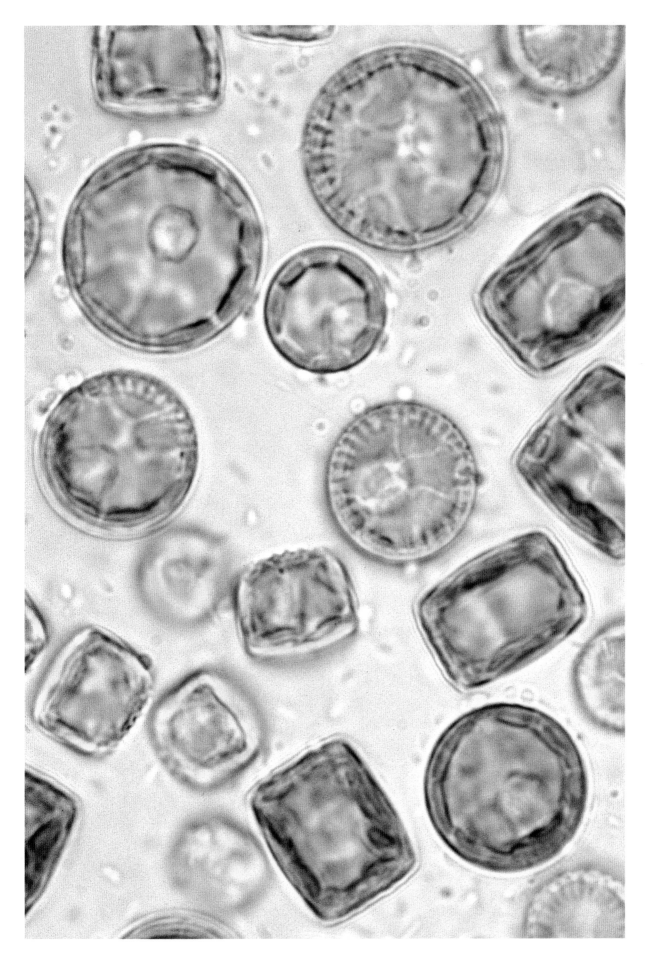

presence of a marginal ring of spines and in other features of the valve markings. In fig. 215 the cell in valve view (lower cell) shows the ring of marginal spines, striations radiating from the central region and the irregular but rounded outline of the discoid chloroplasts (also seen in the upper cell). The upper cell which is in girdle view also shows that the two valves are of different shape. The lefthand valve is convex, as is seen more clearly in fig. 216. The righthand valve appears flat but actually is centrally concave. The two valves of the species of *Cyclotella* seen in fig. 209 are alike, both being slightly convex.

The same cells of *Stephanodiscus* seen in fig. 215 are shown at a more central plane in fig. 216. It can now be seen that the discoid chloroplasts are plano-convex in shape, the plane side being closely adpressed to the cell margin. The round body in the centre of the cell is the nucleus.

Very fine, long threads arise from the margins of both *Cyclotella* (fig. 214) and *Stephanodiscus*. These threads are not siliceous, in the species shown they can be over 150 µm long.

Most of the colonial freshwater centric diatoms are filamentous. Some like the *Ellerbeckia* (fig. 210, previously a species of *Melosira*) form tight filaments and others, like the *Cyclotella* of fig. 211, loose ones. The latter forms filaments of several or even up to about 100 cells at the beginning of a period of increase, although single cells also are present. As the population grows, filaments become shorter and shorter, and fewer and fewer, until virtually the whole population consists of single cells. *Ellerbeckia* always is filamentous.

Fig. 213. *Cyclotella*. Cell (36 µm diam.) wall markings. **Pn**.

Fig. 214. *Cyclotella*. Cells (circa 10 µm diam.) dried to show the fine, non-siliceous threads (setae), the full length of which can be 10-15 times the diameter of a cell. **P**.

Figs 215, 216. *Stephanodiscus*. Below, valve (65 µm diam.) and above, girdle views; cells in fig. 216 at a lower plane of focus (note central nucleus in valve view) than in fig. 215 (note short spines). The groups of minute (c. 1.5 × 0.5 µm) cells in the background are parts of colonies of a Cyanophyte, *Aphanothece* (p. 196).

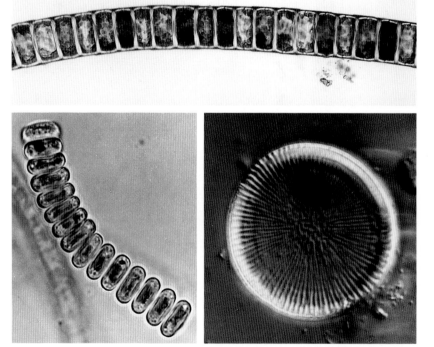

Fig. 210. *Ellerbeckia* (named after the authors' home, Ellerbeck). Girdle view (× 256).

Fig. 211. *Cyclotella*. Girdle view (× 1600).

Fig. 212. *Ellerbeckia*. Valve view (× 1025). **N**.

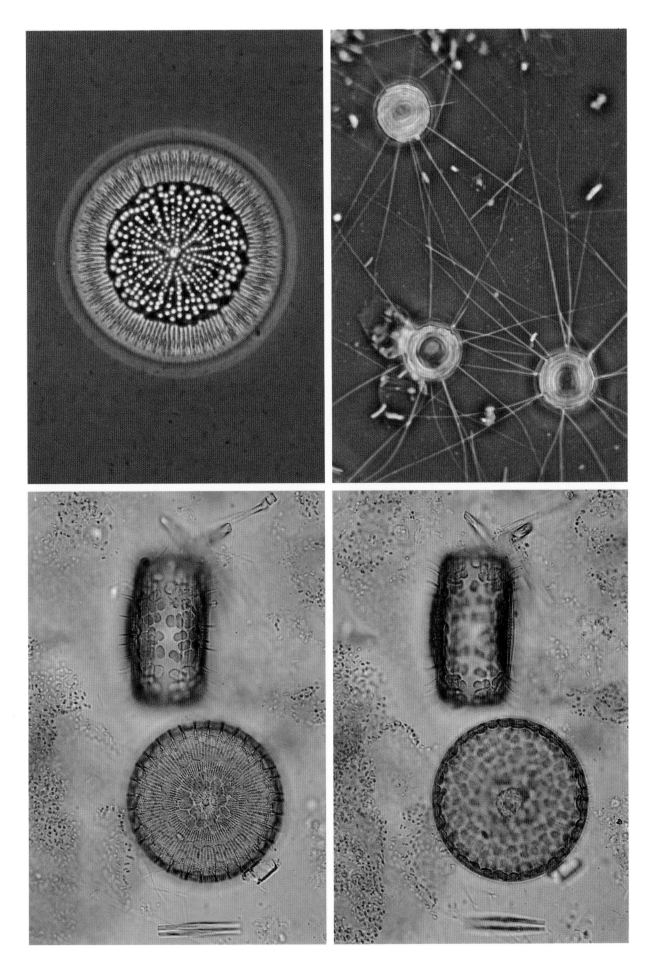

The next sixteen genera shown (figs. 218-240) are pennate diatoms.

Pinnularia (figs 207, 208, 218) is a common diatom living on mud or other substrata. There are about 200 species. In the species shown there is a chloroplast lining the girdle face (fig. 208) of each valve and extending about one third of the way across the valve face (fig. 207). The round bodies present are oil globules. The greyish protoplasm forming a central bridge in fig. 208 harbours the nucleus. The horizontal lines (striations) running down the length of each side of the valve are chambers within the siliceous wall. Seen in the empty cell of another species (fig. 218) each striation has in it a central light oval area. This "light spot" marks an internal opening in the chamber. Such empty cells or walls devoid of the cell contents are needed in order to study the valve markings and so to identify most species. Usually it is necessary to remove all the organic matter by chemical treatment in order to obtain such cells or parts thereof free of living or dead cell contents.

When a *Pinnularia* is looked at in valve view two centrally placed lines, one originating from each end of the valve, converge towards the centre of the diatom cell where they end in a pinhead-like swelling. These lines can be seen faintly in fig. 207 and clearly in fig. 218 as well as in the genera depicted in figs 219-221. They form the main part of a structure called a raphe which involves slits in the silica wall. The raphes in figs 218-221 all run down the centre of the long axis of the valve.

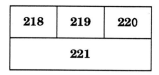

Fig. 218. *Pinnularia*. Valve view, only part shown. Width at centre 30 μm. In this and figs 219-221, the organic cell contents have been removed to show wall markings and raphes more clearly. **Pn**.

Fig. 219. *Didymosphenia*. Valve view (125 μm l.). **P**.

Fig. 220. *Didymosphenia*. Central part of valve (40 μm br.) at a higher magnification. **Pn**.

Fig. 221. *Stauroneis*. Central part of valve (44 μm br.). **Pn**.

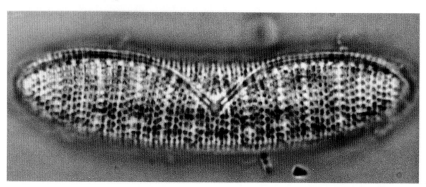

Fig. 217. *Epithemia*. Valve view, note elongate V-shape of raphe (× 1600).

In other genera they are more-or-less markedly, even wholly, displaced towards the margins, as in the *Epithemia* shown (fig. 217). These displacements may be accompanied by torsions or other changes in the simple "box-like" structure of the cell.

Raphes are only present in certain genera of the pennate diatoms, they are absent in centric diatoms. Some pennate diatoms have a raphe on each of the two valve faces as in *Pinnularia*, others only on one valve. Diatoms possessing a raphe can move smoothly or jerkily over a surface.

Since movement of all but one or two diatoms is restricted to species with a raphe, it must be caused

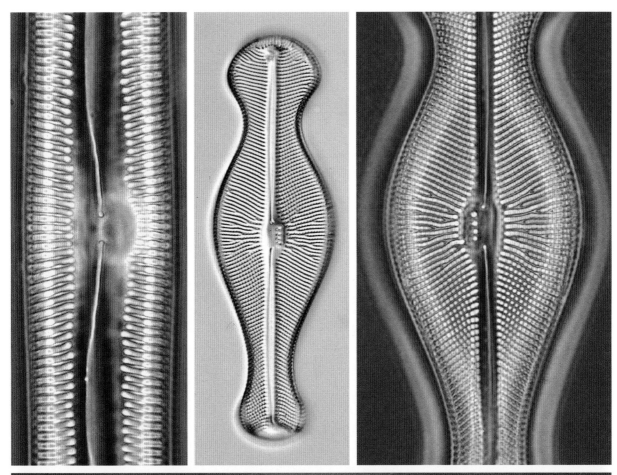
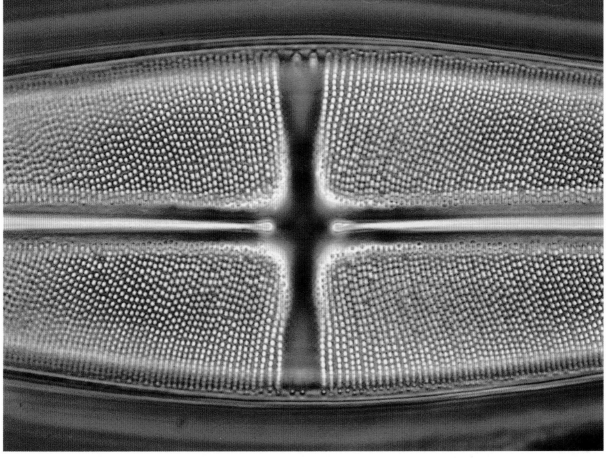

directly or indirectly by the protoplasm in this fissure but it is still uncertain exactly how movement is accomplished. Like some desmids, a very small number of diatoms move by means of the excretion of mucilage. This movement is very slight compared to that of diatoms with raphes.

The raphes and valve markings of *Didymosphenia* and *Stauroneis* are seen in figs 219, 220 and 221 respectively. *Stauroneis* is common in similar habitats to *Pinnularia*. *Didymosphenia*, as will be seen later (fig. 257), lives attached to a surface by a long stalk. It favours rocky places where there is plenty of water movement, such as turbulent rivers and the wave washed shores of lakes. Large numbers of cells occur together forming greyish masses which by eye can be mistaken for some other kind of organism.

Surirella (figs 223, 224) has boat-like cells. They may be wider at one end than the other and the wall can be developed into prominent strutted wings (fig. 224).

Gyrosigma (fig. 225) has more or less markedly S-shaped cells. Species of *Gyrosigma* and *Surirella* are common on soft deposits such as fine silt or mud.

Cymbella (figs 226, 227) can be free-living but usually is attached to something, commonly by a mucilage stalk or the cells live in a tube of mucilage (fig. 227). The tube-living species shown has at various time been included in a separate genus, or subgenus and because it has other characteristic features apart from living in tubes the old name *Encyonema* has been resurrected.

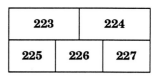

Fig. 223. *Surirella*. Valve view (169 μm l.).

Fig. 224. *Surirella*. Girdle view (158 μm l.).

Fig. 225. *Gyrosigma*. Valve view (132 μm l.). N.

Fig. 226. *Cymbella*. Valve view (166 μm l.).

Fig. 227. *Cymbella* (*Encyonema*) cells living in a mucilage tube (valves to 34 μm l.). N.

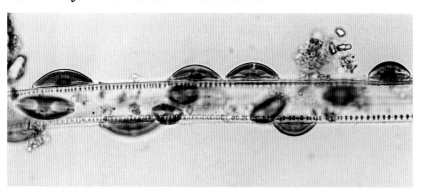

Fig. 222. *Amphora*, on a dead cell of *Nitzschia* (× 640).

Amphora (figs 222, 379). The valves are so arched (like a gable) and the raphe system so placed that the cell usually is seen in girdle view together with the parts of the two valves bearing the raphes. However, these characteristic generic features cannot be seen in the living cells shown at the focal plane and magnification concerned. *Amphora* is common on sand grains, stones, plants and other algae. The species shown in fig. 222 characteristically lives on a certain species of *Nitzschia* (*N. sigmoidea*), although it is sometimes found on other large diatoms (e.g. *Surirella*). Incidentally, *Nitzschia* is a genus in which the raphe is displaced from the central position referred to earlier. The markings along the edges of the *Nitzschia* cell show the position of the raphes.

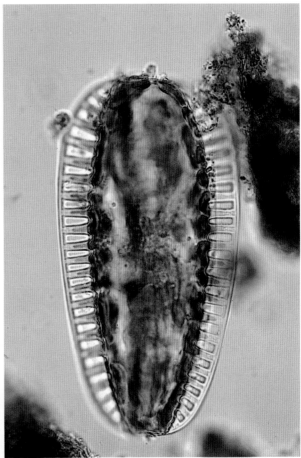
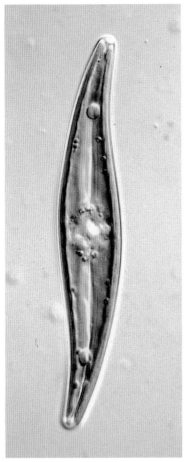
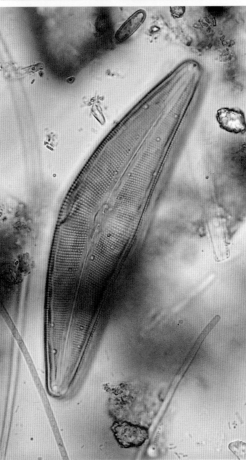
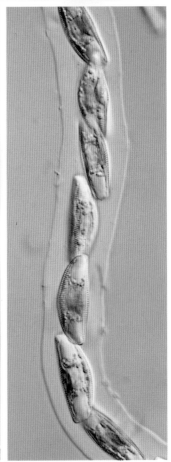

Synedra species (fig. 229) often are attached to a surface by a blob of mucilage, here to a filament of *Mougeotia* (p. 58). In girdle view the cells are rectangular but in valve view usually are fusiform (spindle-shaped) in outline. Unlike the previous pennate diatoms, the cells of *Synedra* do not have raphes and so cannot move.

Gomphonema (figs 228, 230, 231) cells have raphes on both valves but spend most of their lives attached to something either by a mucilage pad (fig. 230) or by a stalk (fig. 231) which can be branched to form a tree (fig. 228). The species of *Gomphonema* in fig. 230 is attached to a filament of *Vaucheria* (p. 108) and that in fig. 231 to *Oedogonium* (p. 62). The oval bodies in the *Vaucheria* are chloroplasts but those in the unhealthy cells of *Oedogonium* are starch grains. *Gomphonema* cells are wider at one end than at the other and wedge-shaped in girdle view.

Achnanthes (fig. 232) has cells which are bow-shaped (arcuate) in girdle view. The fine mucilage stalks by which this species is attached to an unidentified algal filament have been made clear by staining them. *Achnanthes* belongs to a group of pennate diatoms in which only one of the valves of the cell has a raphe.

Fig. 229. *Synedra*. Cells (to 138 μm l.) attached to a filament of *Mougeotia*.

Fig. 230. *Gomphonema*. Cells (to 41 μm l.) attached to a filament of *Vaucheria*.

Fig. 231. *Gomphonema*. Note stalks. Cells attached by stalks to a moribund filament of *Oedogonium*. Paired cells 34 μm l.

Fig. 232. *Achnanthes*. Cells attached to unidentifiable algal filament; stalks stained mauve by the dye brilliant cresyl blue (cells circa 20 μm l.).

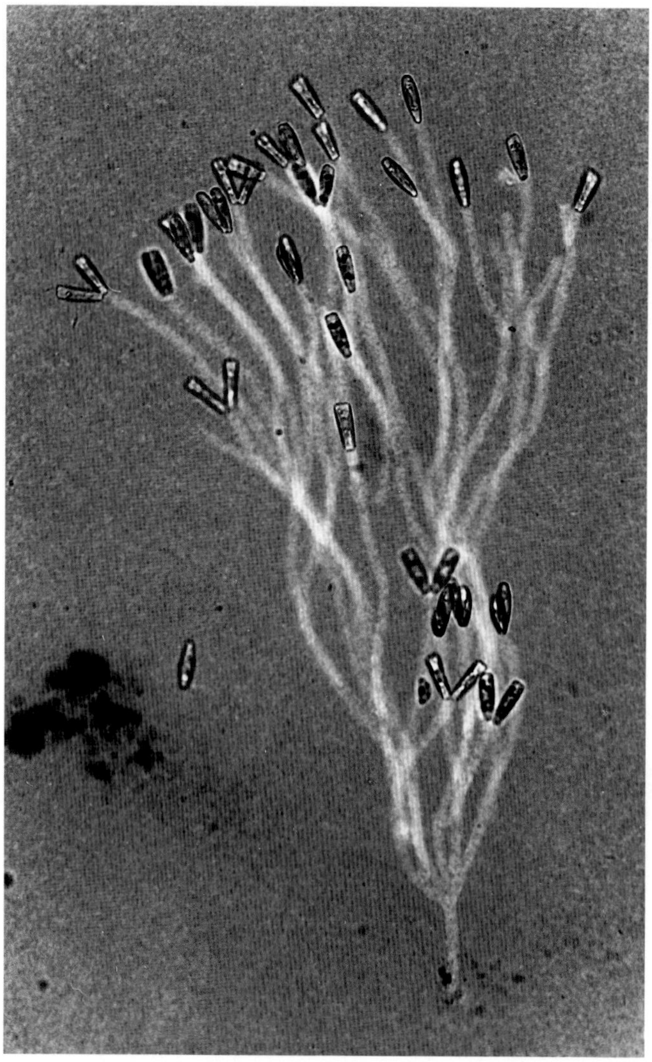

Fig. 228. *Gomphonema*. Tree-like colony of cells on long stalks (× 256). I.

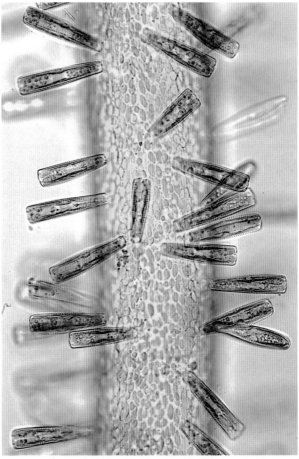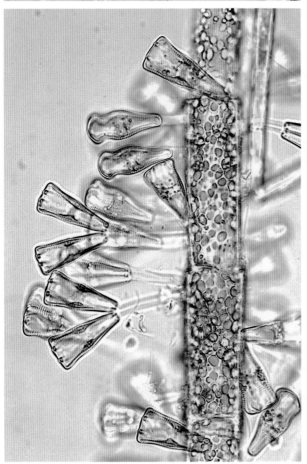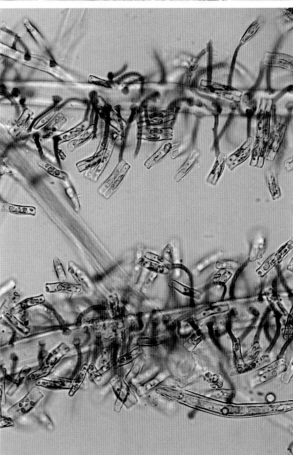

Nitzschia (figs 222, 235) is the second largest genus of freshwater diatoms as well as being common in the sea. Nearly all the species live as solitary individuals in benthic habitats, only a minority being planktonic. The cells of the small species shown (fig. 235) sometimes group together to form stellate colonies.

Meridion forms fan-shaped colonies (fig. 236). The cells lack raphes and are wider at one end than at the other.

In *Navicula* (fig. 237) most of the species are more or less boat-shaped and virtually all of them live on but not attached to a surface. The picture shows a rich population living on mud. Also present are threads of the Cyanophyte *Oscillatoria* (see p. 222). Both valves have raphes. The genus *Navicula* used to consist of some 2000 species but recently it has begun to be split up by placing certain groups of species into separate genera. A major reason for so doing is the view that its taxonomy should not be based solely on the form and structure of the siliceous wall. As a result, certain old generic names, which fell into disuse when what might be called silicon taxonomy reigned preeminent, have come back into use. This is not the reason why the *Navicula* in figs 233 and 234 has recently been placed in an old, but recently resurrected, genus called *Craticula*. The distinction between *Navicula* and *Craticula* lies in certain anatomical and structural features. The figures show the two chloroplasts which line the girdle sides of many naviculoid diatoms. Since figs 232 and 233 are valve views, only the upper or lower edges of the chloroplasts are seen. Also visible, across the centre of the cell, is the protoplasmic bridge which contains the nucleus. The two large oil globules to each side of the nuclear bridge are nearly always present in this, the commonest freshwater species of *Craticula*.

Fig. 235. *Nitzschia*. Stellate colony (57 μm diam.).

Fig. 236. *Meridion*. Curved (fan-shaped) colonies (cells 12-19 μm l.).

Fig. 237. *Navicula*. A rich population moving over mud (cells 65 μm l.). Filaments of *Oscillatoria* also present.

Figs 233, 234. *Craticula* (× 1065).

Fig. 233. At level of protoplasmic bridge containing the nucleus.

Fig. 234. At level of the raphe slits.

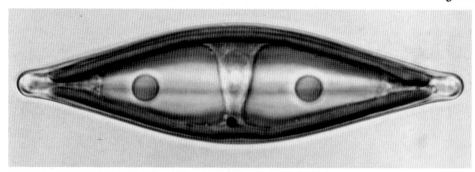

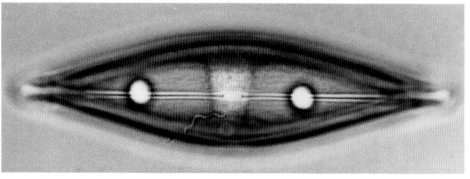

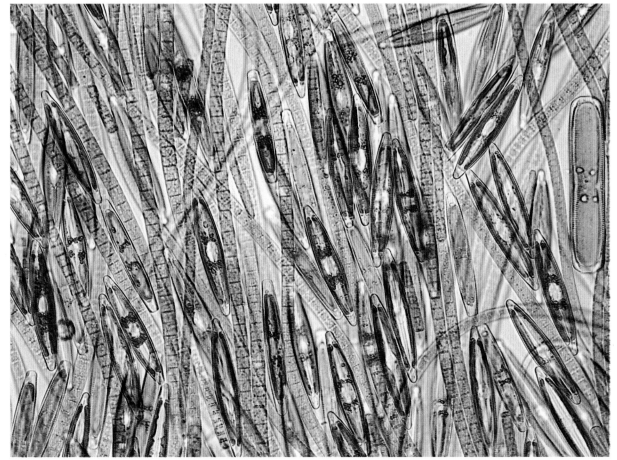

The pennate diatom genera so far depicted consist wholly or predominantly of forms living on or attached to a surface, though a large number can move over the substratum. The next group of species is planktonic. Something has been said about plankton earlier (p. 14). All freshwater diatoms have a positive rate of sinking. The shape of many of them may be considered as adaptations to ensure that their rate of sinking will be slow. The *Cyclotella* in fig. 209 and the *Stephanodiscus* in figs 215, 216 are planktonic. In the pictures, the cells appear to be without any structural feature which would reduce the rate of sinking of its short cylindrical body. However, there are very fine threads extending (fig. 214) from the margin of the cells, a feature common to many centric diatoms of similar shape. These threads are not composed of silica but of organic matter and they can be removed without harming the cell. Such threads on cells of a species in an allied genus were found to decrease the rate of sinking of cells by about 50% compared to those cells from which these threads had been removed.

Tabellaria contains species which are attached to or entangled among underwater objects (fig. 238) and those which are planktonic. The planktonic species shown (fig. 239) has stellate, parachute-like colonies which can be considered as an adaption to reduce the rate of sinking.

Fig. 238. *Tabellaria*. Zig-zag chain of a usually non-planktonic species (×460).

Asterionella (figs 240, 246-249) has similar colonies to those of the *Tabellaria* in fig. 239. Also, like *Tabellaria*, sometimes the cells are not in stellate groups but joined end to end in zig-zag colonies. The colourless blobs of mucilage joining one cell to the next can be seen in figs 247, 248.

Asterionella is one of the commonest planktonic algae. It is a nuisance in many reservoirs because it is a very effective blocker of filters. In very large numbers and so often when concentrated it gives off a smell of geraniums.

Asterionella was discovered as a result of the cholera, typhoid and other waterborne epidemics in London during the last century. The cause was the contamination of the public water supplies with faecal and other foul matter. A doctor, A. H. Hassall, made pioneering studies on London's water supplies, in the course of which he made microscopical observations and encountered the little star which he called *Asterionella formosa*. The words *Asterionella* and *formosa* together mean little star of beautiful form, hence it can be said that Hassall found beauty amongst filth.

Fig. 239. *Tabellaria*. Eight-celled stellate planktonic colony (106 μm diam.). **P.**

Fig. 240. *Asterionella*. If the cell at 7 o'clock reproduced, it would start a second ring of cells at a higher level than the first. Central space circa 8 μm. **P.**

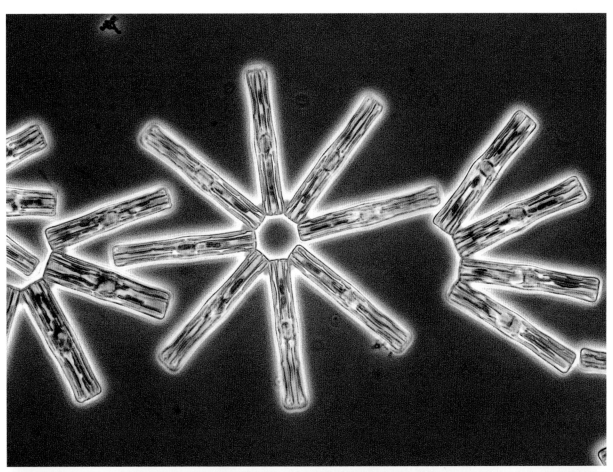
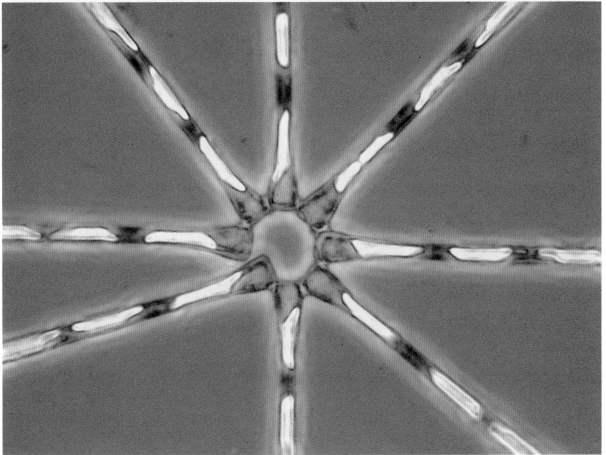

Fragilaria (fig. 243) also contains both planktonic and non-planktonic species. The species shown is a common plankton diatom. Its elongate cells are joined centrally to form colonies like double-sided combs.

The *Synedra* illustrated in fig. 244 has cells of similar shape to those of the *Fragilaria*. The cells are solitary except for a time after cell division when they are in pairs. More elongation of a given cell shape will retard the rate of sinking. If, for example, the cells of the *Cyclotella* (fig. 209) were pulled out into the long cylindrical form seen in the *Fragilaria* and *Synedra*, then it is reasonable to assume that the rate of sinking would be less than that of its actual box-like cells, irrespective of the added influence of the presence or absence of the threads mentioned earlier.

A few planktonic species of *Synedra*, have much longer and finer rod-shaped cells than the one shown, reaching half a millimetre or more in length. Like the two previous planktonic pennate diatoms *Tabellaria* and *Fragilaria*, some species of *Synedra*, as has been seen (fig. 229), live attached to a surface.

Rhizosolenia (figs 241, 242, 245) is a centric diatom which has an elongated cylindrical cell with a spine on each valve. The main length of the cell is provided by the girdle side. The chloroplast is small and occupies only the central region of the cell. It is one of the three or four diatoms whose cells contain contractile vacuoles. If contractile vacuoles have the function suggested on p. 82, then the thousands of other diatoms have another means of controlling their water and salt relations.

Fig. 243. *Fragilaria*. Length of colony, 103 µm. P.

Fig. 244. *Synedra*. Longest cell, 172 µm. P.

Fig. 245. *Rhizosolenia* (*Urosolenia*). Length of cell, 162 µm. P.

Figs 241, 242. *Rhizosolenia* (*Urosolenia*) (× 1600). P. In fig. 241, a contractile vacuole (at 12 o'clock – arrow) is visible. It then contracts and becomes invisible (fig. 242).

In fig. 241, one of these contractile vacuoles (indicated by the small white spherical area arrowed), can be seen in the "open" position. In fig. 242, it has closed and so now has "disappeared" from view.

The few freshwater species of *Rhizosolenia* recently have been split off from the many marine ones and given the name *Urosolenia*.

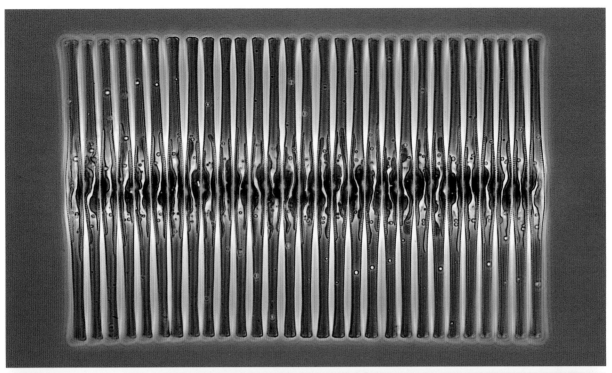
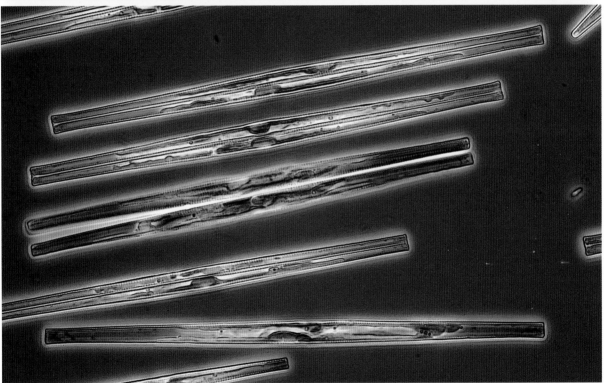
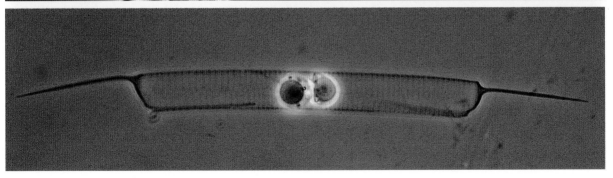

Cell size

Figs 246-249 illustrate a situation common to all but a very few diatoms, namely reduction in size. The kind of rigid siliceous box in which most diatoms live causes the average size of the cells in a population to decrease unless the normal course of multiplication can be altered. In the next section the common method for reversing this diminution in size is shown. To understand why the mean size of the cells in a population will decrease, so long as the usual process of cell division continues, we can, for simplicity, forget that the two cell valves are joined together by girdle bands (p. 118) and consider the cell to be a simple box, as it appears to be in figs 207 and 208.

Multiplication is by binary fission. Of the two daughter cells so produced, one keeps the wall of the larger valve, that is the top of the box and forms a new wall to fit inside it. Hence, this new cell is the same as the cell it arose from i.e. the mother cell. The other daughter cell arising from this binary fission retains the wall of the smaller valve of the mother cell and forms a new wall which fits within it like the bottom of the original "box". Now, we have two new cells, one the same size as the mother cell and one smaller than the mother cell.

When these two daughter cells divide, three of the four cells now present will be smaller than the fourth which retains the size of the original mother cell. Moreover, one of the cells of this second generation will be smaller than the smaller of the two cells of the first generation. As cell multiplication continues so more and more cells smaller than the original cell will arise and the smallest cells will become ever smaller. For example, by the tenth generation there are 1024 cells (the first generation has two cells, the second four and so on). Only one of these 1024 cells is the same size as the cell originating this population and there will be various numbers of cells of ever decreasing size. If this process continues, cells of such small size will arise that they cannot continue to grow normally or at all. The small cells of *Asterionella* in fig. 248 have not reached such a small size. They can grow as fast as the larger cells. Curiously, despite their viability, such small cells are rarely seen in nature though almost always produced in populations cultivated in the laboratory. One reason for this, though not an explanation for their absence in nature, is that such cultured populations do not have a method of arresting the diminution of size. Since *Asterionella* must have such a method in order to exist in nature, it would be true to say that the present methods of culture are imperfect because they do not permit the production of the kinds of cells which can return the diatom to its maximum size. Such cells are called auxospores (see later).

246	247
248	249

Figs 246-249. *Asterionella*. P.

Fig. 246. Note yellowish-brown chloroplasts and central nuclear area (appearing coloured dark-blue). Cells 75 µm l.

Fig. 247. Colony of shorter cells (22 µm l.).

Fig. 248. Colony of very short cells (7 µm l.) and one of "normal" size.

Fig. 249. Upper colony with four short cells consequent on a misdivision. Normally produced cells 62 µm l. Note that the lower colony has two superposed "rings" of cells, actually it is a helix.

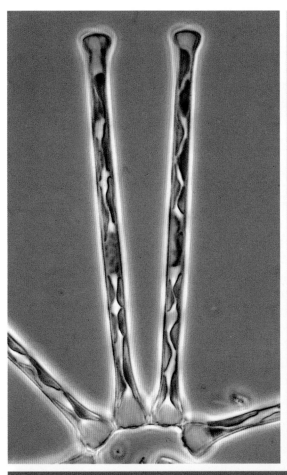
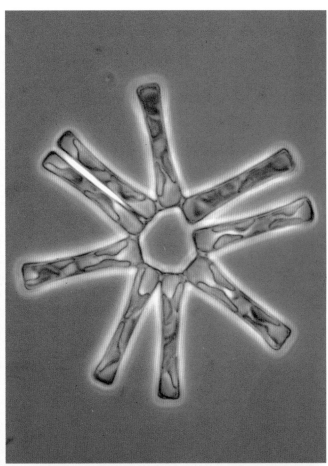
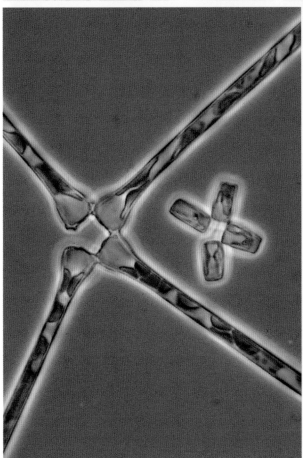
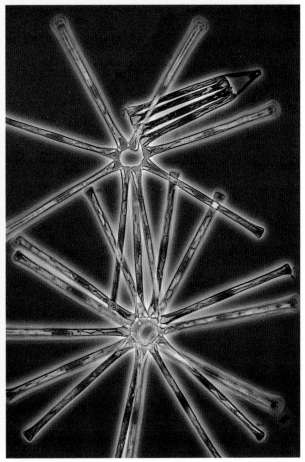

137

In any one cell division the smaller of the two daughter cells will only be shorter and thinner than the other by a fraction of one micrometre. Consequently, these two cells appear to be the same size. It follows that, provided reproduction takes place normally, differences in cell sizes in a population arising from a single cell will only become noticeable a considerable number of generations later. For example, the cells in a single colony normally will all appear to be of the same size (see figs 210, 239, 243, 246, 247) because they are the product of too few generations for reductions in size to be visible.

In the upper colony of *Asterionella* in fig. 249 all but four of the cells follow the system of appearing to be the same length. The group of four cells at 1 o'clock are markedly shorter than the rest. This "sudden" shortening has arisen due to faulty division, a condition which can occasionally occur in diatoms. It is another curious fact that such faulty divisions are very rarely seen in natural populations, as is illustrated in fig. 249, but are common in cultured ones.

Asterionella is a very common diatom and has so often been studied that many millions of colonies must have been observed. However, as yet, nobody has seen how this diatom overcomes the problem of decreasing size. An increase in the average cell size of a population at certain times has been recorded by several workers. This increase has been produced by the appearance of longer cells than were previously present. It thus seems likely that cell enlargement is also by some kind of auxospore formation.

Auxospores and sexual reproduction.

In fig. 250 there are two apparently swollen cells in one of the filaments of the centric diatom *Melosira*. What has actually happened is that the cell contents have enlarged and pushed apart the two halves of the siliceous wall. These cells, the auxospores, will produce a new filament of large cells. In fig. 251 the righthand filament is close to or within the size range in which auxospores may be formed. Below or above this range of size, auxospores are not formed. Hence, cells which have reached a size below this range are doomed to form yet smaller cells until some of their descendants are too small to continue to exist. The large-celled filaments in figs 251, 252 have recently arisen from auxospores. The cell at the top of the lefthand filament (fig. 251) has a teat-like tip lying within a colourless cap. The latter is one valve of the cell which produced an auxospore which, in turn, gave rise to this large celled filament. The post-auxospore filament (fig. 252) shows the intricate pattern of the chloroplasts of this *Melosira* species.

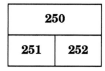

Figs 250-252. *Melosira.*

Fig. 250. Filament (14 μm br.) with two auxospores (37 μm br.).

Fig. 251. The wide filament (34 μm br.) has recently developed from an auxospore. Note "cap" (valve of pre-auxospore cell) on teat end of filament.

Fig. 252. Another filament (34 μm br.) developed after formation of an auxospore. Note shapes of chloroplasts in surface view.

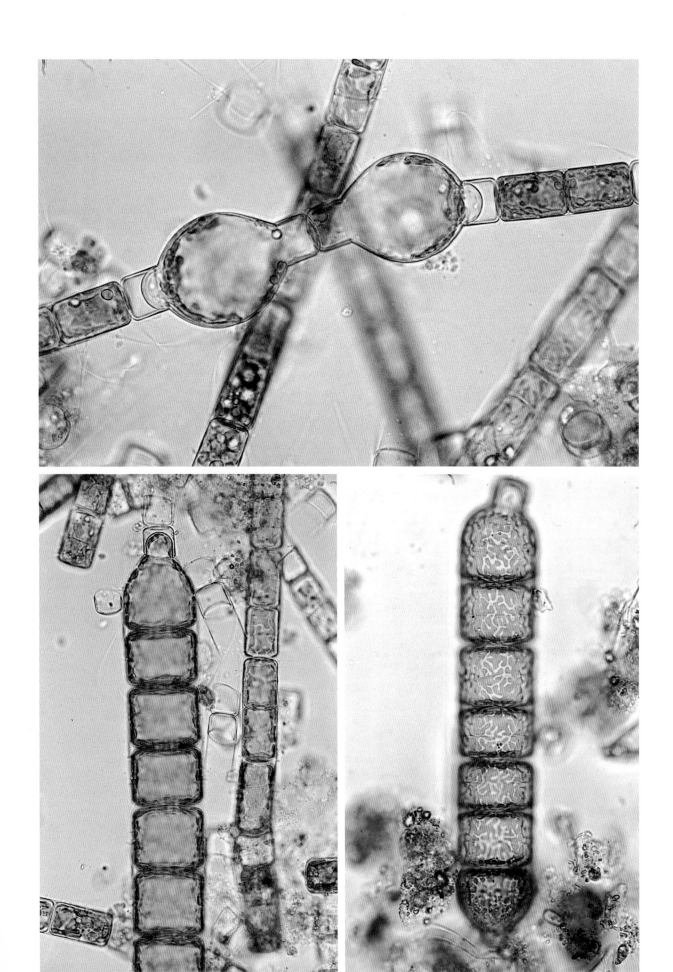

139

In fig. 255 there are filaments of various diameters belonging to another centric diatom, *Aulacoseira* (previously called *Melosira*). In fig. 256 a small filament of this diatom has a large cell on its end. This has arisen from an auxospore and will produce a wide-celled filament. Auxospores are at first "naked" in that they are not enclosed in a siliceous "box". The auxospores of the *Melosira* and *Aulacoseira* are spherical and before they divide to initiate a new filament of maximal size they surround themselves with a siliceous wall. In fig. 256, part of the markings on the silicified wall can be seen to the right and left of where it is still attached to one valve of the cell from which the auxospore arose. In centric diatoms, auxospore formation usually is a consequence of a sexual union. A minute sperm enters a female cell through an already present or specially produced gap in its siliceous wall. The fertilised cell, that is the zygote, then enlarges into an auxospore. The fact that a few bilaterally symmetrical supposed pennate diatoms are in fact centric ones became clear when their method of sexual reproduction was discovered.

Pennate diatoms do not produce eggs and sperms though they have a variety of types of sexual reproduction connected with the production of auxospores. Also, like some centric diatoms, auxospores are formed in some species without sex being involved. In fig. 257 two cells of *Didymosphenia* (p. 126) are paired within a mucilage envelope prior to sexual reproduction. The cells always pair "head-to-tail". Cells of *Didymosphenia* usually live attached to a substrate by a mucilage stalk, though the cells can free themselves and move. Since, in *Didymosphenia*, one gamete is on a stalk and the other is not, it seems that the latter actively seeks out the former.

Each cell in figs 253, 254 has produced two auxospores. No gametes are visible, hence auxospore formation seems to be asexual. However, this might not be so, because some pennate diatoms have an extreme form of interbreeding in which division of the protoplast or nuclei within a cell is followed by sexual union. Such sex does cause some redistribution of genes but no new genetic material is present in the zygote. When two separate individuals are gametes, their union mixes two separate packs of genes, i.e. it is crossbreeding. Without knowledge of what happened before these auxospores arose, it is impossible to say if sex was or was not involved.

Fig. 255. *Melosira* (*Aulacoseira*). Filament widths 5-10 µm.

Fig. 256. *Melosira* (*Aulacoseira*). Recently formed auxospore (15 µm br.) with partial formation of a silicified wall.

Fig. 257. *Didymosphenia*. Long (109 µm l.) and short (78 µm l.) cells, paired prior to sexual reproduction. Note, stalk of larger cell. **I.**

Figs 253, 254. Auxospores of a pennate diatom (? *Cymbella*) and beside them the valves of the cell from which they have arisen Both × 640. **N.**

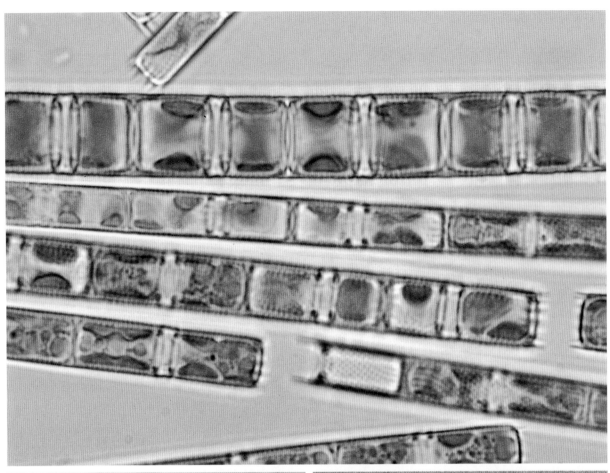
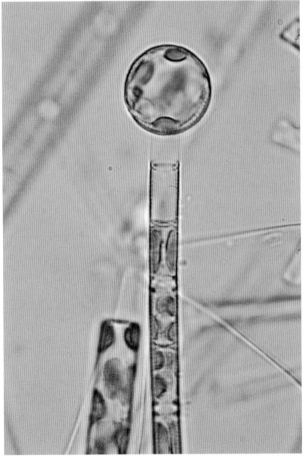
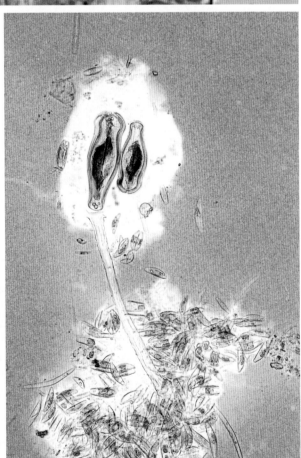

Fossil diatoms, industrial uses, drowning, artistry.

The silica wall of diatoms is sparingly soluble except in highly alkaline solutions such as those of sodium or potassium hydroxide or, among acids, in hydrofluoric acid. Hence, their remains can accumulate on the bottoms of lakes and seas. Eventually, after thousands or millions of years these diatomaceous deposits may become land when a waterbody fills up or is raised up by earth movements. The "mountains" in the distance in fig. 258 in fact are deposits of fossil diatoms present in Washington State, U.S.A. Such large deposits usually are of marine origin (fig. 259). Apart from the large amount of marine diatomaceous deposits which have been raised up in geological time, there are vast expanses of diatomaceous ooze on the bottom of some parts of the oceans which one day may become land. Some idea of the amount of fossil remains of diatoms which can be preserved in the bottom deposits of lakes can be given by estimating the production of the major planktonic diatoms in Windermere. Though this is England's largest lake (volume 3.14×10^8 m^3), it is not large on a world scale. During the last 50 years some 20,000 tons of diatom silica have sedimented to the bottom of the lake. The major part of this is made up of the remains of *Asterionella*. The vastness of the number of siliceous walls of *Asterionella* preserved in the deposits can be realised by the fact that an average-sized teacup (200 ml) full of Windermere water supports the growth of some two million cells of *Asterionella* each year, most of which eventually will enter its sediments. Some cells of course are washed out of the lake especially during periods of heavy rainfall and flooding. At a conservative estimate, over a million metric tons of diatom silica have been incorporated into Windermere's bottom deposits since the end of the last ice age.

Diatom silica is not absolutely insoluble and, for example, in sinking through great ocean depths all but the most heavily silicified walls may be dissolved. Mechanical damage or breakage too will increase the rate of loss because of the larger surface exposed to solution.

The purer the diatomaceous deposit, the whiter its colour (e.g. fig. 258); many freshwater deposits are brownish, reddish or greyish in colour. The common name for diatom rich material is diatomite, though the older German name Kieselguhr may be used sometimes. Diatomite is an important industrial product. In 1988 about two million tons were used. The USA was the largest producer. The known reserves of diatomite amount to about two billion tons so that there is unlikely to be any shortage in the foreseeable future. In 1988 about 69% of the world's diatomite production was used for filtration processes and 16% as fillers. The value of diatomite lies in its resistance to heat and chemicals, its porosity and

Fig. 258. Professor W. T. Edmondson of the University of Washington in front of diatomite "mountains" (Washington State, U.S.A.).

Fig. 259. Cells from a marine commercial diatomite (largest cell 325 μm diam.).

lightness. Though the specific gravity of diatomaceous silica is over twice that of water, its porosity gives it a high surface area to weight ratio. The large surface area makes it a valuable absorbent material. Apart from how pure the diatomite is, its component species with their varying structure affect the uses for which a given deposit is suitable. Roughly, low grade diatomite is mainly suitable for fire or heat resistant purposes or as a filler. Higher grade material can be used for filtration, absorption and various chemical processes. Diatomite has been used for a long time. The Greeks and Romans used it as a building material, for example for the dome of St. Sophia in what was then Constantinople. Later, goldsmiths and silversmiths used it in secret formulae for polishes and it has been used for this purpose in toothpowders which now have been replaced by toothpaste. It was also added as an adulterant to flour in times of shortage. In more recent times, Alfred Nobel of Nobel peace prize fame, used it to stabilize the nitroglycerine in dynamite. This was a commercially important advance, for previously dynamite had an unfortunate propensity to explode before it was meant so to do. Diatomite no longer is used for this purpose.

The major uses of diatomite for some time have been and still are for clarification and filtration of substances such as beer, wine and sugar, for resistance to and insulation from heat and fire in furnaces; in lightweight constructional materials such as bricks and blocks; as a filler in paints and lacquers; for polishing metal and as an absorbent of gases or noxious materials. A survey we made of the patent literature between 1973 and 1986 gives some idea of recent and future trends in the use of diatomite. 29% of the patents were for chemical processes (largely catalytic ones); 12% for special cements, concrete and other constructional materials; 10% for thermal insulation, 8% for paper, plastic sheets and board; 7% for absorbent materials, 7% for filtration processes and 6% for agriculture, mainly to prevent caking in fertilisers and so to assure their even spreading in use.

If it is not obvious but possible that someone has died from drowning, then the presence of diatoms in the lungs of a corpse can produce valuable evidence. Some caution however, is necessary for diatoms have been found in the lungs and other tissues of people who have not died from drowning. Those who work with diatomite are particularly likely to have diatoms in their tissues. They also are in the same danger as those exposed to other siliceous particles which are easily blown about (e.g. asbestos).

The love of microscopists for diatoms has led to such artistry as the floral arrangement seen in fig. 260. Bearing in mind that a pattern such as this is produced by selecting, separating and pushing the diatoms into place under a microscope, one can only marvel at the patience,

Fig. 260. Artistic arrangement of diatoms (diam. of "flower" 537 µm) by Klaus Kemp. **D.**

dexterity and steadiness of hand that can produce such artistry.

In spite of the resistance of the walls of diatoms to dissolution, they have the relatively short fossil history of about 150 million years. It is thought that they existed long before this although no traces have been found. If this surmise is correct, it is a surprising situation because fossils of relatively delicate algae without siliceous walls are known from much earlier times.

Diatoms as environmental indicators

Freshwater diatoms are important indicators of the nature of the water in which they are found and consequently their remains in the sediments can be used to indicate past conditions in lakes. Further, the nature of the condition in a lake is determined by its surroundings. Therefore, diatoms and other fossils can be used to deduce past changes in the environment, for example changes in sea level, since fresh, brackish and sea waters have different types of diatom population.

They can also be used to indicate past changes in the nutrient status of a lake. The diatoms that dominate the plankton of nutrient-poor (oligotrophic) waters are different in kind, number or both together from those typical of nutrient-rich (eutrophic) lakes. Hence, the courses of natural or man-made eutrophication (p. 12) can be followed.

The vast majority of lakes in the world were formed or reformed during the last ice age. The movement of the ice scoured out the bottoms of previous lake basins or produced new lake basins so that when the ice age ended millions of new lakes arose. The ice age ended some 10 to 15,000 years ago, the time depending on how far north or at what altitude the area was. The diatoms and pollen in the deposits of postglacial lakes tell us much about how the environment has changed since then. Diatoms are important in elucidating the acidification of lakes by rain. The effects of the industrial revolution on many buildings, notably those made of limestone, have long been obvious. In more recent times the effects of this industrially produced acid rain on lakes and rivers, especially on fish and their food organisms, has caused great concern. The onset and development of the acidity of these waters can be traced by studying the changes in the diatom flora in cores taken from them. Diatom populations are affected by the pH of the water, some only grow well in acid waters, others in alkaline ones and yet others show no clear relationship to the pH of natural waters. Using this knowledge, the past pH variations in the water of a lake can be determined within quite a small margin of error.

In order to use the remains of diatoms and other organisms in studying the past, there must be some method of dating that past. In dating lake deposits and so the diatoms within them, the methods used include the decay of radioactive elements, the magnetisation of particles and direct counting of layers of different colour or constitution.

Radioactive elements can be used to measure time because they decay at a constant rate. The rate at which a radioative material decays can be expressed by its half-life. This is the time taken for half of its atoms to decay. A much used radioactive isotope of carbon, carbon-14 (^{14}C), has a half-life of about 5570 years. With so long a half-life it can be used for dating back to about 30,000 years ago. However, its rate of decay is too slow to be used to date very recently laid down deposits. Lead-210 (^{210}Pb) and caesium-137 (^{137}Cs) are suitable. Lead-210 with a half-life of 22 years, can be used for about 150 years back from the present. Lead-210 is a decay product of radon-226 (^{226}Rn; half-life 3.8 days), a gas issuing from the earth and well known for the health danger it poses to people living in houses built over certain kinds of rock. Caesium 137 (^{137}Cs; half-life 30 years) is one of the fission products of atomic bombs and so with a recorded history of fall-out from 1945 to a maximum in the sixties and a decrease thereafter as testing bombs in the atmosphere decreased. Now a new marker has been provided by the Chernobyl disaster but, as with atomic bomb markers, it does not have the worldwide distribution that carbon-14 does.

Changes in the horizontal and vertical directions of the earth's magnetic field can be used for chronology if suitable particles are present, notably those of magnetite. If a particle becomes magnetised at the time of reaching or entering a deposit on the bottom of a lake it becomes magnetised in the direction of the earth's field. This field is known to change with time. There are historical records for over 400 years and older changes can be discovered by comparing magnetic data with other chronological methods (e.g. carbon-14). In a relatively small number of cases, layers of different structure or composition can be found in lake deposits which are not subject to disturbance after formation. These layers can be counted to determine age just like tree rings can, for they too represent changes in growth, between autumn and winter on the one hand and spring and summer on the other. These variations in the seasonal growth of organisms can produce visible colour changes in a lake sediment or other changes detectable by laboratory analysis. An example, not uncommon in sheltered calcareous lakes, is a whitish layer marking summer and a darker layer marking winter. In summer the active photosynthesis of plants and algae produces a deposit rich in calcium and magnesium carbonates, so giving it a lighter colour than in winter when there is little photosynthesis and so no precipitation of carbonate.

Chapter Six

The Chrysophytes

The possession of siliceous coverings to cells is not confined to diatoms, even some amoebae have them, for example *Paulinella* (p. 252). However, the type of wall enclosing diatoms is not seen in any other organism. Likewise, some Chrysophytes have a characteristic and unique type of siliceous coating to their cells and most of them produce a special type of silicified spore or cyst, we use the words synonymously.

Mallomonas (figs 261-266, 269) is a single-celled free-swimming flagellate, the body of which is covered by siliceous scales and spines, so giving it a hairy appearance (figs 261, 269). The scales overlap one another, like tiles on a roof (figs 263, 265, 266). This overlapping is also just visible around the edge of the specimen in fig. 269. Each silica scale usually has a spine attached to it, which may possess a serrated distal portion (fig. 262). Scales and spines are produced separately inside the cell. One may marvel at how they are then extruded, the scales arranged over the cell surface and each spine attached to a scale.

The taxonomy and so identification of species of *Mallomonas* is based on the diverse structure and arrangement of the scales and spines. As with the cell walls of diatoms, examination by electron microscopy is necessary in order to reveal and elucidate the detailed fine structure of these scales and spines.

The possession of two flagella of unequal size is a feature of Chrysophytes and by electron microscopy it can be seen that one of them, the longer, bears fine hairs. The size difference between the two flagella ranges from slight to very great. Some Chrysophytes appear to have only a single flagellum but it is now known that most of them actually have two. *Mallomonas* exemplifies this very well. It possesses a single functional flagellum enabling it to swim through the water. Such a flagellum can be seen as an undulating thread (pointing north) arising from the apex of each of the two cells in fig. 261. A second flagellum is also present but it is reduced to a rudimentary knob, so small that it is difficult to see by light microscopy.

The otherwise identical genus to *Mallomonas*, namely *Mallomonopsis* has two functional flagella. Since flagella length is the only difference between them, some authorities do not accept that *Mallomonopsis* is a separate genus.

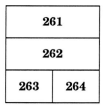

Figs 261-264. *Mallomonas*.

Fig. 261. Note rigid spines and wavy single flagellum; lefthand cell 38 μm l. **P.**

Fig. 262. Scale and spine (35 μm l.). **P.**

Fig. 263. Overlapping scales and spines; cell (50 μm l.). **P.**

Fig. 264. Cyst (30 μm diam.) containing large leucosin globule and surrounded by the siliceous cell envelope. **N.**

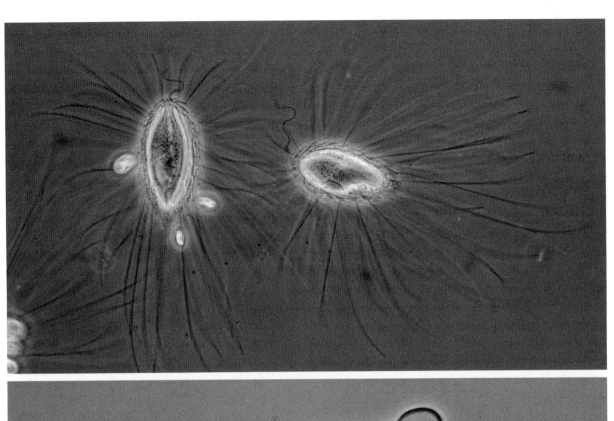
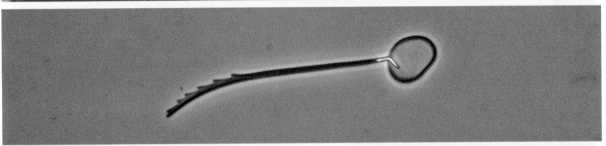
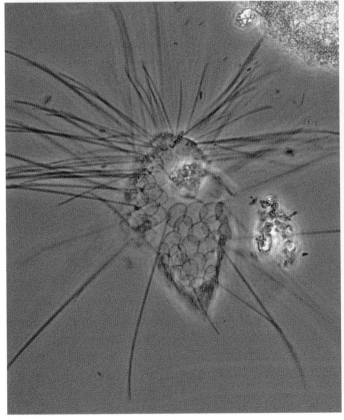
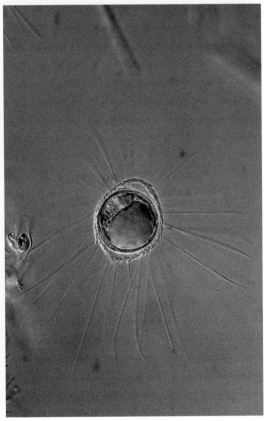

Like diatoms, the chloroplasts of *Mallomonas* (figs 261, 269) and most other Chrysophytes are brown or yellow in colour. However, the cells of several genera lack chloroplasts and live on dead organic matter, that is they are saprophytes, or ingest other microorganisms including other algae. Even photosynthetic species may also live saprophytically or ingest other organisms. A single cell may be capable of all three modes of nutrition. Thus, such a cell is living like a plant, fungus or animal and illustrates the fact that there are no absolute distinctions between such kinds of organism.

Species of *Mallomonas* are present in almost all kinds of freshwater, in the plankton, among weeds and near the bottom provided there is sufficient light for photosynthesis.

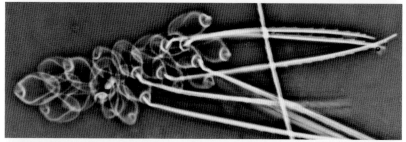

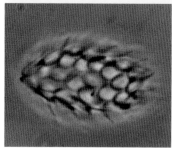
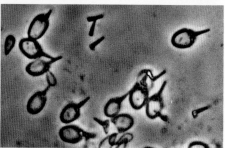

Figs 265, 266. *Mallomonas* (× 1600). **P.** Scales and parts of spines of two other species.

Figs 267, 268. *Synura* (× 1600). **P.** Fig. 267. Coating of scales; fig. 268, scales. Note short spines.

Fig. 269. *Mallomonas* (47.5 μm l.). Note chloroplast.

Fig. 270. *Synura*. Part of a globose colony. Cells (circa 13 μm l.); note the short spines.

Fig. 271. *Synura*. An elongate colony (150 μm l.). **P.**

Synura (figs 267, 268, 270-275) also has overlapping siliceous scales and spines, though the latter, if present, are short (figs 267, 268, 270, 271). The cells are united into motile colonies by their stalk-like ends (fig. 272). The colonies usually are globose (figs 273, 274) but may be of rather irregular shape and, in extreme cases, elongate (fig. 271). Some authorities consider the elongate *Synura* to belong to a different genus. *Synura* cells have two flagella of somewhat uneven length (fig. 275). As in other biflagellate Chrysophytes the two flagella beat in different ways and have structural differences which are difficult or impossible to see except by electron microscopy.

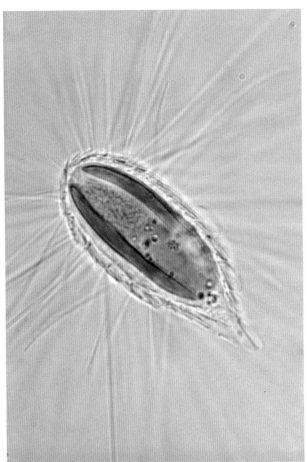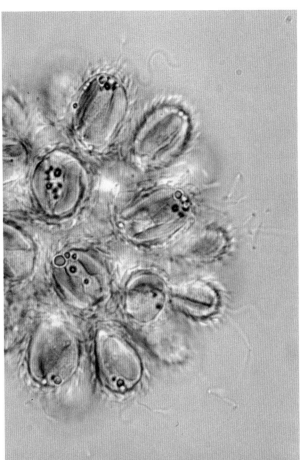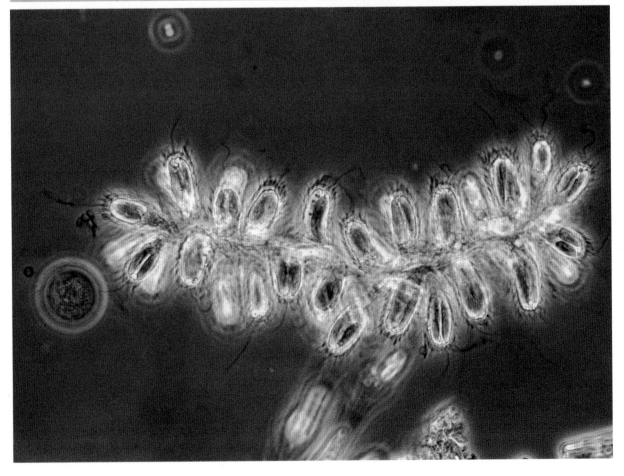

A rare alga of similar structure, called *Syncrypta*, has cells without siliceous scales and spines and so looks like the *Synura* in figs 272-275 which presents a "smooth" outline. Indeed, we thought at first that the alga in these pictures was *Syncrypta*. The scales of this species of *Synura* are so delicate that they only become clearly visible by light microscopy at the highest power of magnification and using phase-contrast lenses. The siliceous spores of *Mallomonas* (fig. 264) and *Synura* (figs 281, 284) are considered later.

274
275

Fig. 274. *Synura*. Colonies (75-105 µm diam.) with very delicate and here invisible coating of scales; easily mistaken for *Syncrypta* (see text).

Fig. 275. *Synura*. Squashed colony (120 µm l.). Note flagella. **P**.

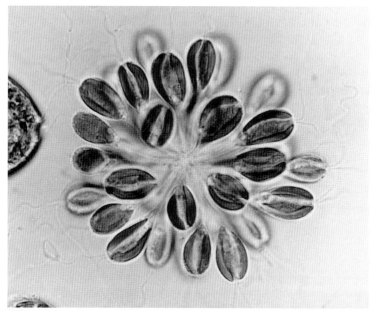

272
273

Fig. 272. *Synura*. Small colony, note stalks to cells and lack of visible scales and spines. However, they are present and this is not a species of *Syncrypta* (× 1008).

Fig. 273. *Synura* among filaments of the Cyanophyte *Oscillatoria* and a few cells of a naviculoid diatom (× 256).

Species of *Synura* are common in the plankton and also in weedy, shallow waters, sometimes close to the bottom and so among algae living on the mud surface (e.g. fig. 273).

Several Chrysophytes (e.g. *Dinobryon*, *Uroglena*) impart a fishy odour and taste reminiscent of cod liver oil to drinking water when abundant, thus causing problems for the water industry. Relatively small numbers of *Synura* can be a nuisance in this respect. Indeed, it is said to be the worst offender of all.

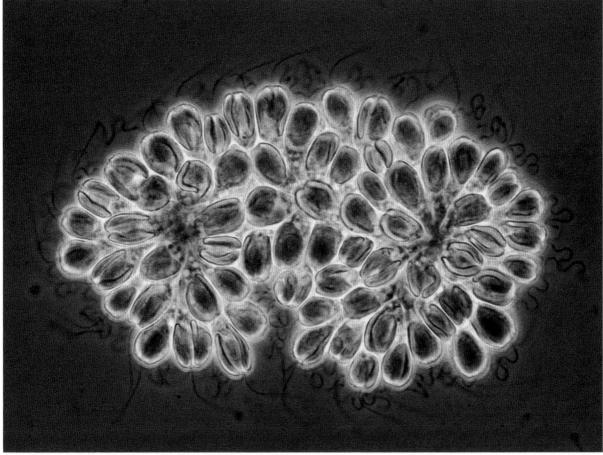

The cells of *Dinobryon* (figs 276-279, 282, 283) are not covered by siliceous structures. They live in vase-like cases composed of organic matter. The cell is attached to its case by a filiform basal elongation (fig. 278). In some species the cells are solitary but in the common species shown the cases are united to one another to form bushy or dendroid (tree-like) colonies. The colony swims in a rather slow but graceful manner. Each cell has two flagella of different lengths as can be seen in a cell on the righthand side at about 3 o'clock in fig. 278 and the upper cell in fig. 282.

It is remarkable that such an ungainly group of cells, each in a separate container, can swim in what one can call a purposeful manner. For example, the colony will swim towards a source of light, provided that it is not too bright. Each cell has a red eye-spot which constitutes part of a direction finding apparatus. An eye-spot can be seen at the apex of the two cells in fig. 282 and many of the cells in fig. 278. In the latter figure, taken under phase-contrast illumination, the eye-spot appears as a small rounded yellowish body.

The cells of *Dinobryon* reproduce by binary fission, dividing longitudinally. One of these two daughter cells remains in the mother cell case. The other daughter cell moves to near the top of the case, attaches itself to it and produces a new case. The colonies are formed by a repetition of this process. Hence the bushy colony in fig. 278 has arisen by successive cell reproduction from the single cell at its base. A newly produced cell can also swim out of the mother cell's case and start a new colony.

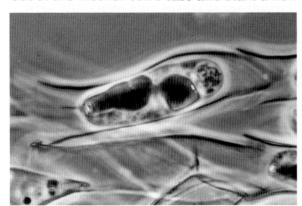 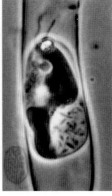

Figs 276, 277. *Dinobryon*. Two cells with basal vacuoles containing bacteria (× 1600). **P.**

Fig. 278. *Dinobryon*. Height of colony 194 μm. The apically situated eye-spots appear yellowish in phase-contrast.

Dinobryon is very common in the plankton of lakes or pools which are relatively poor in nutrients and not very alkaline. Such waters or pools are common in mountainous and other regions with poor and usually more or less acid soils. Recently, it has been shown that *Dinobryon* ingests bacteria (figs 276, 277). Since plant nutrients are not abundant in the waters in which *Dinobryon* lives, bacteria can be an important source of nutriment additional to that obtained through photosynthesis.

The cells of *Dinobryon* can pass through periods unfavourable for growth in the type of cyst or spore which is peculiar to Chrysophytes (fig. 283).

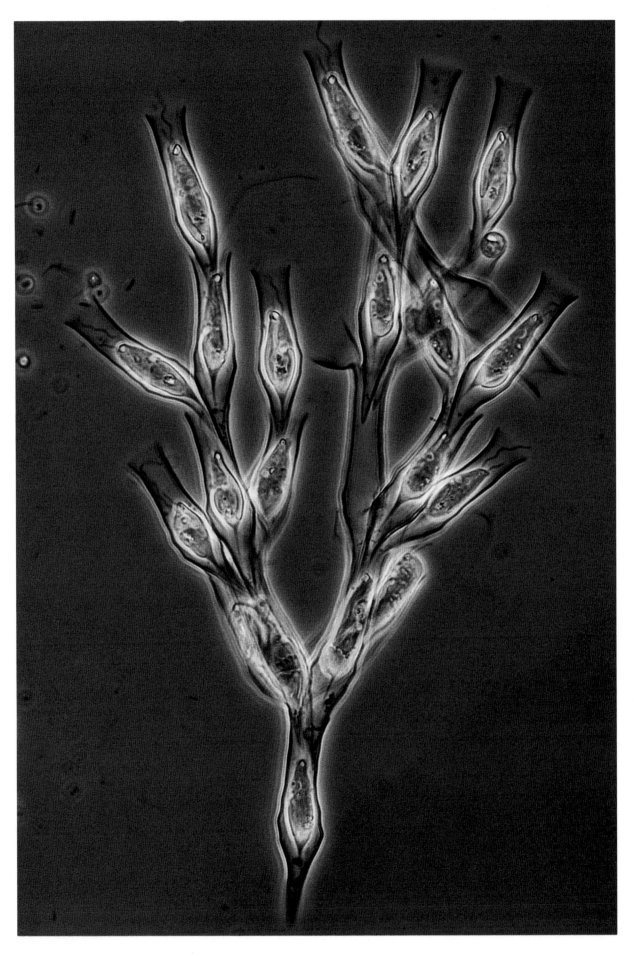

The spores of *Dinobryon* are located on the branched colony (figs 279, 283) and formed within a round case.

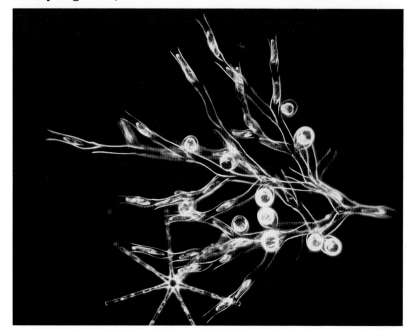

Fig. 279. *Dinobryon* (× 256). **D**. Colony with spores (round bodies). Also present, colony of *Asterionella*.

The spore consists of a cell enclosed in a siliceous wall which is only perforated at one place. The plugged pore of two *Dinobryon* spores can be seen in fig. 283. The pore lies at the bottom of a siliceous tube or spore collar and is closed by a hemispherical plug of organic matter. Plugs of spores of *Uroglena* and an unknown Chrysophyte are seen in figs 280, 285. The "caps" on the spore tubes in fig. 283 are temporary investments. At the time of germination the organic matter breaks down and the cell or cells within squeeze through the pore and swim away.

The spores of silicified genera such as *Mallomonas* and *Synura* are formed inside the coating of scales and spines. The overlapping of the scales of *Mallomonas* is seen in fig. 264 and of *Synura* in figs 281, 284. The spore of the *Synura* in fig. 284 has not yet produced a fully silicified wall, whereas the mature spore in fig. 281 has done so. In the lower part of a spore there is a large hyaline body which represents a carbohydrate reserve of food called leucosin or chrysolaminarin. It is most clearly visible in the *Mallomonas* spore (fig. 264).

Figs 282. *Dinobryon*. Two cells (10 μm l.) in their cases. The oblong colourless body containing small globules under the lower cell is a fungal parasite (*Rhizophydium*). The white body (out of focus) on top of the case of the upper cell is a non-photosynthetic flagellate.

Fig. 283. *Dinobryon* spores (15 × 13.5 μm). **N**.

Fig. 284. *Synura*. An immature spore (17.5 μm diam.), not yet fully silicified.

Fig. 285. Spore (12.5 μm diam.) of an unknown chrysophyte.

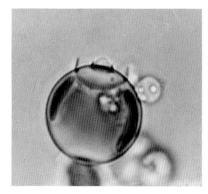
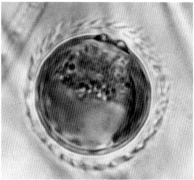

Fig. 280. *Uroglena* spore (× 1600). The white body on the spore (right hand side at 2 o'clock) is a fungal parasite.

Fig. 281. *Synura* spore (× 1600). Note cell's scales and spines around spore.

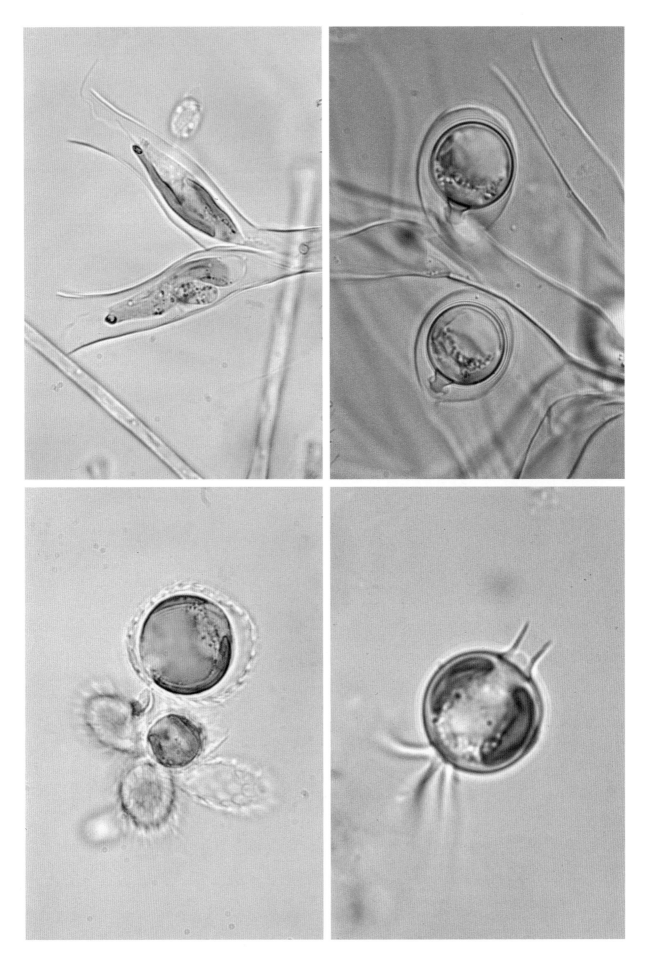

As with the auxospores of diatoms, spore formation in Chrysophytes may follow sexual union, an example of which is seen in a species of *Kephyrion* (fig. 286). Cells of *Kephyrion*, like those of *Dinobryon* live in a vase-shaped case. In fig. 286 only the empty cases are left attached to a spore, formed as a result of the sexual fusion of their protoplasts.

Species of *Kephyrion* are common in the plankton but easily overlooked because of their small size (nearly all are less than 10 μm long or broad). There is, or there appears to be, only one flagellum when examined by light microscopy but electron microscopy may well reveal that a second flagellum or a rudiment thereof is present as in *Mallomonas*. If a second flagellum has been observed already by light microscopy, then Chrysophytes otherwise identical with the genus *Kephyrion* are placed either in a genus called *Pseudokephryion* or one called *Kephyriopsis*.

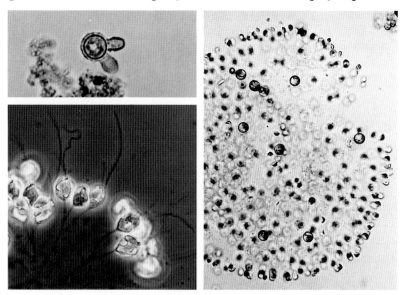

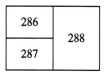

Fig. 286. *Kephyrion*. Sexual reproduction. Note spherical zygote and empty cases of the two gametes (× 1025).

Fig. 287. *Uroglena*. Cells and, at the bases, threads which with others form a system of connections reaching back to the centre of the colony (× 640). P.

Fig. 288. *Uroglena*. Spores, (the spherical bodies) in a colony (× 256).

Uroglena (figs 280, 287-290) forms colonies which are globose, like most of those of *Synura*. However, the cells are "naked" and do not have a coating of siliceous scales and spines. The colonies usually contain many more cells than those of *Synura*. The cells in young colonies are at the ends of a system of thin and fine (fig. 287), or broad and conspicuous dichotomously branched threads arising from the centre. These threads may be lost or become very difficult to see as the colonies increase in size or age. The whole colony is mucilaginous. Each cell has two flagella of very unequal length (fig. 290) and a prominent anterior red eye-spot (fig. 289). Spores (fig. 280) occur in the colony (fig. 288). *Uroglena* is prevalent in the plankton of similar waters to those favoured by *Dinobryon* and is a very common cause of fishy tastes and odours. It too ingests bacteria.

Cyclonexis (fig. 295) is not a common alga. The elongate "naked" cells are grouped close together in a ring-shaped colony which rotates as it swims. *Cyclonexis* typically occurs in ponds and small pools.

Figs 289, 290. *Uroglena*.

Fig. 289. Colony of medium size (137 μm diam.). Note red eye-spots.

Fig. 290. Part of squashed colony. Note flagella and eye-spots (yellowish in phase-contrast). Cells 9.4-13.7 μm l. P.

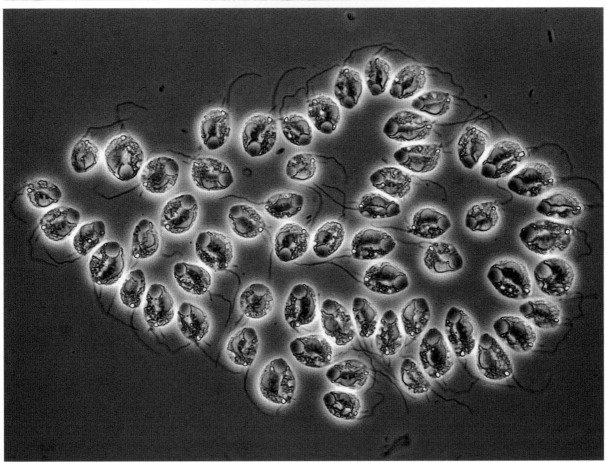

Chrysophytes also include unicellular and colonial non-motile and a few filamentous species.

Chrysopyxis (fig. 296) is a unicellular, non-motile form in which, like *Dinobryon*, the cell lives within a vase-like case. The case is attached to a surface. The species shown nearly always lives on the filamentous Chlorophyte *Mougeotia* (p. 58). It is attached to the *Mougeotia* by a fine thread encircling the filament.

In *Hydrurus* (figs 291-294, 297) large numbers of cells are embedded in slippery mucilage to form branched colonies (figs 291, 292) which are large enough to be seen by the unaided eye. Any cell can develop a flagellum and swim away. The shape of the motile cell (figs 293, 294) is unusual, even allowing for a certain degree of plasticity which it may undergo. Typically it is triangular in outline with more-or-less elongate angles.

Fig. 295. *Cyclonexis*. Colony, 49 × 44 μm.

Fig. 296. *Chrysopyxis* (case 7.5 μm br.), attached to a filament of *Mougeotia*. N.

Fig. 297. *Hydrurus*. Part of a branch of the alga. Cells circa 9 μm l. or br.

Fig. 298. *Phaeodermatium*. Part of a typical crustose specimen. Cells circa 2.5-4.5 μm l.

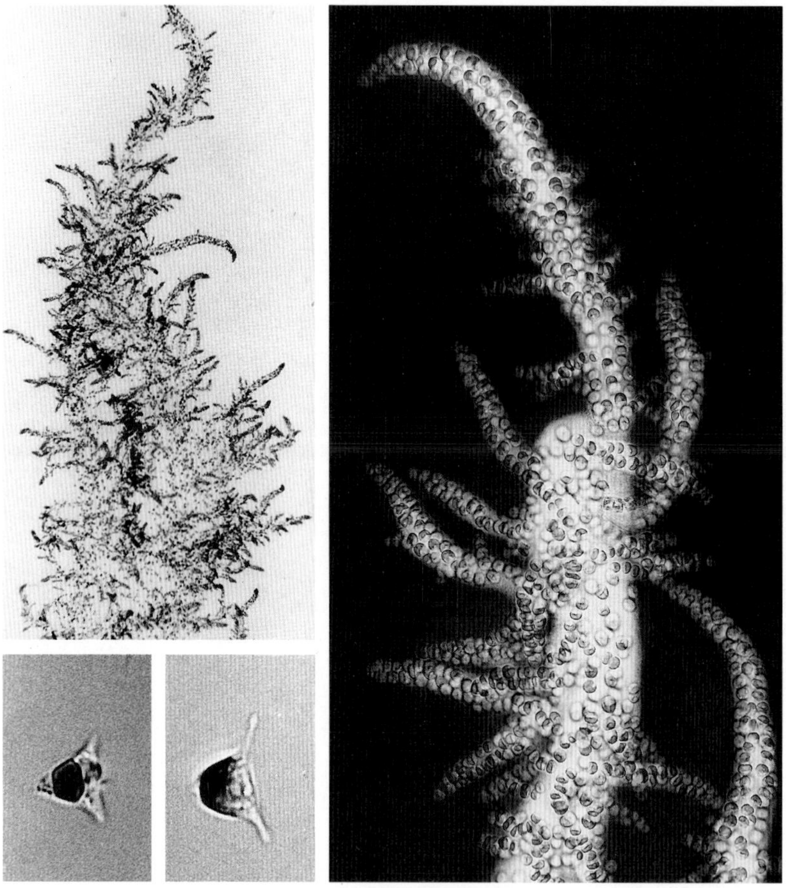

Figs 291-294. *Hydrurus*.

Fig. 291. Part of a colony (× 28).

Fig. 292. Cells embedded in mucilage (× 160). I.

Figs 293, 294. Free-swimming cells (× 900).

Hydrurus is one of the few algae which can be recognised by smell. Like many other Chrysophytes it is a very delicate alga and, unless kept cold, often is difficult to get back to the laboratory or home before it has begun to disintegrate. When collected and put in a bottle it rapidly produces a powerful pungent odour, which in this case can be a precursor of death. However, the characteristic smell is not necessarily a sign of impending death. It may be present or absent before collection and when *Hydrurus* is very abundant in a stream the distinctive odour sometimes can be detected well before the stream is reached.

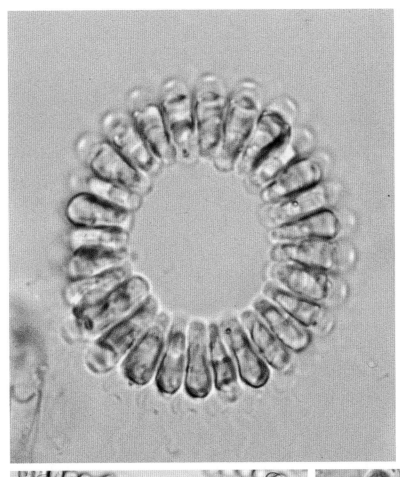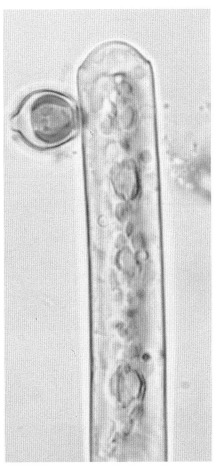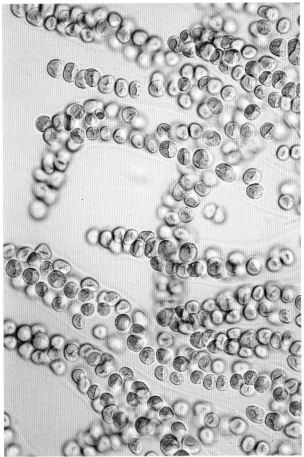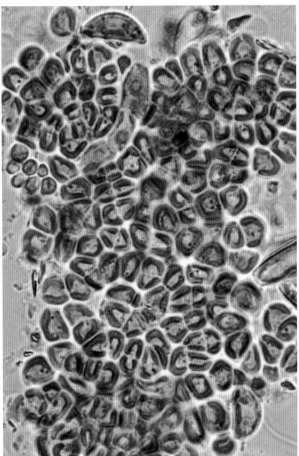

Hydrurus is not a widely distributed alga but not rare, indeed often abundant in favourable places. Such habitats are streams or rivers in mountainous districts, especially during the colder months of the year. When the temperature rises much above 10°C, some cells swim away, others form Chrysophycean cysts, many probably die and the feathery colonies break up and disappear.

Phaeodermatium (fig. 298) also lives in streams and rivers, thriving in the colder months of the year and usually disappearing in summer. It is commoner than records of its occurrence suggest because it is not easy to see or collect. It forms thin (mainly one cell thick) filamentous crusts on rocks and stones. The cells are so closely appressed to the substratum and to one another that their filamentous arrangement often is not easy to see. In fig. 298, the radiating filamentous nature of the *Phaeodermatium* crust is seen clearly only at the top of the picture.

Like *Hydrurus*, any cell may develop a flagellum and swim away. Moreover, the shape and size of the motile cell appears just like that of a motile *Hydrurus* cell. With its similar habitat and seasonal occurrence, similar motile cells and cysts, it has been suggested that *Phaeodermatium* is a growth form of *Hydrurus*. Extra support for this belief is the fact that the filaments can disintegrate and the crust change into a mass of separate cells embedded in mucilage. However, the development of long branched mucilaginous colonies like those of *Hydrurus* has not been seen; nor does it have the very characteristic smell of *Hydrurus*. In the stream running through our garden *Phaeodermatium* can be so abundant, even in February, that the stones are slippery to walk on. By about June it has disappeared. However, in 45 years *Hydrurus* has never been seen in this stream. This strongly suggests that *Phaeodermatium* and *Hydrurus* are not growth-forms of one and the same alga. Even so, this is not proof, for it could be that there is something in our stream habitat that inhibits the development of the *Hydrurus* stage.

Algae like *Phaeodermatium* which form crusts can be both difficult to see even when abundant and difficult to remove undamaged. Further, *Phaeodermatium* typically lives in fast-flowing water so that what is scraped off is likely to be lost. A method for collecting *Phaeodermatium* unharmed, as well as many other algae growing closely adherent to a surface, is to use microscope slides for them to grow upon. These slides can have holes in them or be in some kind of frame. They are suspended in the water and attached by strong, non-decomposable string (e.g. terylene) to an immovable object on the bank of a river or lake. They need to be left there for a fortnight or more to give organisms time to develop from any spores which may attach themselves to the slide. If a definite alga is being targeted, then the kind of habitat and time of year must be taken into account. In our northern England, if *Phaeodermatium* is sought, the slides should be put in during autumn and examined from January onwards because the alga only flourishes in cold water and starts active growth very early in the year.

The motile cells of *Hydrurus* and *Phaeodermatium* neither live in a case like *Dinobryon* nor are covered by siliceous scales like *Mallomonas* but, like those of *Cyclonexis*, they are "naked", that is with no wall or special outer integument other than the membrane around the protoplast. Single free-swimming cells of this kind are common in this group and the largest genera are *Ochromonas* and *Chromulina*.

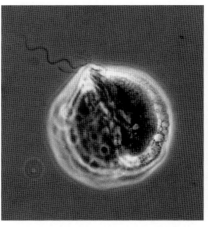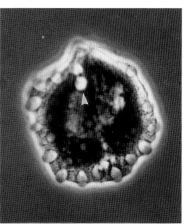

Figs 299, 300. *Ochromonas*.

Fig. 299. Note, (at 11 o'clock) the two flagella of unequal length (× 640). **P**.

Fig. 300. Note the balloon-shaped vesicles around most of the cell. The circular white body (arrowed) is a contractile vacuole (× 860). **P**.

Ochromonas (figs 299, 300) has two flagella, one shorter, often much shorter, than the other and, as in other Chrysophytes, they are of different fine structure and motion. In the species shown, which is discoid in shape, there are numerous balloon-shaped vesicles just below the surface, even bulging out to distort its outline. When the cells are irritated or harmed artificially, thick strands shoot out of these vesicles; whether such threads are ejected under natural conditions is not known. Their function and so that of the vesicles from which they arise are not understood. It has a spiny spore (fig. 586).

Chromulina only differs from *Ochromonas* in having one flagellum but, as has been mentioned, the fact that only one flagellum is detected by light microscopy does not mean that another minute or rudimentary one is not present. In the past, the two major lines of evolution in the Chrysophytes were thought to arise from uniflagellate (e.g. like *Chromulina*) or biflagellate (e.g. *Ochromonas*) ancestors. Today, an apparently uniflagellate Chrysophyte would need examination by electron microscopy to check that a second flagellum is not present. Hence the validity of the genus *Chromulina* is uncertain. About 90 species of *Chromulina* have been described and about 80 of *Ochromonas*. Species of both genera are commonest in small bodies of water, among weeds and near muddy or boggy deposits in both small and large waterbodies.

A feature of the cells of many "naked" Chrysophytes is that they can change in shape, the degree to which they do so depending on the inherent properties of the species concerned and the environmental conditions. Cells of the *Ochromonas* described above have been seen to produce thread-like extensions (pseudopodia, see Glossary) when kept in the laboratory. A variety of other Chrysophytes may do so at times. These protoplasmic threads are not permanent features of these species but there are genera which characteristically live in this state.

Chrysamoeba (figs 301-304) has discoid cells when living on a substratum. The cell is of irregular outline, largely because of protoplasmic extensions, varying in number and position, which radiate out from it. At the base they are relatively broad and conical in shape, narrowing into long, thin threads. These extensions of the protoplast are called rhizopodia. In the species shown, there are small granulate thickenings along the rhizopodium giving it a beady appearance. The rhizopodia can change in length or even

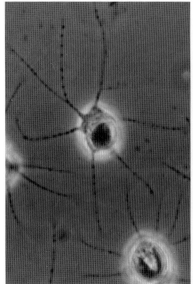
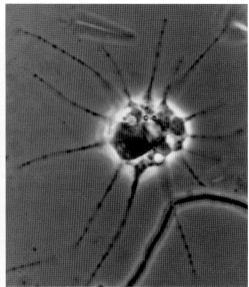
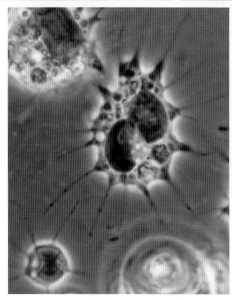
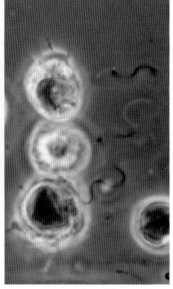

301	302
303	304

Figs 301-304. *Chrysamoeba* (× 1600). P.

Figs 301-303. Rhizopodial stage, cells of various sizes.

Fig. 304. At bottom, a cell which has become motile. Note waving flagellum on righthand side (at about 3 o'clock). The two cells (out of focus) above it are also flagellate.

be retracted and new ones can develop. However, all such changes are very slow in this non-swimming stage. The cells may be solitary or in groups. The species seen here apparently has no flagellum and for this reason would in the past have been called *Rhizochrysis* because *Chrysamoeba*, as originally described, had a flagellum. As in other cases mentioned, electron microscopy has shown that a flagellum always is present, though not always long enough to be detected by light microscopy. Further, again as in other cases, there are in fact two flagella, the second usually being little more than a rudiment. The cell can produce a long and so prominent flagellum when it transforms into a free-swimming flagellate (fig. 304). In this process, the cell retracts most or all of its rhizodia and become globose.

Chrysamoeba is not common but its rarity partially stems from the difficulty of finding it in collected material. It lives on mud and other substrate which are disturbed in collection. In addition, the cells may become motile and then look just like a species of *Chromulina*. The best method for finding *Chrysamoeba* and many other fascinating algae is as follows. Place the sample containing mud or other detritus in a container, leave to stand for a day or so in a cool place. Decant as much of the water as possible without disturbing the deposit at the bottom. Mix this residue and pour into one or more shallow dishes (e.g. Petri dishes). Place microscope coverglasses on the deposit or water above it. From time to time carefully remove a coverglass, place it on a microscope slide and examine. Coverglasses can be examined from time to time for weeks, even months, afterwards. The dish should be kept in low light, never in sunlight, and is best covered by something transparent (e.g. the Petri dish lid) to prevent dust falling on the coverglasses. *Chrysamoeba* may also appear on slides placed in the water, as described on p. 162.

Chrysophytes and lake history

The siliceous parts of Chrysophytes (scales, spines and cysts), like the cell walls of diatoms, are preserved in the material laid down on the bottoms of lakes. Hence, they too can be used to elucidate past environmental conditions. However, fossilized remains of Chrysophytes are much less valuable than those of diatoms. The siliceous scales and spines are small and so not as easy to find as diatoms. Further, they can be so delicate that they are lost sooner or later. The spores are better fossil material because their silica is more massive so that they are easier to find and unlikely to dissolve or disintegrate. On the other hand, while the fossilized scales of *Mallomonas* and *Synura* almost always can be related to known species, most of the Chrysophyte spores present cannot (e.g. the spore in fig. 285).

Chapter Seven

The Haptophytes (Prymnesiophytes)

The name Haptophyte has now been replaced by Prymnesiophyte in order to comply with the internationally agreed rules of botanical nomenclature. We keep the name Haptophyte because it refers to the haptonema, an organelle peculiar to this group and present in both the genera illustrated.

Prymnesium (figs 306, 307) is unicellular and motile. It has two flagella of similar length and in between them a short, filiform appendage which looks like a third flagellum (fig. 307). This is the haptonema. The brown cells when alive (fig. 306) are somewhat elongate. The cell in fig. 307 has become spherical due to the flattening needed to photograph it in a very thin film of water in order to show the minute haptonema. The surface of the cell is covered by scales which are too small and diaphanous to be seen except by electron microscopy. They are not composed of silica but of organic matter. Cell multiplication is by binary fission. The species of *Prymnesium* shown grows best in moderately or weakly brackish water. If the water is rich in nutrients very large populations of *Prymnesium* can arise. Moreover, if other environmental conditions are right, *Prymesium* produces an extremely powerful toxin which preferentially affects organisms with gills, such as fish and certain molluscs. Catastrophic mass deaths of fish can then occur. In hot, dry countries where the water so often is salty, *Prymnesium* can cause serious losses in inland fish farms. The flourishing farming of fish in Israel was seriously threatend by *Prymnesium*, until scientists devised simple methods of detecting its presence and controlling its abundance.

Chrysochromulina (figs 305, 308) is a common freshwater, planktonic alga. The majority of species are marine and at least one of them is harmful to other organisms. The freshwater species shown is not harmful even when present in large numbers (e.g. ten million or more cells per litre). Like *Prymnesium*, the cell has two flagella of similar length and its surface is covered by fine scales. The haptonema is much longer than the flagella (fig. 305). When irritated mechanically or chemically it immediately

Fig. 306. *Prymnesium*. Longest cells circa 10 μm.

Fig. 307. *Prymnesium*. Minute haptonema lying between the cell's two flagella. Cell (9.5 μm diam.) has rounded off. **P.**

Fig. 308. *Chrysochromulina*. Cells killed, in life to 8 μm l.; central haptonema coiled, when extended up to 80 μm l. Fixed in a solution of iodine in potassium iodide. **P.**

Fig. 305. *Chrysochromulina*. Chemically treated specimen, showing the two flagella and long haptonema (× 1940).

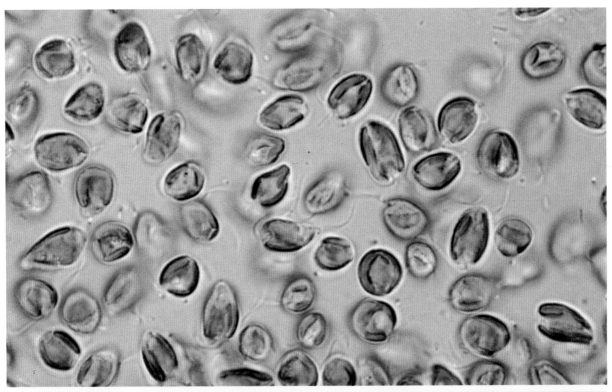
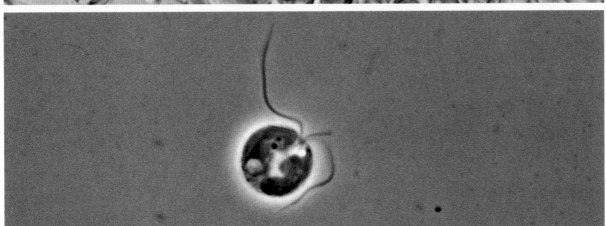
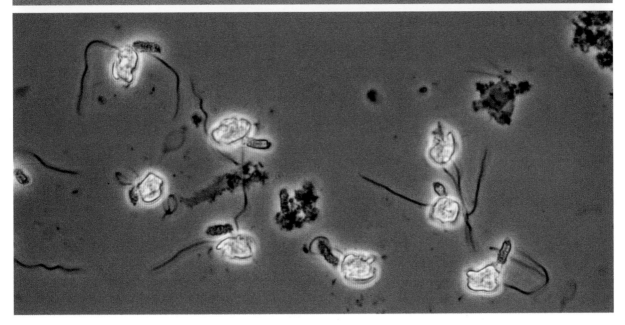

coils up into a tight spring. This has happened to the material in fig. 308 as a result of treatment with a solution of iodine in potassium iodide.

Until recently, the function of the haptonema was unknown. Now, for a marine species it has been shown that it's haptonema can collect and concentrate small microorganisms into a bolus which is applied to the cell's surface and finally incorporated into it. Whether this also occurs in the freshwater species has not yet been established. It also remains unclear whether the very short haptonema of *Prymnesium* can function in the same manner.

In a few freshwater Haptophytes the scaly covering to the cell is calcified. These scales are called coccoliths. Coccolith Haptophytes are common in the marine plankton, especially in tropical seas. Some deep sea oozes are rich in coccoliths and some which were so in the past are now land, forming chalk cliffs and chalky soils.

The Raphidiophytes

Gonyostomum (figs 309-312) is the commonest freshwater member of this small group of unicellular, flagellate algae. The species shown is widespread and sometimes abundant in small lakes, pools and ponds, usually in water rich in humus and so often yellow or brownish.

The cells are bright green to yellowish-green in colour and of rather varied shape. The commonest form, when swimming, is top-shaped (fig. 309) but they can also be ovoid; the body is more-or-less strongly flattened (compare figs 309 and 310). There are two flagella arising in an apical pit.

When swimming, one flagellum is directed forwards and the other curves backwards over the cell (fig. 309). The former drives the cell forward and the latter seems to act solely as a rudder. There are numerous discoid chloroplasts (fig. 311). Just below the cell surface – there is no cell wall – are numerous narrow, refractive rod-like bodies (fig. 312) called trichocysts. Most of them may be congregated near the apex or base of the cell or be distributed relatively evenly over it. If the cell is irritated or harmed, the trichocysts eject mucilaginous, slimy threads. It is uncertain to what extent such bodies are extruded under natural conditions, and their function is unknown.

In Sweden, there has been a great increase in the abundance of *Gonyostomum* during the last 40 years. *Gonyostomum* is especially common in acid waters (e.g. pH 5.5-6.5) and many waters in Sweden have been greatly affected by acid rain in the same period, leading to a serious deterioration in their aquatic life. Therefore, it is likely that the increase in *Gonystomum* is related to acid rain but there is no firm evidence that this is so. Nevertheless, massive growths of *Gonystomum* have ruined popular bathing places. Swimmers have come out of the water covered in slime and suffered severe skin irritation. Similar troubles have arisen since about 1980 in Finland.

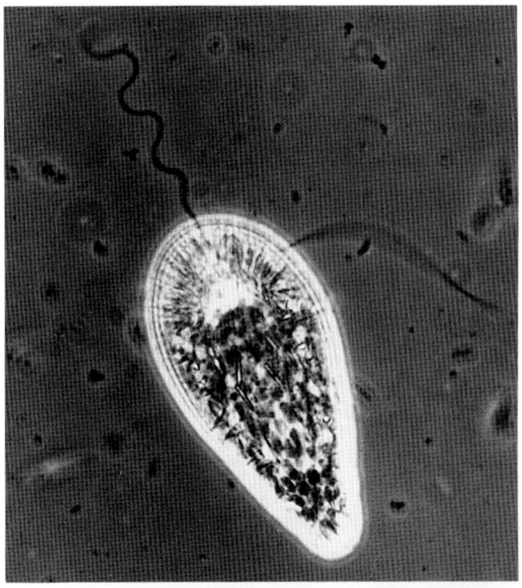

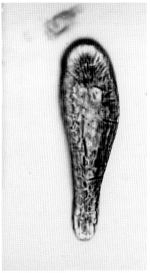

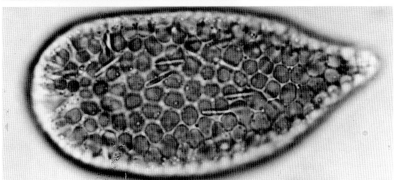

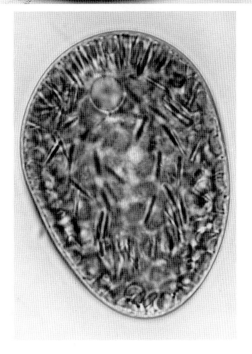

Figs 309-312. *Gonyostomum.*

Fig. 309. A free-swimming cell seen from the broadside. The white area below the apex marks the apical pit from which the two flagella arise (× 640). **P.**

Fig. 310. A cell seen from the narrow side (× 640).

Fig. 311. Numerous discoid chloroplasts. The rod-shaped bodies are trichocysts, the distribution of which is more clearly shown in fig. 312. Here a contractile vacuole (round ring) is also visible. Both × 1440.

The Cryptophytes

Cryptophytes are of virtually universal occurrence in freshwater and often abundant in plankton. They are also common in brackish water and in the sea. They are an important source of food for small invertebrates (fig. 471).

Most Cryptophytes are yellow or brown in colour, some are brownish-red and a few red, blue, blue-green or colourless. To a certain extent, their colour depends on their physiological state so that, for example, reddish species can also be more brown than red. All are unicellular.

Cryptomonas (figs 313-315) is the largest genus. The cells are flattened. Here they are seen from a broad side, in section they would be oval, ellipsoid or oblong. They are often more or less twisted. The cells have two flagella of slightly uneven length (fig. 313). An invagination, called the gullet, extends down from near the apex of the cell. This gullet is lined by an array of small bodies called trichocysts or ejectosomes (fig. 314). The latter name is based on the fact that under certain rather ill-defined conditions they eject fine threads. The earlier name of trichocyst was based on the belief that the threads they produce catch microorganisms or deter animals which might otherwise feed on them, as is the case of the trichocysts of certain animals. There is no good evidence that the threads of Cryptophytes do serve either function.

The cells of *Cryptomonas* can become non-motile and live in clumps (fig. 315) held together by mucilage. This is the so-called palmelloid stage (cf. *Euglena*, fig. 161). They can also round off and form a wall (fig. 315, righthand side at 1 o'clock). Such cells or cysts are also resting spores.

Chilomonas (fig. 316) is like *Cryptomonas* but lacks chloroplasts. It is common in water rich in decaying plants or other suitable sources of organic matter. The granules in the cells are starch grains. Starch often is so common in cells of *Cryptomonas* and other photosynthetic genera that it masks the structures visible in the nearly starch-free cells seen in figs 313-315.

In the past, colour was used as a generic character. Thus brown species belong to *Cryptomonas*, red or reddish ones to *Rhodomonas* and blue or blue-green ones to *Chroomonas*. Recent work suggests that such generic distinctions are unjustified.

Investigations utilising electron microscopy and molecular biology have led to a theory that Cryptophytes arose (presumably long ago) from a flagellate organism, which incorporated into itself a non-flagellate organism. The latter is thought to share a common ancestry with the Rhodophytes (Chapter 9). In Chapter 8, p. 180, a Cryptophyte incorporated into a Dinophyte and in Chapter 11, p. 252, Cyanophyte-like bodies acting as chloroplasts in diverse organisms are encountered. Such kinds of union may have played an important role in evolution (see p. 253).

313	314
315	316

Fig. 313. *Cryptomonas*. Note flagella. Cell 25 μm l., fixed in a solution of iodine in potassium iodide. P.

Fig. 314. *Cryptomonas*. Note lines of ejectosomes (trichocysts). Cell 52.5 μm l.

Fig. 315. *Cryptomonas*. Clump of cells in mucilage (palmelloid stage). Note cyst (round cell, 16 μm diam.).

Fig. 316. *Chilomonas*. Cells (21-25 × 5.9 μm); note grains of starch.

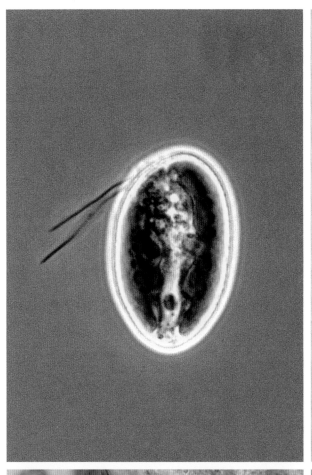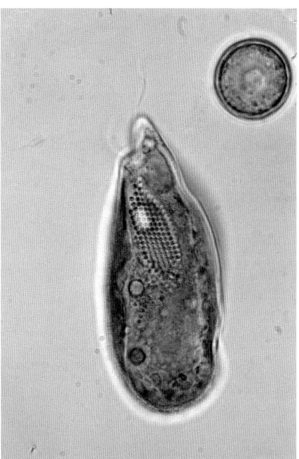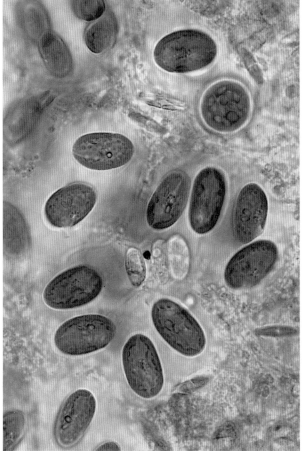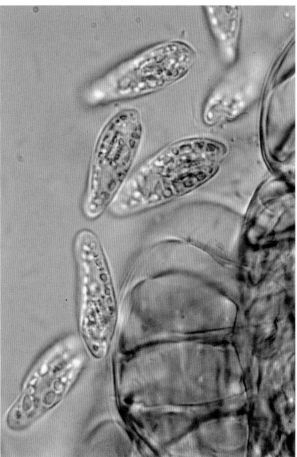

Chapter Eight

The Dinophytes

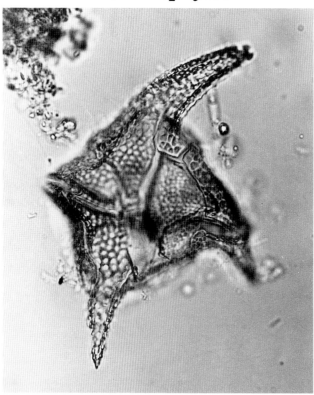

Fig. 317. *Ceratium*. A species with a curved apical horn. Note transverse and longitudinal grooves, also cell wall plates (× 640).

Most Dinophytes are free-swimming, unicellular organisms. A few are non-motile unicells and a couple are filamentous, in each case reproducing by motile dinophycean spores which are of such a characteristic structure that they cannot be mistaken for other algae. Some Dinophytes are "colourless" parasitizing animals or feeding on a variety of microorganisms (see later). The "coloured" photosynthetic forms usually are yellow or brown but reddish and blue-green species are known.

Dinophytes are common in freshwater but find their greatest diversity in the sea. They are the cause of "red tides" and shellfish poisoning. Many marine organisms, notably corals, contain symbiotic Dinophytes the photosynthetic products of which are essential for their existence. Though there are no "red tides" in freshwater at least one Dinophyte produces toxic substances and red waterblooms are known to occur.

Ceratium (figs 317-325, 332) belongs to the major group of Dinophytes, the Dinoflagellates. It is common in the plankton of lakes and some species are often abundant in those which are relatively rich in plant nutrients such as phosphates and nitrates. In such lakes it is often accompanied by Cyanophytes (fig. 318), a group considered later (Chapter 10).

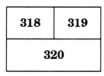

Fig. 318. *Ceratium* among filamentous Cyanophytes. Upper central cell circa 212 µm l.

Fig. 319. *Ceratium* (50 µm br.), showing long, external flagellum. Above, part of a colony of *Dictyosphaerium*. **P.**

Fig. 320. *Ceratium* dominated plankton (cells circa 170-200 µm l.) **D.**

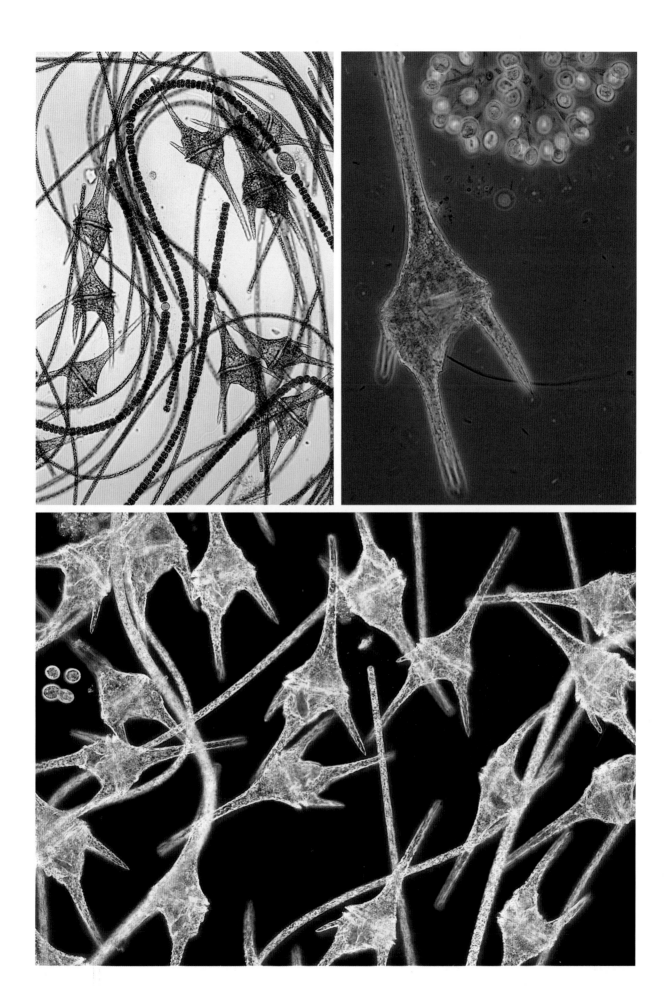

The strongly flattened body is extended into three (fig. 317) or four (fig. 319) rod-like portions which commonly are called horns. The longest horn is an extension of the anterior end of the cell.

The cell is divided into two halves by a transverse groove running a little more than half way round the cell (figs 317, 322). This groove contains a flagellum. Despite being in a groove, the wave-motion of this flagellum drives the cell forward. A second flagellum extends beyond the cell (fig. 319) and arises from a broad longitudinal groove in the lower half of the cell. This longitudinal groove connects with the ends of the transverse one (see figs 317, 322 lefthand cell). It is this arrangement of grooves with their attendant flagella which gives the Dinoflagellates and the motile spores of other Dinophytes their unique appearance.

The cell wall is composed of a set of plates which themselves are composed of a network of parts (figs 317, 324). Genera with such a plated type of wall often are said to be "armoured" Dinoflagellates. This kind of wall structure also occurs in the next genus *Peridinium*. The number and arrangement of these plates is used to distinguish the genera of such "armoured" Dinoflagellates.

The cell reproduces by dividing in an obliquely longitudinal fashion (figs 321, 323). Each half then has one or more of the mother cell's horns and in its further growth produces additional new horns and so becomes a copy of the mother cell.

322	323
324	325

Figs 322-325. *Ceratium*.

Fig. 322. Cells (58-64 μm br.), seen from two sides.

Fig. 323. Cell (diagonal width 81 μm) in process of division.

Figs 324, 325. Cysts (greatest lengths 100 and 144 μm respectively) of two species.

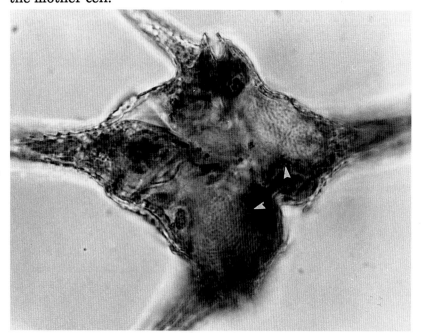

321

Fig. 321. *Ceratium*. Dividing cell, with two nuclei (arrowed); note rod-shaped chromosomes in each of them (× 1008).

In fig. 321 the large oval body in each daughter cell is its nucleus. Both nuclei contain many rod-shaped bodies, the upper (at 1 o'clock) more clearly than the lower, partially obscured nucleus (at 6 o'clock). These bodies are the chromosomes. They are the bearers of the genetic material essential for the working of the cells and organism and for

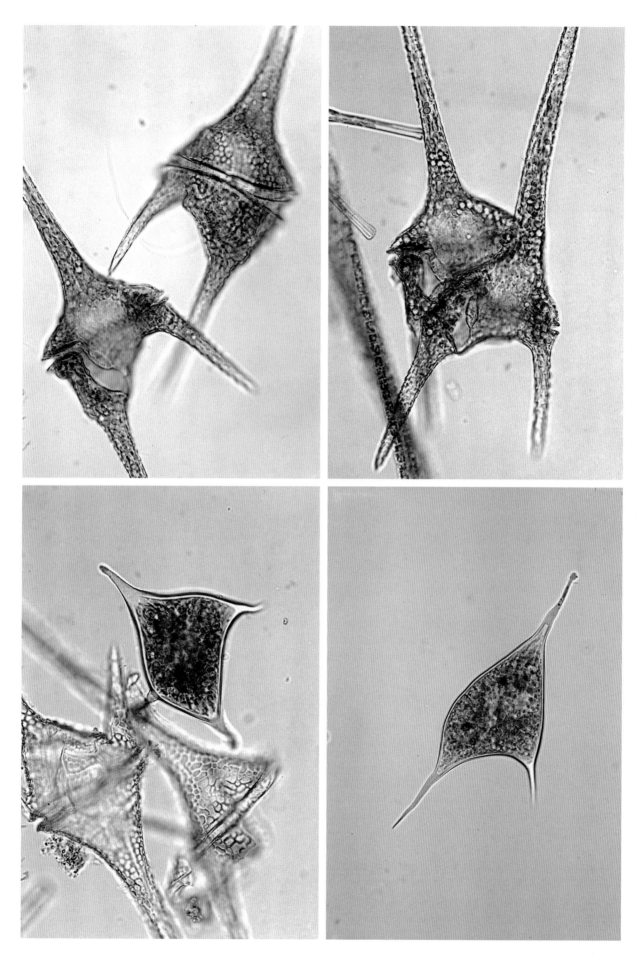

the transmission of hereditary features. This visible state of the chromosomes in nuclei which are not dividing, the so-called "resting" nucleus (it is far from resting!), is a striking feature, peculiar to Dinophytes (see also figs 333, 335), virtually all of which can be recognised by their nuclei alone. In other algae, chromosomes are not visible in living cells, except sometimes when their nuclei are dividing.

Like most Dinoflagellates, *Ceratium* has yellow or brown coloured chloroplasts. Apart from photosynthesis it appears that *Ceratium* may ingest other microorganisms but it is uncertain how commonly this happens in freshwater.

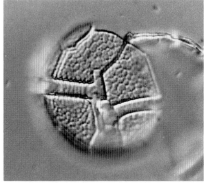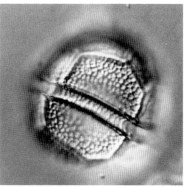

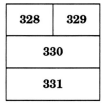

Figs 326, 327. *Peridinium* (× 820). N.

Fig. 326. The side (ventral) with the longitudinal groove (lower half of cell) and its junction with the horizontal one.

Fig. 327. The opposite side (dorsal) with the transverse groove.

Peridinium (figs 326-331) has more globose cells than *Ceratium* and no horns, though some species have spinous extensions to the wall. The number and arrangement of the cell wall's plates differ from those in *Ceratium*. As with *Ceratium* and other Dinoflagellates, the cell is divided into an upper and lower half, delimited by the transverse groove, and the two sides are different. One side (fig. 326) has a broad groove running down the lower half of the cell. This is the groove from which the long external flagellum seen for *Ceratium* in fig. 319 arises. At its apex, the longitudinal groove joins the ends of the transverse one which girdles most of the cell and contains the second flagellum. The opposite side of the cell (fig. 327) has no longitudinal groove, only the transverse one.

Some of the plates composing the cell wall are seen in figs 326, 327 and 331. The plates are divided up into an angular meshwork and demarcated from one another by linear areas which lack the reticulate ornamentation of the plates themselves.

Not all armoured Dinoflagellates have a type of cell division like that of *Ceratium*. In *Peridinium* the cell can divide into two within the plated cell wall (fig. 330). The daughter cells at first lack the plated wall and look like cells of *Gymnodinium* (fig. 338). In the species shown (fig. 330) mucilage is produced during cell division; remnants of the plates of the mother cell can be seen on the surface of this mucilage. *Peridinium*, like *Ceratium*, is common in both inland waters and the sea.

Figs 328-331. *Peridinium*.

Fig. 328. Cell (45 μm diam.) seen from side with the longitudinal groove.

Fig. 329. Cell (54 × 63 μm) seen from opposite side to fig. 328 and hence only transverse groove visible.

Fig. 330. Divided cells (plus envelope of mucilage 131 × 100 μm). I.

Fig. 331. Cysts (37 & 31 μm). Note fragments of cell walls, and the plates of which they are composed.

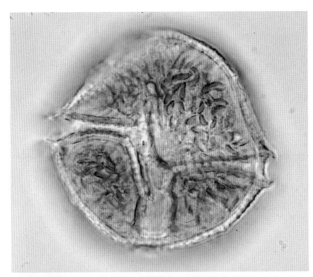
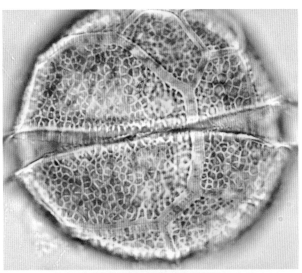
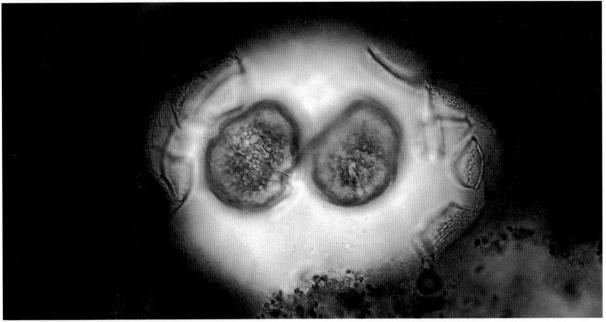
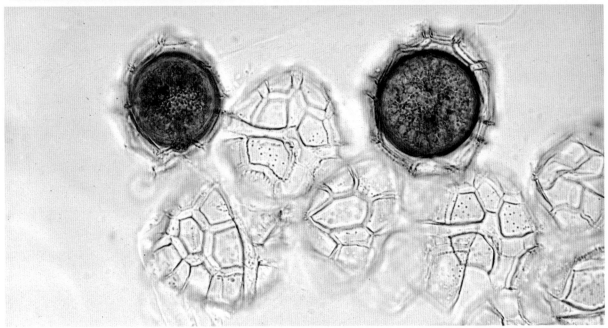

Cysts

Ceratium, *Peridinium* and many other Dinoflagellates form cysts. The cyst is formed inside the cell wall which eventually breaks to release it.

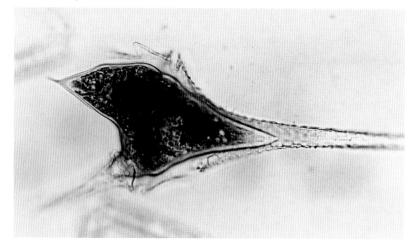

Fig. 332. *Ceratium*. Cyst not yet shed from mother cell (× 640).

In *Ceratium* (figs 324, 325, 332) the number and length of the spinous extensions to the cyst reflects those of the horns and the general shape of the cells within which they were formed. Two cysts of *Peridinium* still contained within their original armoured cell walls, as well as several broken empty specimens are present in fig. 331.

The cysts of *Ceratium* and many other Dinoflagellates are also resting cells. They are surrounded by a thick membrane and contain abundant food reserves. In some cases the involvement of a sexual process has been reported.

In temperate and northern regions the cysts of *Ceratium* usually appear in late summer or early autumn at the end of a summer period of population growth. They sink rapidly to the bottom of the water concerned and remain there until the water begins to warm up next spring or early summer when they germinate. It follows that the number of spores formed each year and the proportion of these spores which survives the winter, and death through fungal parasitism (see later) or ingestion by animals, is a very important factor in determining the success of the population which develops in the succeeding summer. Spores can survive for more than a year but not many last more than a few years. Why some spores do not germinate in the succeeding year is not known for certain. However, if they are buried in the mud and silt at the bottom of a lake, they may not be sufficiently warmed the next year to bring about their germination. There is as yet no evidence that even if spores die their walls will remain undecomposed for long periods of time in freshwater. However, in marine deposits, fossil remains of Dinoflagellate spores are relatively common and can be valuable stratigraphic markers in oil exploration. Fossil coccoliths (p. 168) can also be valuable stratigraphical markers.

The usual fate of organisms ingested by Dinoflagellates is their death and digestion. The *Peridinium* in fig. 340 has ingested three microorganisms which have undergone so much digestion that they are unidentifiable. Their remains fill more than half the cell and the amount of food ingested in relation to the size of the *Peridinium* is remarkable. This *Peridinium* lacks chloroplasts. In old terminology it certainly is an animal since it both swims, captures and "eats" its food. In figs 333, 335 the large nucleus and chromosomes of the same *Peridinium* are especially clear. However, in the latter case there is a body with a thick wall and shiny content in the *Peridinium* and in fig. 336, two such bodies are present. These bodies are the resting spores of an undescribed protozoan which preys on *Pseudocarteria*, a Chlorophyte differing from *Chlamydomonas* (p. 80) in having four flagella and many contractile vacuoles. The cells are entered and their chloroplasts devoured. The remains of such a cell can be seen in fig. 334 and here too, two resting spores of the predator are present. Whether the *Peridinium* ingests algal cells plus the contained resting spores of the predators or whether only the latter after their release into the water remains unknown.

333	334
335	336

Figs 333, 335, 336. A "colourless", holozoic *Peridinium*.

Fig. 333. Note large Dinophyte nucleus (× 1600).

Figs 335, 336. The *Peridinium* cells have ingested one and two resting spores of a protozoan (× 1600).

Fig. 334. Two spores of the protozoan in a cell of the Chlorophyte *Pseudocarteria* which it infests (× 1008).

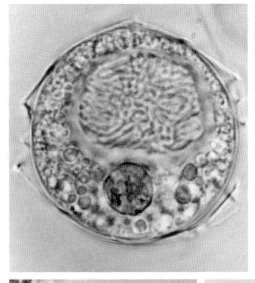
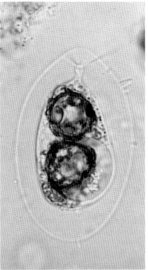
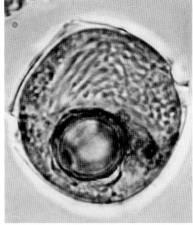
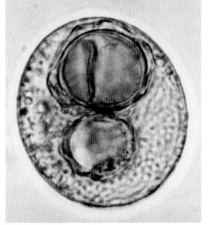

Gymnodinium (figs 338, 339) lacks the cell wall plates of the preceding two genera. Most species are photosynthetic and brown in colour (fig. 338). A very small number of species of *Gymnodinium* are blue-green or blue in colour (fig. 339). Several lack chloroplasts and live by ingesting other microorganisms (e.g. like the *Peridinium* in figs 333-335, 340). It is known that some of the blue-coloured species are dual organisms in that inside a *Gymnodinium* which lacks chloroplasts is one, or possibly, more cells of a Cryptophyte with blue-coloured chloroplasts. The Cryptophyte which is now living within the *Gymnodinium* may no longer possess all the organelles of a free-living cell and indeed be represented by little more than its chloroplasts. The *Gymnodinium* has, as it were, domesticated the Cryptophyte or the bits of it which can carry out the process of photosynthesis.

How has such a dual organism originated? Presumably by a *Gymnodinium* catching and ingesting the Cryptophyte and in one case it has been shown that colourless (chloroplast-less) *Gymnodinium* cells do ingest cells of a blue-coloured Cryptophyte. The Cryptophyte cannot escape from the *Gymnodinium* nor can it reproduce. In laboratory experiments it has been impossible to grow the *Gymnodinium* without supplying it with cells of the Cryptophyte. Presumably, all blue-coloured species of *Gymnodinium* are also such dual organisms. Some may have ingested a Cryptophyte long ago and, though other parts of the Cryptophyte cell have been lost, the chloroplasts remain as functional organelles which can multiply when the host *Gymnodinium* does so. In fact, we now have the evolution of an independent unitary organism derived from the incorporation of one organism into another. It may be that other kinds of algae or other organisms have arisen in analogous ways. This possible evolutionary process is considered further on pages 252-253. The blue-green Cryptophyte is called *Chroomonas*.

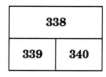

Fig. 338. *Gymnodinium*. Cells (largest to 30 × 26 μm) with red eye-spots.

Fig. 339. *Gymnodinium*. Cell (32 × 27 μm) containing one or more blue Cryptophytes which act as its chloroplasts.

Fig. 340. *Peridinium*. The same species as in figs 333-336, here containing partially digested remains of 3 microorganisms (probably algae). Cell 35 × 30 μm.

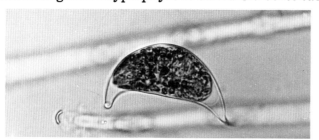

Fig. 337. *Cystodinium* (× 640).

Cystodinium (fig. 337) is one of several genera which are non-motile except when reproducing. Some live attached, often to an alga, others are free-floating. In the species shown, the cells are lunate with spinous poles. There are numerous brown chloroplasts. Reproduction is by zoospores, usually 2-4 in number, looking just like cells of a *Gymnodinium*.

Species of *Cystodinium* are quite common in small bodies of water such as boggy pools and ponds.

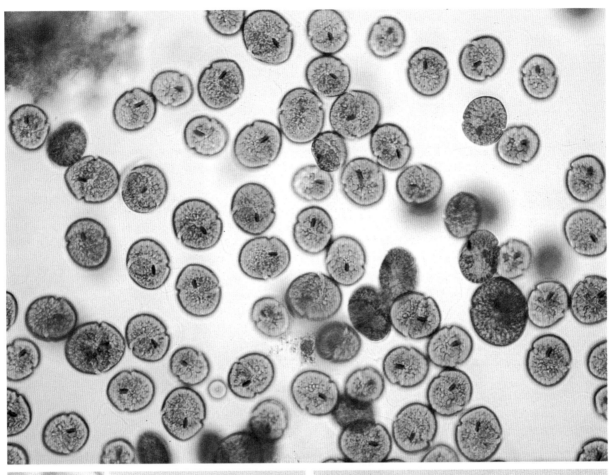
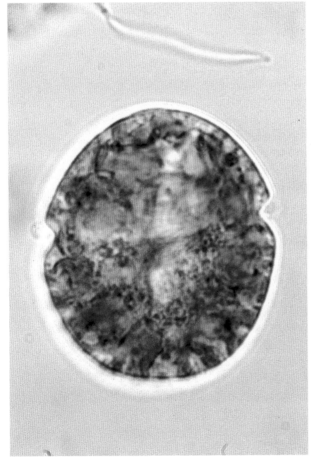
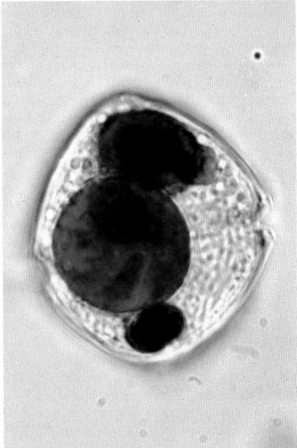

Chapter Nine

The Rhodophytes

Most Rhodophytes or Red Algae are marine. Many are macroscopic rather than microscopic and with their red or reddish colour are easily recognised as Rhodophytes. These red seaweeds are common on the seashore, either between tidemarks or in deeper water. Freshwater Rhodophytes are fewer in number and so form a less prominent part of the flora of inland waters. Further, few are red in colour. Nevertheless, freshwater Rhodophytes are of widespread occurrence, especially in rivers and streams.

Porphyridium (fig. 341) is one of the few unicellular Rhodophytes. However, it often is visible to the naked eye because the cells live in gelatinous masses of a blood or wine-red colour. The cells contain a single chloroplast, the main mass of which occupies the central regions. From this central mass, lobes pass out towards the cell surface, so giving the whole chloroplast a somewhat stellate appearance. The cells can move by a kind of slow gliding motion effected by the excretion of mucilage. The only known method of reproduction is by cell division.

Porphyridium can grow on salty or non-salty soils. The species shown is particularly common in greenhouses and at the base of walls and trees.

An example of a Rhodophyte which is as red as any red seaweed is seen in fig. 342. This alga, *Hildenbrandia*, is growing on an old-fashioned ginger beer bottle found on the bottom of a lake. Its normal habitat is as a thin crust on rocks and stones in running water. It nearly always is found in shady stretches or, if in the open, on the shaded side of rocky places. In lakes, it is likely to be most common in deep water.

The vast majority of Rhodophytes are filamentous. The crusts of *Hildenbrandia* (fig. 342) contain threads a few cells high which are tightly appressed to one another. Many Rhodophytes have main shoots composed of one or more filaments from which diverse kinds of short branches arise. The filaments, particularly those of the shorter branches, can be so tightly packed together that they appear to be part of a sheet or tissue of cells rather than an aggregation of filaments.

Lemanea (fig. 343) is an example of a Rhodophyte whose filamentous construction is not obvious from its general appearance. Yet its structure basically is the same as that of the next alga shown (*Batrachospermum*) where the filamentous nature of the plant is easy to see.

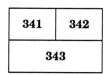

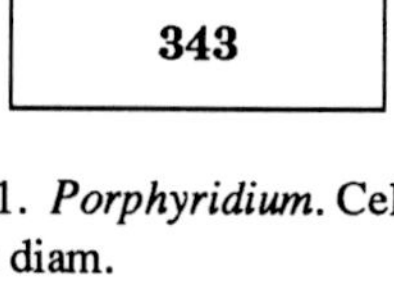

Fig. 341. *Porphyridium*. Cells 5-7 μm diam.

Fig. 342. *Hildendbrandia* growing on an old-fashioned ginger-beer bottle retrieved from the bottom of a lake.

Fig. 343. *Lemanea*, circa life size.

Lemanea is a stiff bristle-like branched or unbranched plant, somewhat similar to coarse black or dark olive-green horsehair. Close inspection will show that the plant has small swellings at intervals along its length. However, the apparent swellings in fig. 343 are tufts of another Rhodophyte growing on it. The plants commonly are 10 cm or more in length and grow in bunches. There is no other freshwater alga looking like it and this, combined with its relatively large size, make it easily recognisable.

Lemanea is common in fast flowing, even torrential parts of rivers. In summer, it may disappear or seem to do so. It has a second stage in its life-cycle composed of single, branched filaments, only visible in the mass. This kind of filament is shown later (*Rhodochorton*).

Batrachospermum (figs 344-358) consists of branched, beady threads reaching several centimetres in length (fig. 345). On the basis of its beady appearance to the naked eye, it has been given the common name of the frogspawn alga. This is something of a misnomer since frogs' eggs are distributed randomly in a formless mass of jelly. A better name would be toadspawn alga because the common toad releases its eggs in strings which are more like the beady threads of *Batrachospermum*. The degree of beadyness varies from species to species.

The structure of the *Batrachospermum* plant represents a simple type of the filamentous construction typical of Rhodophytes. There is a main axis composed of a filament containing the largest cells from which arise secondary axes (fig. 346) with a similar type of cell. These primary and secondary axes are potentially of unlimited growth and so length. From both main and secondary axes bunches of richly branched short filaments arise (figs 346, 348, 358). These branches are of limited growth in length. The bushy short branches arise at regular intervals along the main axes, giving the plant its characteristic toadspawn appearance. In many species there are further kinds of filaments. For example, from each node where the short lateral branches arise, filaments may grow up and down the cells of the central axis (fig. 344), even to the extent of covering it completely. Similar filaments at the base of the plant can attach it to the substratum.

The mucilage in which the plant is embedded often is inhabited by a variety of microscopic algae (fig. 346); in this instance numerous small diatoms.

Figs 345-348. *Batrachospermum*.

Fig. 345. Spread out in a dish (diam. 83 mm).

Fig. 346. Long (main) and short (lateral to 312 μm l.) branching system; note many small algae living in the surrounding mucilage, the larger of which can be seen to be diatoms.

Fig. 347. Sexual reproduction. Four male cells (5.5 μm br.) attached to an elongate female cell situated at the apex of a branch (centre, righthand side of picture). P.

Fig. 348. Tufts of short branches. The very dense ("bird's nest") balls of branches (250 μm diam.) arise after sexual union and later liberate spores.

Fig. 344. *Batrachospermum*. A cell from a main axis hidden by a covering of filaments arising from a node (× 256).

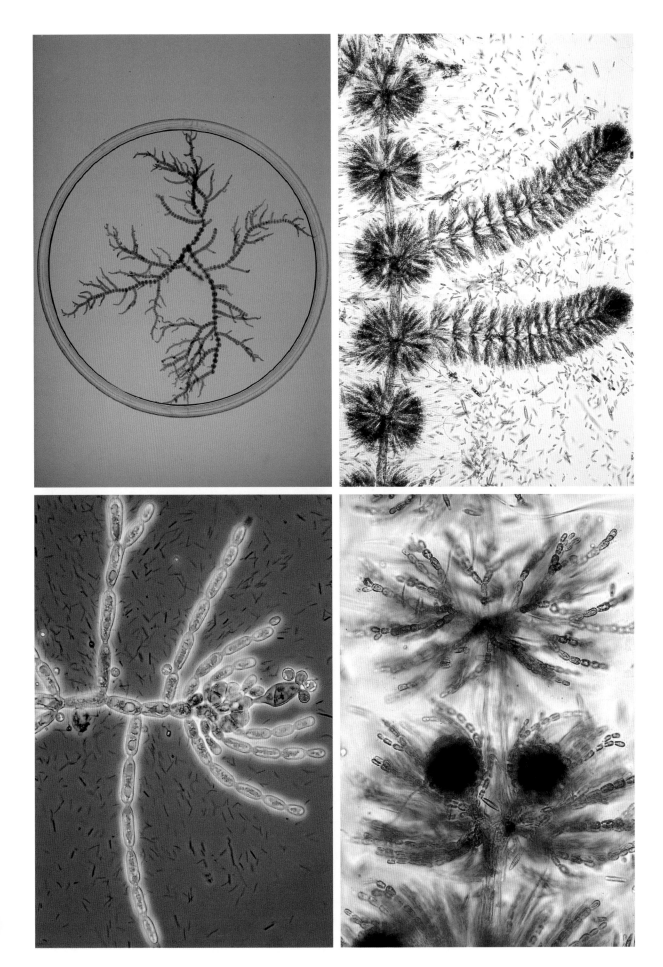

As mentioned, the basic construction of *Lemanea* is similar to that of *Batrachospermum*. However, there is a much greater differentiation between the main central thread and its branches. The short, outermost filaments are so tightly appressed to one another that what appears to be an outer skin-like tissue is produced and this gives the plant its firmness. It is more difficult to see sexual reproduction (see below) in *Lemanea* than in *Batrachospermum* because of this dense outer surface.

The bunches of short branches of *Batrachospermum* often contain even denser masses of branched filaments looking like a bird's nest (figs 348, 355). These dense masses arise from sexual reproduction. The female cell is located at the end of a very short branch (figs 347, 353, 354). It usually looks quite different from the terminal cells of neighbouring branches. In many Rhodophytes the upper part of the female cell is smaller than the lower part, though it may be longer. In this *Batrachospermum*, the upper part is both longer and broader than the lower, from which it is demarcated by a short neck-like constriction. The female cells can be difficult to find among a mass of lateral branches. In fig. 354 two female branches are present in the V-shaped region between the main filament and the bunch of lateral branches (at one o'clock). They are markedly shorter than the asexual filaments below them. On top of each female cell is a spherical male cell. In fig. 348 the two "bird's nests" of branches arising after sexual unions are in the same positions; one on each side of the main filament in the V-shaped space between it and the bunch of short lateral branches.

The male cells also arise terminally, often in clusters (fig. 350). They are small and globose and after liberation (fig. 352) depend on water movements in order to reach a female cell. In fig. 347, four male cells are attached to the end of a female cell. The content of only one male cell will enter the female cell and its nucleus fuse with that of the female cell; it appears that the contents of the lowermost male cell are entering the female cell. In fig. 353, a single male cell has made contact with the tip of the female cell which its content will soon enter, judging from the loss of the wall on the female cell where it is in contact with the male cell. Fertilisation takes place in the basal part of the female cell.

From this sexual union arises the dense mass of filaments referred to as the bird's nest stage (fig. 348). This stage reproduces by asexual spores which are formed singly on the ends of its filaments (fig. 356). The spores are non-motile and resemble those produced by the male cells. These spores germinate (fig. 357) to form microscopic structures composed of branched filaments, like those of the next genus *Rhodochorton* (e.g. fig. 364). This is called the *Chantransia* or chantransioid stage for reasons explained later.

Figs 349-355. *Batrachospermum*.

Fig. 349. A terminal female cell.

Fig. 350. Globular male cells at ends of branches.

Fig. 351. A terminal male cell.

Fig. 352. The cell content of the male has just been liberated.

Fig. 353. A female cell with a male on its apex.

Fig. 354. In the V between a cell of a main axis and a bunch of lateral branches (righthand side), two short female branches with a male cell on the apex of each female cell (arrowed).

Fig. 355. Part of a plant with many "bird's nest" stages (black globular bodies).

Figs 349-353 (× 1008); fig. 354 (× 256); fig. 355 (× 20).

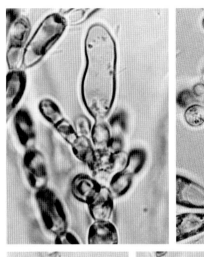
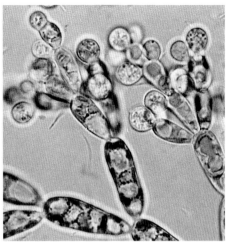
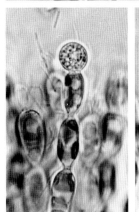
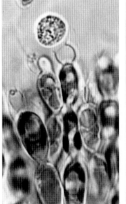

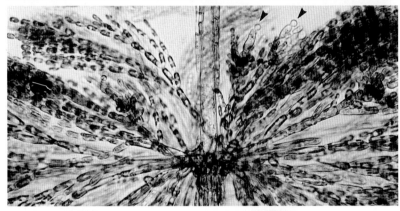

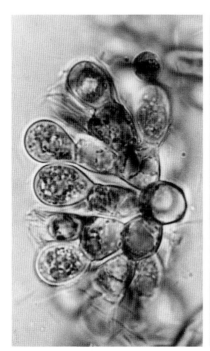
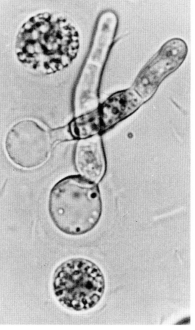

Figs 356, 357. *Batrachospermum*.

Fig. 356. Spores forming at ends of branches of the "bird's nest" stage (× 1380).

Fig. 357. Two spores which have germinated (short filaments) and two which have not (× 1600).

The chantransioid plants in time grow into the macroscopic, "toadspawn" form of *Batrachospermum*. "In time" may mean quite a long time, because the juvenile form can reproduce itself by its asexual spores which are similar in appearance to the spores from which this stage arose. These spores are called monospores. The adult plants of a few species of *Batrachospermum* can also produce such monospores. Therefore there are three kinds of body on such a plant which look similar enough to be confused with one another.

So, the life-cycle of *Batrachospermum* contains three kinds of plant or three generations. The "toadspawn" plant (fig. 346) whose cells contain a single set of chromosomes until sexual reproduction takes place, as a consequence of which the "bird's nest" growths of tightly packed filaments arise. This second generation (fig. 348) grows on the toadspawn stage and its cells contain a double set of chromosomes, one from the female and the other from the male nucleus. This generation produces spores (fig. 356) which germinate (fig. 357) to form the chantransioid stage (third generation) which has filaments like those of a *Rhodochorton* (e.g. fig. 364). One of the branches of the chantransioid stage develops into a new toadspawn stage and somewhere – it is not always clear exactly where – during this process the number of chromosomes per cell is halved. This is a complicated life cycle though no more complex than the life cycles of many other Rhodophytes or indeed many other eukaryotes. It ensures a new mix of genes in the process of doubling and then halving the chromosome number, just as in a variety of ways this arises in us and other eukaryotes.

Fig. 358. *Batrachospermum*. A somewhat greenish coloured species (cells of main axis up to 23 μm br.). Note a few diatoms living in the mucilage.

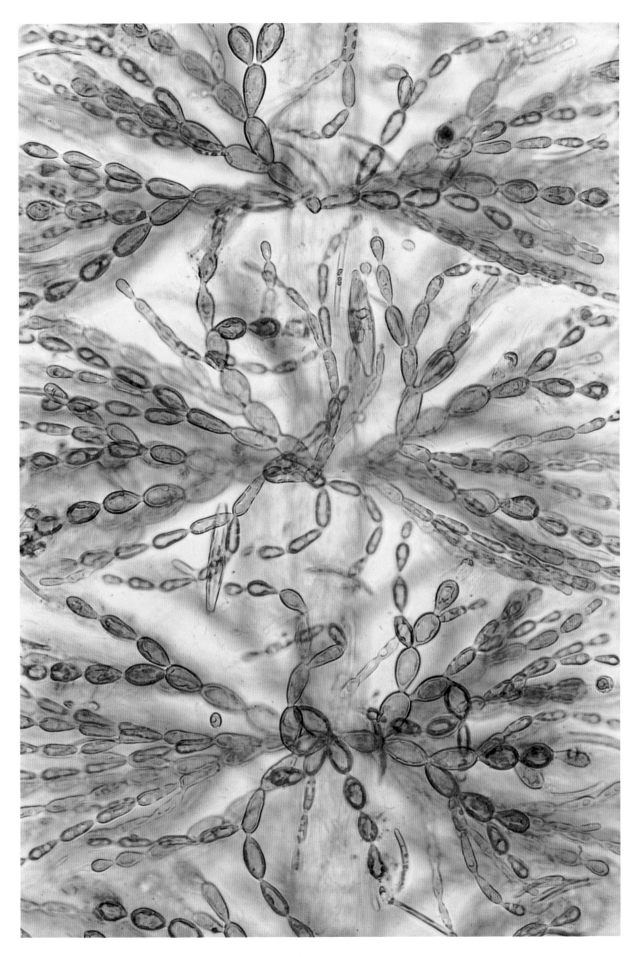

Batrachospermum is a common alga, notably but not always in running water. It is very variable in colour both from species to species and in relation to the environmental conditions. It can be shades of violet, yellow or brown and a group of species are greenish (fig. 358) or bluish-green but neither the green of Chlorophytes nor the blue-green of the next group, the Cyanophytes. *Batrachospermum* does not live in saline water.

Rhodochorton (figs 359-370) also varies considerably in colour. Common colours are reddish brown, vinous red, olive green, violet-red or violet. Unlike *Lemanea* and *Batrachospermum*, there is no such marked differentiation into main, secondary or tertiary axes. It consists mainly of simple, branched filaments (figs 364, 368).

The common type of reproduction is by asexual, unicellular spores, the monospores, produced at the apices of the branches (fig. 365). Branches bearing monospores usually are short and arise in small clusters (fig. 361).

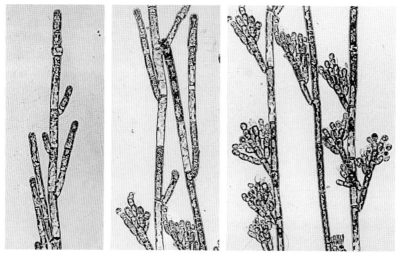

| 359 | 360 | 361 |

Figs 359-361. *Rhodochorton* (*Audouinella*). Progressive series from apex to lower parts of a plant bearing clusters of monospores (× 160).

Two other methods of reproduction exist but are not easy to find. One of these is sexual, like that of *Batrachospermum* but no dense mass of filaments develop after fertilisation, hence there are no "bird's nests".

The second type of reproduction is found on plants of similar construction to those bearing sex organs or only monospores. However, instead of sex organs they form sporangia (tetrasporangia) containing four spores (tetraspores) arranged tetrahedrally (fig. 369). Each tetraspore possesses a single set of chromosomes and so germinates to produce a potentially sexual generation.

Thus, the life-cycle of *Rhodochorton*, though basically the same as that of *Batrachospermum*, differs in its tetrasporic generation. One generation is similar to that of the "toadspawn" generation of *Batrachospermum* in that its cells have a single set of chromosomes until sexual organs are produced and the female cell is fertilised. These male and female cells normally develop on separate plants. From the fertilised female cell (i.e. the zygote) there arises a generation which, as in *Batrachospermum*, grows on the

| 362 | 363 |
| 364 | 365 |

Figs 362-365. *Rhodochorton* (*Audouinella*).

Fig. 362. Filament masses on a rock.

Fig. 363. Pink and green coloured cells (9-12.5 μm br.).

Fig. 364. Branching system.

Fig. 365. On lefthand side three developing monospores, the largest (12.5 μm diam.) is almost mature.

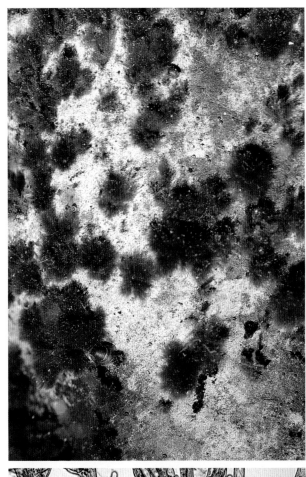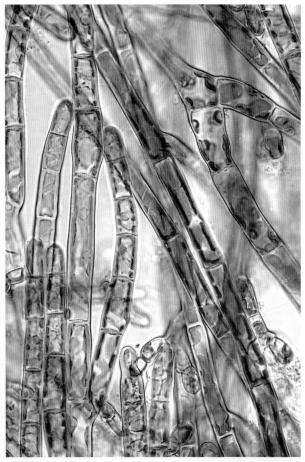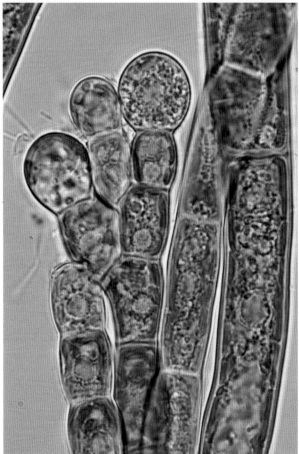

sexual plant but does not produce such a tight mass of filaments as in the "bird's nest" stage of *Batrachospermum*. Spores from this stage germinate to form a generation of similar structure, to the chantransioid generation of *Batrachospermum* and so to the previous, sexual generation in *Rhodochorton* itself. Unlike *Batrachospermum*, filaments of this *Rhodochorton* generation do not

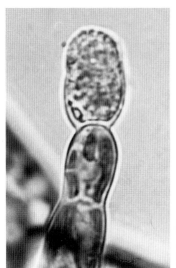
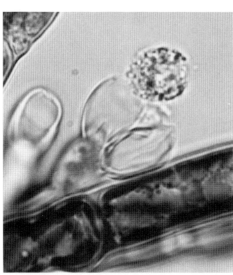
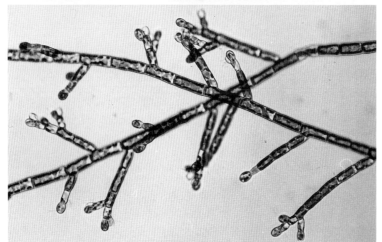
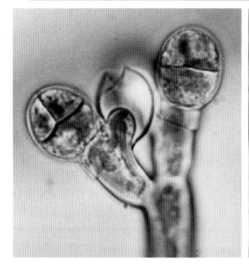
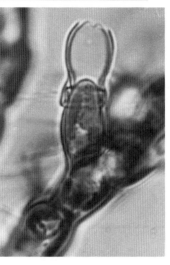

Figs 366-370. *Rhodochorton*.

Fig. 366. A monospore issuing from a sporangium (monosporangium) (× 1750).

Fig. 367. A liberated monospore (× 1780).

Fig. 368. A tetrasporic plant (× 285).

Fig. 369. Two tetrasporangia beginning to divide into tetraspores (cross-walls visible). In the centre is an empty tetrasporangium, with a cell growing into it so that proliferation may take place (× 1780).

Fig. 370. An empty tetrasporangium, the two walls indicate that proliferation has occurred (× 1780).

transform into the next generation, that is back to the sexual generation. Instead, they liberate spores (tetraspores) in the production of which the chromosome numbers are halved. The plants so arising are potentially sexual, so closing the cycle of life. The word "potentially" is used because both generations with one or two sets of chromosomes can reproduce for a prolonged period solely by monospores. These spores are more important in increasing plant numbers than are the sexual or tetraspore plants but, of course, do not alter the genetic constitution of the plants arising from them. The restriction of the sexes to separate plants is not peculiar to *Rhodochorton* or even all its species.

There is argument about the correct name of this genus. *Audouinella* seems to be preferred to *Rhodochorton* at present. We retain the name under which the life history of the plant shown was elucidated. An older name for a variety of similar plants is *Chantransia*. It is known now that most of the *Chantransia* species are stages in the life histories of other genera such as *Lemanea* and *Batrachospermum*. Hence, the name for these stages is the *Chantransia* or the chantransioid stage. The *Chantransia* stage of *Batrachospermum* has been mentioned and its monospores which, when present, are the same as the monospores of *Rhodochorton*. Since a chantransioid stage of *Batrachospermum* or *Lemanea* can reproduce itself and be indistinguishable from *Rhodochorton*, it may be a long time before a certain identification is possible, for example by seeing a chantransioid plant develop into a young plant of *Batrachospermum* or *Lemanea*. An example is the *Rhodochorton* in our garden stream. Chantransioid plants have grown under a small waterfall for some 45 years. No plants of *Batrachospermum* or *Lemanea* have ever been seen in the stream during these years. Therefore, it seemed unlikely that the plant growing under the waterfall was a chantransioid stage of another Rhodophyte. The view that the plant was neither a *Lemanea* nor a *Batrachospermum* was substantiated when we found the tetraspore stage which is present in *Rhodochorton* (and many other Rhodophytes) but not in *Lemanea* or *Batrachospermum*.

This species of *Rhodochorton* (or *Audouinella*) is common in rivers. The majority of *Rhodochorton* species are marine. The tufts of filaments on the *Lemanea* plants mentioned earlier and seen in fig. 343 probably belong to our *Rhodochorton* but the reproductive bodies which could prove this belief were not found in the material examined.

It will have been noted that only non-motile spores have been mentioned. There are no free-swimming, flagellate stages in the life history of any Rhodophyte. In this they are like the next group the Cyanophytes but in cell structure and methods of reproduction they are very different.

Chapter Ten

The Cyanophytes (Cyanobacteria: blue-green algae)

It was realised long ago that Cyanophytes stand apart from all other algae. Over a hundred years ago there were those who allied them with bacteria. Certainly there was something odd about their "nuclei" and "chloroplasts". Indeed, it soon became clear that there were no structures like the nuclei of plants and animals or the chloroplasts of photosynthetic plants or, as might be said, of ordinary, typical plants. At usable magnifications obtainable with light microscopes (circa 1500 times), it seemed that hereditary (genetic) material was present in some kind of structures akin to the chromosomes of plants and animals but they were not assembled in a body like a nucleus. The same was true of the photosynthetic apparatus which was not confined within one or more chloroplasts.

Electron microscopy revealed that the cell structure of Cyanophytes was bacterial in nature and it and other types of modern biochemical and physiological investigations have revealed other features in common with bacteria. It became obvious that whereas there were, as we have seen, no absolute differences between animals and plants there were such between both of them and bacteria and related organisms.

As was mentioned in the introduction, the living world, as we now know it, consists of two groups or kingdoms of organisms which are given the names, prokaryotes (also spelt procaryotes) and eukaryotes (also spelt eucaryotes). The former includes bacteria and their allies such as Cyanophytes, the latter animals, plants and fungi. Among the prokaryotic features of Cyanophytes are the absence of a nucleus, and so also nucleolus, and of chloroplasts. In other algae, all of which are eukaryotes, the main genetic (hereditary) material, such as the chromosomes, is enclosed in a body (organelle) separated from other cell bodies by its own membrane (e.g. figs 47, 86). The same applies to the chloroplast, which contains the photosynthetic apparatus and a certain amount of genetic material of its own.

Today, Cyanophytes commonly are called Cyanobacteria in scientific texts; in the media and popular press they are referred to as blue-green algae. As yet, only some Cyanophytes have been characterised and classified by purely bacteriological methods. The only comprehensive oversights of the group are those based on a botanical and to a large extent pre-electron microscopical basis, that is, these works are floras. The most recent revision of the group, which is in the form of a series of articles, includes many changes from these floras. It follows botanical nomenclature and is based on these past floras, though taking into consideration modern knowledge. Cyanophytes are still found in all modern books about algae and so both in the botanical and bacteriological literature.

In such a pictorial introduction, it might have been convenient to exclude what is said above, which may seem rather confusing to the layman, but it would not have been scientifically correct. Moreover, it is necessary to point out that Cyanobacteria have two features not found in other bacteria but characteristic of plants. They contain the photosynthetic pigment chlorophyll a and release oxygen in photosynthesis. Indeed ecologically it is not unreasonable to say that many Cyanobacteria are plant-like and they are always likely to be included in any discussion of the plant plankton (phytoplankton).

We have used the name Cyanophytes in place of Cyanobacteria or blue-green algae to correspond with those given for the other groups of algae (e.g. Chlorophytes, green algae; Rhodophytes, red algae).

Cyanophytes may be blue-green in colour but can also be more blue than green, reddish, violet, yellow or brown. They are not the grass green of Chlorophytes. The green of their chlorophyll is masked by or blended with blue and red photosynthetic pigments or by yellow or brown coloured walls or sheaths external to the protoplasm. The cell contents of many species vary in colour under different environmental conditions. This variation is caused by alterations in the proportions of the photosynthetic pigments. Cyanophytes include unicellular, colonial and filamentous forms. There are no flagellate stages. Cyanophytes are a very ancient group, they or something closely similar are known from rocks over a billion years old. As probably the first photosynthetic organisms releasing oxygen in the process, they so altered the primitive, virtually oxygen-free atmosphere that the way was opened for the evolution of the living world we now inhabit. Today, they are of universal occurrence in fresh and salt water, from the antarctic to the arctic, in permanently frozen lakes and in hot springs, in soil, deserts and tundra.

Cyanothece (figs 373, 374) is unicellular. Its species previously were included in the genus *Synechococcus*, the cells of which usually are markedly smaller. It has one method of reproduction, a method common to many unicellular bacteria. The cell divides into two (binary fission) transversely. A cell starting to divide is seen in fig. 374. A cross-wall has begun to grow inwards, rather like the closing of a camera's diaphragm. As the new wall grows, so the cell constricts more and more, eventually separating into two equal parts.

Species of *Cyanothece* are common in fresh and salt waters. The cells of some of its species, together with species of *Synechococcus* and of a third genus (*Synechocystis*), are minute (circa 1 µm long or broad). These small organisms until recently largely have been overlooked or, if seen, mistaken for other bacteria because they appeared to be colourless. However, in a few cases, they occurred in such large numbers that the water appeared blue-green or, when they were cultured it became clear that they were Cyanophytes. The recent realisation that such minute Cyanophytes are an important part of both the marine and freshwater plankton has been caused mainly by the use of microscopes and stains which reveal the photosynthetic pigments as fluorescent colours. Then these algae shine out as minute coloured specs against a black background. The taxonomy of Cyanophytes still is mainly botanical but bacteriologists are erecting a rather different system. Until these two systems fuse, the taxonomy of these minute species will remain confused as is that of *Cyanothece* and *Synechococcus*.

Figs 373, 374. *Cyanothece*.

Fig. 373. (45 × 22.5 µm).

Fig. 374. (42 × 22.5 µm). Early stage in cell division.

Fig. 375. *Chroococcus*. Four-celled colony (50 × 65 µm). N.

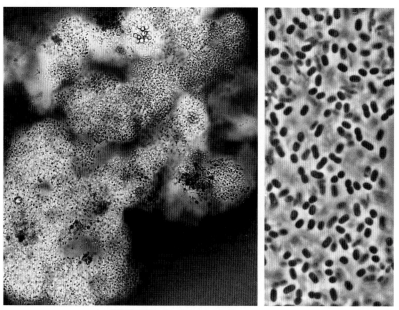

Figs 371, 372. *Aphanothece*.

Fig. 371. A colony (× 256). I.

Fig. 372. Cells, recently divided ones still in pairs (× 1600). P.

Aphanothece (figs 215, 216, 371, 372) is colonial. The cells are like those of *Synechococcus* but in irregular masses held together by mucilage as in *Microcystis* (p. 202). Like *Synechococcus*, different species may have cells of markedly different size. In the planktonic species shown, the cells are of similar size (c. 1-2 µm long and 0.5 µm broad) to the small species of *Synechococcus* referred to above. Because large numbers of cells can be present, the colonies themselves often are easily seen. Nevertheless, even many-celled colonies of this Cyanophyte can appear to be colourless and so misidentified. When their colour can be detected it may be pinkish or blue-green.

The cells of *Chroococcus* (figs 375-377) usually are hemispherical and in pairs, with their straight sides facing one another. These paired cells can occur in small groups

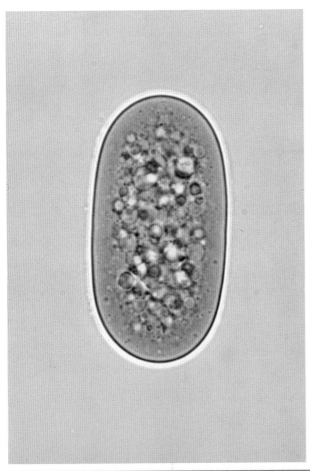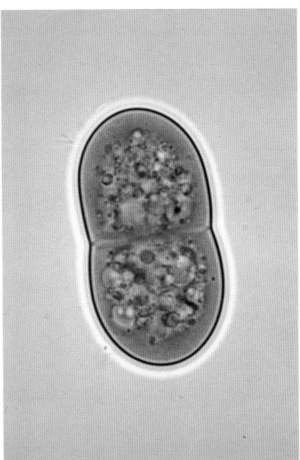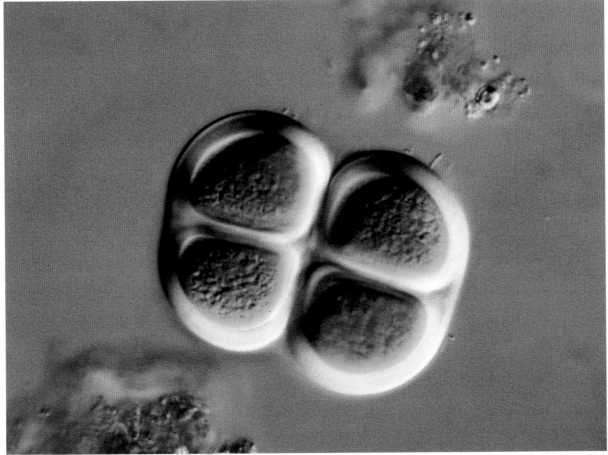

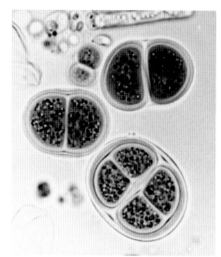
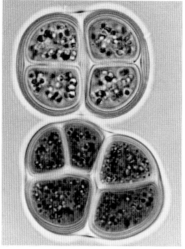

Figs 376, 377. *Chroococcus*.

Fig. 376 shows two and four-celled colonies; in fig. 377 one cell of a previously four-celled colony has divided. Fig. 376 × 765; fig. 377 × 1008.

embedded in mucilage which usually is firm and homogeneous or stratified. *Chroococcus* is particularly common in bog pools and ponds, and on wet rocks.

Merismopedia (figs 378, 379) forms tabular colonies one cell thick. The cells generally are grouped in fours and these groups are in more or less regular transverse and longitudinal rows. The number of cells in a colony can range from a few to thousands. *Merismopedia* is common on soft underwater substrata. Small colonies such as that in fig. 378 can move over a surface, albeit very slowly. In general, the larger the colony, the slower it moves and it seems that the largest colonies, those composed of thousands of cells, are non-motile. How these gliding movements are carried out is not known. Certain species of *Synechococcus* and *Cyanothece* also can move but it is as yet unknown whether all species can do so.

The two colonies of *Merismopedia* shown come from the surface of the same sample of mud and so were living at the same depth in the water and therefore apparently receiving the same quality and the same quantity of light. Nevertheless they are of different colour. As has been mentioned, many Cyanophytes can vary the colour of their cells by altering the proportions of green, red and blue photosynthetic pigments in them. Such colour changes usually depend mainly, if not wholly, on variations in the quantity of blue and red pigments present. It was suggested that such colour changes in nature are reactions to the alterations in the colour of light as it passes through a column of water. Naturally, such colour changes are related both to the colour of the water itself and the kind and amount of suspended matter in it. For example, the colour of the light passing through many freshwaters becomes more and more green with depth. It was suggested that the colour of those algae which can produce significant changes, such as Cyanophytes and Rhodophytes, will change with depth to compensate for the change in the spectral quality of light because some of its

Figs 378, 379. *Merismopedia*, with violet and green coloured cells (c. 3 μm br.) respectively. Large diatom in fig. 378, *Pinnularia* (215 μm l.) and small diatom in fig. 379, *Amphora* (38 μm l.).

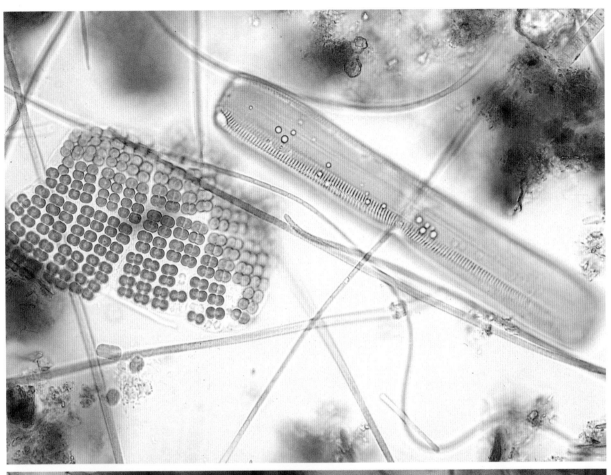

199

wavelengths have been more strongly absorbed than others. The name chromatic adaptation was given to this theory. Some experiments support this theory, for example a Cyanophyte may be more blue in colour in red than in green light. However, this theory does not seem to explain the different colours of the two *Merismopedia* colonies (figs 378, 379) found at the same depth in the water. It is known that races or forms of the same species of Cyanophyte differ in colour under light of a given spectral composition and that some change colour more readily than others and yet others do not change colour at all. A common effect of reduced irradiation is that the total amount of the photosynthetic pigments increases. With algae such as diatoms, with a pigmentation mix that lacks blue and red components and whose colour is almost wholly determined by the brown photosynthetic pigment present, there is little or no possibility for chromatic adaptation. However, it is a common experience to find that diatoms living in poor light are browner than those exposed to high light. To summarise, colour changes in algal cells may be related to the spectral quality of the light they receive or to how bright it is or to nutritional factors, and how readily colour changes or to what degree it can change depends on the alga's genetic make-up.

Chamaesiphon (figs 380-382) is a unicellular Cyanophyte common on rocks, stones and other objects in rivers and lakes. Its cells are cylindrical, ovoid or pyriform. It reproduces by one or more transverse divisions of the apical part of the cell which "bud off" as spherical spores. The spores are liberated when the apex of the cell ruptures.

Fig. 382. *Chamaesiphon* growing on *Rhodochorton* (*Audouinella*). Longest cell (terminated by a spore, 51 μm l.).

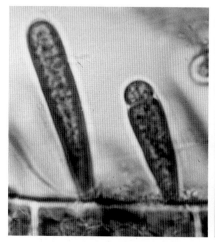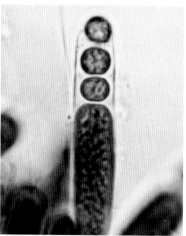

Figs 380, 381. *Chamaesiphon*, one cell releasing a single spore and another, three spores. Both × 1600.

The species shown is common on filamentous algae. Here it is growing on the Rhodophyte *Rhodochorton* (p. 190). All stages in growth and reproduction are present. In fig. 380 a spore is developing and in fig. 381 three fully-formed spores are present, the uppermost protruding from the ruptured cell apex. In fig. 382, spores which have recently become attached to *Rhodochorton* and all stages in subsequent development can be seen.

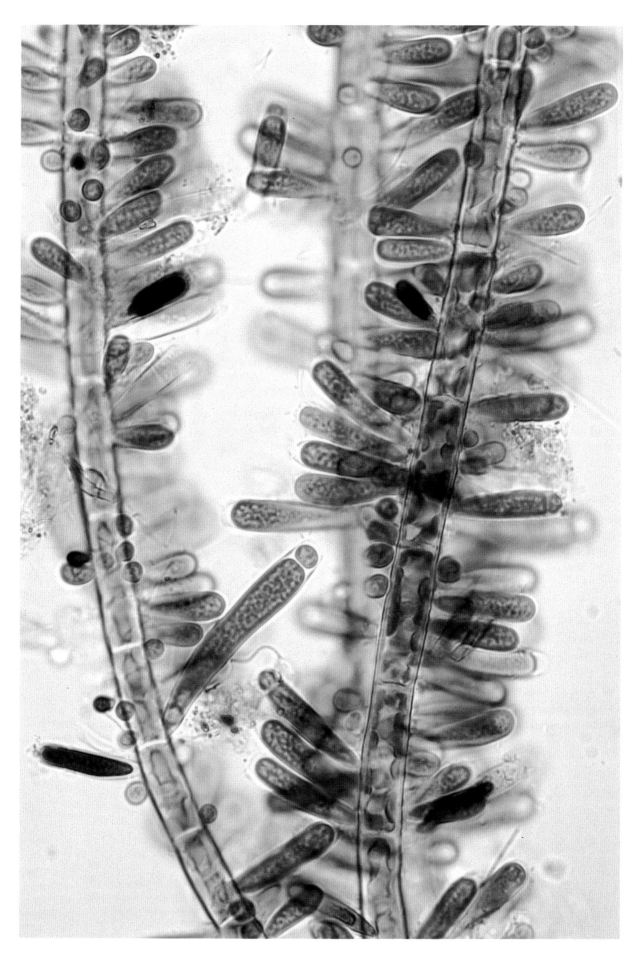

201

Microcystis (figs 383-385). Numerous small, globose, gas-vacuolate cells are embedded in mucilage to form colonies of varied shape and size. The largest colonies are just visible to the naked eye and often have one or more holes through them. Figure 383 is a collection of individual photographs of *Microcystis* which illustrates the variability in colony form and in places includes some artistic licence. All the top row of figures as well as the second and fourth figures from the left on the next line and the "duck" below represent the shapes of single colonies as they were found in nature. Two colonies form the "mushrooms" (cap and stalk), the central "man" (head, body and legs). The "giraffe" comprises four different colonies and the body of the "monster", three.

Gas vacuoles and flotation

Microcystis and many of the other Cyanophytes pictured contain structures called gas vacuoles. The word vacuole is a misnomer because in fact the "vacuoles" consist of aggregations of hollow, cylindrical bodies with conical ends. These gas vesicles are too small to be seen by light microscopy. Similar gas-filled cylinders are found in some other prokaryotes but not in any eukaryote. Recently the word aerotope has been proposed to replace gas vacuole; it remains to be seen if this will be generally accepted.

If the colonies or filaments of Cyanophytes with these gas vesicles are observed from above, that is by reflected light, they appear blue-green in colour (e.g. figs 10, 421). However, under the microscope (transmitted light) and using low-power bright-field optics the gas vesicles impart a blackish appearance to the cells (e.g. fig. 384, *Gomphosphaeria* and *Microcystis*; figs. 401, 402, *Anabaena*; fig. 411, *Aphanizomenon*; fig. 418, *Gloeotrichia*). At high powers of magnification, the cells appear more speckly due to the presence of these gas-filled cylinders (figs 405, 406, 422). Under phase-contrast illumination the gas vesicles are portrayed as areas of whiteness on an overall blueish background (figs 392, 394, 398, 412). An exception is the *Oscillatoria* (*Limnothrix*) in fig. 424 where the gas vesicles are massed into large gas vacuoles in areas apparently free of other cell contents.

Nearly all Cyanophytes with gas vesicles are planktonic species. If sufficient gas cylinders are present in their cells, colonies of *Microcystis* and other gas-vacuolate Cyanophytes will have a lesser weight or density than the water they are in and so float upwards rather than sink downwards like most other algae (but not all, see pp. 36 and 114). In calm weather, the colonies of such Cyanophytes as the species of *Microcystis* shown (not every species is gas-vacuolate) will rise to the surface of the water to form a superficial scum called a waterbloom in English and water flowers in most other languages (figs 10,

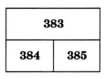

Figs 383-385. *Microcystis*.

Fig. 383. Colonies – a pictorial composition (see text).

Fig. 384. An oval colony (250 × 200 μm) of densely packed cells and an elongate colony with a hole in it, enclosing 7 colonies of *Gomphosphaeria*, many other colonies of which are also present.

Fig. 385. Colony mounted in Indian ink to show the mucilage around it. Cells c. 3-4 μm.

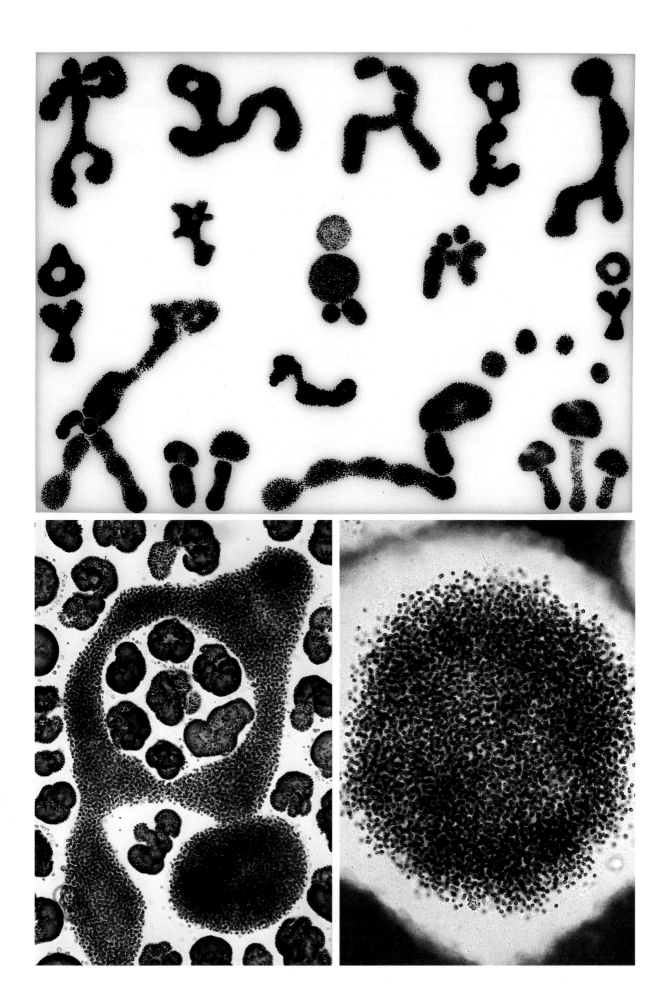

203

421). Commonly, waterblooms include several species of Cyanophytes (fig. 413), though one or more species are likely to predominate; occasionally, the bloom is almost unispecific (figs 391, *Gomphosphaeria*; 403, *Anabaena*; 422, *Oscillatoria*). Often in summer the wind drops in the evening and a still night follows. This gives the right conditions and adequate time for the gas-vacuolate Cyanophytes to float to the surface. Next morning the surface of the water is covered by a striking waterbloom which can give the false impression that these Cyanophytes have increased immensely in numbers overnight. As the land warms up and a breeze develops, the Cyanophytes are dispersed by the consequent water movements. In windy weather, no waterbloom can arise. The chemical environmental conditions favouring the development of large populations of planktonic Cyanophytes are considered on pp. 220-223.

The presence of gas vesicles can be biologically advantageous to the species possessing them. Such Cyanophytes will rise towards the surface and so from any poorly illuminated depth. If wind action mixes them throughout a considerable depth of water, once it drops again they will more or less rapidly return to the surface. Other algae may suffer, for a waterbloom may cut out much of the light below it.

Though rising to the surface ensures better illumination, strong irradiation can be harmful. A waterbloom is exposed to the full blaze of sunshine. The ultraviolet component of light is particularly harmful and can be fatal. This may be one reason why waterblooms can suddenly dissolve into blueish coloured water, the cells having broken up (lysed) releasing their water soluble pigments (fig. 10). Another disadvantage is that a very light breeze will not disperse the colonies in depth but will blow them on to the shore where they will die, often producing an unpleasant scum on the windward side of a lake.

Gas vesicles can collapse if subjected to sufficient pressure, the cells or colonies then will sink. Gas-vacuolate Cyanophytes vary considerably among themselves in the amount of pressure needed to collapse their gas vesicles; that needed to do so in this *Microcystis* is particularly high. It follows, that if wind-induced turbulence in the water drives a Cyanophyte sufficiently deep down into the water, its gas vesicles may collapse and it can no longer float upwards. In some cases, an intracellular rise in pressure may collapse gas vesicles. In photosynthesis, sugars are formed and the more sugar in the cell, the greater the internal pressure (osmotic pressure) within it. This internal rise in pressure obviously is more likely to occur near the surface where photosynthesis will be more active than in less well illuminated depths below. A third possibility of a decrease in buoyancy is that active growth may not be accompanied by active production of new gas vesicles. Then the density of the cells may exceed that of the water and they will sink. Whatever may be the cause of the loss of buoyancy, loss of gas vesicles can be made good by the production of new ones.

The cells of *Microcystis* can contain one or more substances toxic to mammals. Other Cyanophytes also contain toxins. Cases are known from many countries of cattle, sheep and dogs being poisoned by Cyanophytes. There is evidence of human beings becoming ill after taking in water rich in Cyanophytes but there is no certain case of death occurring. Such toxic waters nearly always are very rich in Cyanophyte scums and nobody would drink such water voluntarily, though he might swallow it if he swam in a waterbloom.

The *Microcystis* shown is common in the plankton of waters rich in nutrients and can produce massive waterblooms. The warmer and finer the weather, the more likely it is that waterblooms will arise and the longer they will persist. Some species of *Microcystis* lack gas vesicles and are not planktonic; a few live in wet terrestrial habitats.

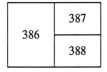

Figs 386-388. *Snowella*.

Fig. 386. Two colonies in a common mass of mucilage (× 640). **I.**

Fig. 387. A small colony showing the mucilage threads arising from its centre (× 860). **P.**

Fig. 388. Cells (× 1665). **P.**

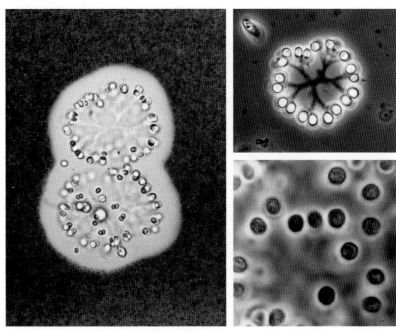

Snowella (figs 386-388) has colonies which are spherical when small, becoming more irregular in shape as they enlarge or divide up into two (fig. 386) or more daughter colonies. The main part of the colony consists of mucilage, a little way below the surface of which there is a layer of small, globose, pale blue, or greyish-blue cells lacking gas vesicles (fig. 388). The cells are joined to the centre of the colony by mucilaginous strands (fig. 387). It is not known whether the mucilage of these strands is the same as that of the rest of the colony but in a more condensed ("jellified") state or differs chemically from it. The same is true of the strands present in the next genus (*Gomphosphaeria*). The strands are not easily seen by bright-field microscopy – a few are faintly visible in fig. 386 – but are clearly visible in phase-contrast (fig. 387). It is important to find out if such strands are present because another genus, *Coelosphaerium*, only differs in their absence and like *Snowella*, is a common planktonic alga.

Gomphosphaeria (figs 384, 389-396) also is colonial. The colonies are globose or somewhat elongate (figs 384, 391) and by volume consist mainly of mucilage. The cells of the species shown contain gas vesicles and hence are speckled black in normal transmitted light (figs 391, 393); in phase-contrast illumination they appear speckled white (figs 392, 394). The cells form a tightly packed layer of irregular outline some way within the surface of the colony (figs 391, 393, 395, 396). Each oval or ovoid cell is near or at the end of a tube of more dense mucilage than the rest (fig. 389). Each tube is part of a branched system of tubes radiating from the central region of the colony (fig. 392).

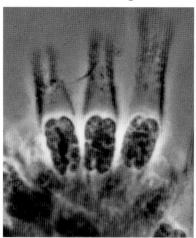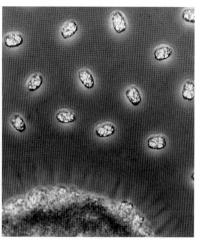

Gomphosphaeria colonies can become very irregular in shape (e.g. fig. 391) as the cells within them multiply. Finally, they break into two or more parts. This and cell division seem to be the only methods of reproduction. However, if a colony is left under a coverslip and the water in it is allowed to evaporate, a stage is reached when the cells shoot out of their mucilage tubes (fig. 390). It is the pressure exerted by the coverslip as the water dries up which causes the ejection of the cells. It has been suggested that ejection of cells in nature could be a major method of colony multiplication. However, even when very large numbers of live colonies are present, for example in a waterbloom, single cells are either very rare or absent.

The inner region of the colony, where there are no *Gomphosphaeria* cells, sometimes contains cells of other Cyanophytes. A mass of small cells can be seen in fig. 393. At higher magnification and with phase-contrast illumination (fig. 394), the cells can be seen to be of identical type to those of the genus *Cyanothece* (figs 373, 374). They do not contain gas vesicles and therefore appear wholly black, thus contrasting with the *Gomphosphaeria* cells also present which appear speckled white in fig. 394. Identical

| 389 | 390 |

Figs 389-390. *Gomphosphaeria*.

Fig. 389. Cells in their mucilage tubes (× 2000). **P**.

Fig. 390. Cells which have been expressed from their tubes by pressure and the edge of the colony they came from (× 720). **P**.

391	392
393	394

Figs 391-394. *Gomphosphaeria*.

Fig. 391. Largest spherical colony, 128 µm diam. **I**.

Fig. 392. Part of a colony squashed to show branched mucilage tubes with cells (3.5-5.7 µm br.) at their ends. **P**.

Fig. 393. Colonies (width of larger 132 µm) containing cells of *Cyanothece* or *Synechococcus* in their central regions. **I**.

Fig. 394. As fig. 393 but at higher magnification. Note dividing cells (c. 2 µm br.). **P**.

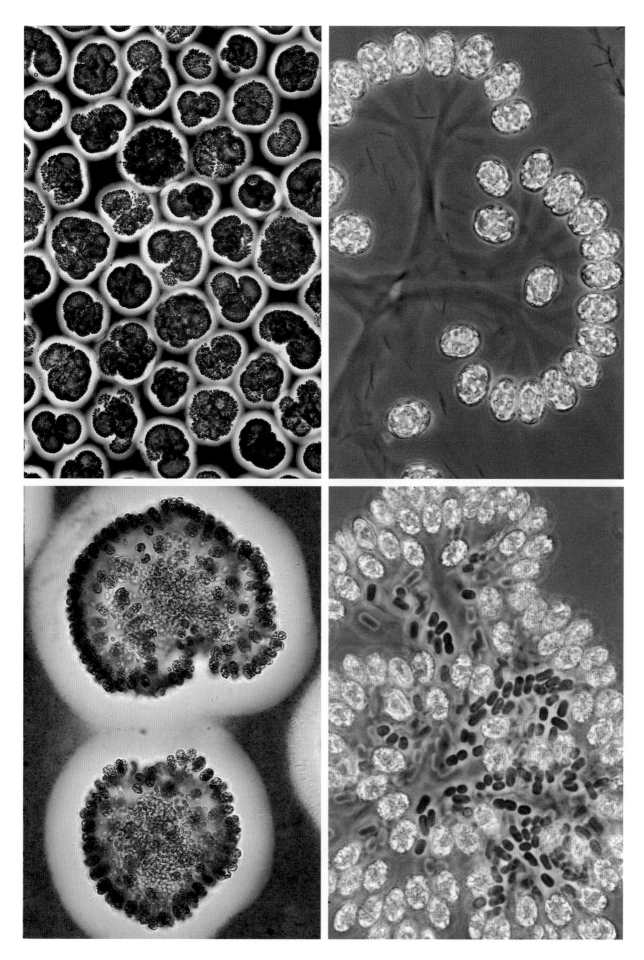

differences in the visual appearance of cells with and without gas vesicles, when viewed by phase-contrast, also occur in the *Anabaena*, fig. 398.

Bacteria also can live in the mucilage. In fig. 395, the colonies have a tightly packed ring of rod-shaped bacteria about half way between the *Gomphosphaeria* cells and the surface of the colony. Yet other organisms may be present, such as the Chlorophyte lying outside the ring of *Gomphosphaeria* cells in fig. 396. This unicellular Chlorophyte is called *Stylosphaeridium* and its cells are attached to those of *Gomphosphaeria* by a very fine thread.

The presence of a variety of microorganisms in the mucilage of *Gomphosphaeria* raises the as yet unanswered question, what are the interrelationships between them and *Gomphosphaeria*? It can be argued that they are simply occupying a suitable space. However, they do not occur in or on the mucilage of all algae producing it but only on a restricted range of species. It is not known whether the Cyanophyte (*Synechococcus*?, concerning taxonomy see p. 196) lives elsewhere; the *Stylosphaeridium* is most commonly found on this species of *Gomphosphaeria*. Nothing is known about the rod-shaped bacterium, so it has no name and its distribution also is unknown. Further, why does the Cyanophyte live within the ring of *Gomphosphaeria* cells and the bacterium occupy a narrow band outside this ring of cells?

This *Gomphosphaeria* is very common in the plankton and so also as a component of waterblooms (e.g. fig. 413). It is not clear whether it is toxic. The species shown used to be placed in the genus *Coelosphaerium* and may be still so named in some works. It now seems probable that in future it will be called *Woronichinia* because it has structural differences from other species retained in *Gomphosphaeria*.

Anabaena (figs 397-405, 407) is filamentous and often looks like a coil or necklace of beads. The filaments are unbranched and can be straight (figs 397, 401), regularly (fig. 402) or irregularly (fig. 405) spirally twisted or twisted and intertwined (fig. 404). In the last case, large numbers of filaments can form loose (fig. 404) or tight tangles (fig. 407) which can be big enough (a few mm) to be seen by the naked eye (fig. 403). In fig. 397, one species has straight filaments and the other spring-like ones. The cells of the species shown contain gas vesicles and are planktonic. The marked change in the appearance of a filament whose cells have lost their gas vesicles can be seen in fig. 398.

Many other *Anabaena* species exist which neither contain gas vesicles nor are planktonic. Recently, they have been transferred to a separate genus called *Trichormus* (see Chapter 11; Symbiosis).

Gas-vacuolate species of *Anabaena* are very common in the plankton and can form or be components of massive waterblooms (fig. 413). Further, most of the species contain toxic substances.

395	396
397	398

Figs 395, 396. *Gomphosphaeria*.

Fig. 395. Colonies (larger colony including mucilage, 204 µm across), each with a ring of bacteria about midway between the outer edge of the mucilage and the *Gomphosphaeria* cells. P, I.

Fig. 396. Cells with attached *Stylosphaeridium* (up to 6 µm l.); both are surrounded by mucilage (size of *Gomphosphaeria* colony 144 × 131 µm). I.

Figs 397, 398. *Anabaena*.

Fig. 397. One species with straight filaments (c. 10 µm br.) and the other with spirally twisted ones. D.

Fig. 398. Four filaments, one without gas vesicles (5.2-6.5 µm br.). P.

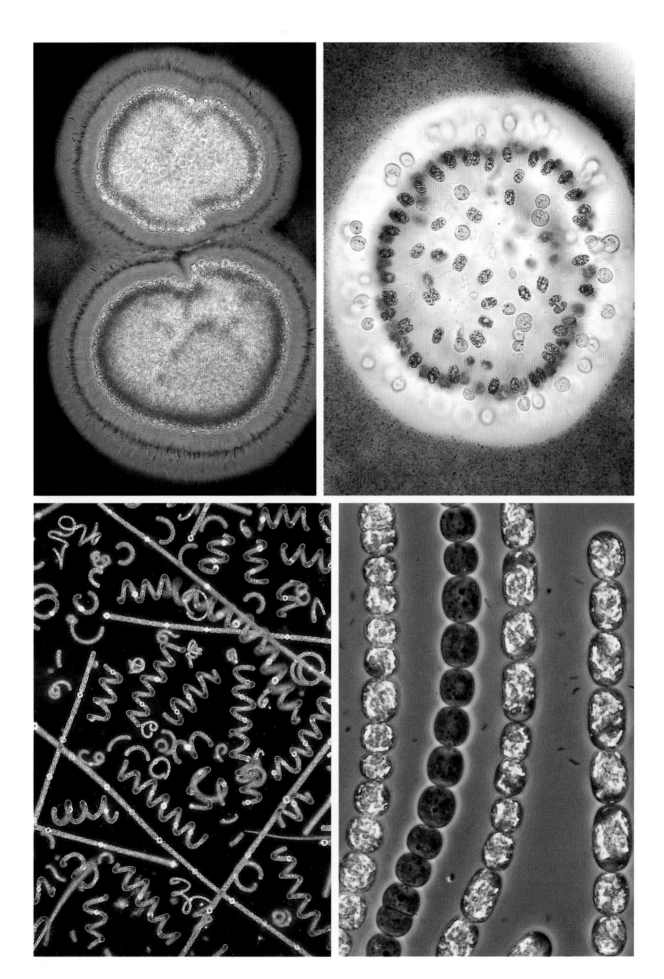

Three kinds of cell can be present in a filament. The commonest is the bead-like vegetative cells. These can be the only cells present (e.g. fig. 398). The vegetative cells only reproduce by dividing into two, so increasing the length of the filament which may eventually break up into two or more pieces. Not always present and always less common are larger, often oval or sausage-shaped cells (figs 399, 401, 402, 405) which are spores. There are also cells without gas vesicles, with homogeneous content and often a visible lump of material at each end; they are called heterocysts (figs 399, 405). Both these types of cell are peculiar to Cyanophytes and so not seen in any other prokaryote, though various bacteria have spores of a different kind to those of Cyanophytes. Spores (also called akinetes) arise by the enlargement of a single cell or of several cells which fuse together to become a single cell. When fully developed they lose most or all of their gas vesicles (figs 402, 404) and sink to the bottom. They germinate to form a new filament (fig. 400), germination starting either before or after rupture of their wall.

399	400
401	402

Figs 399-402. *Anabaena*.

Fig. 399. Filament with heterocyst (11 μm br.) and spore.

Fig. 400. Two mature and a germinating spore (germling 46 μm l.).

Fig. 401 Species with straight filaments and mucilage investment (total width 31-44 μm); spores present in one filament. I.

Fig. 402. Species with spirally twisted filament embedded in mucilage. Note mature spore (28 × 17 μm). I.

Heterocysts and nitrogen fixation

The name heterocyst suggests that the cell is some kind of spore. It is true that their germination to form a new filament has been recorded but only very rarely and some doubt has been expressed whether the cells concerned were fully developed heterocysts. We have seen many thousands of heterocysts but no germination. Because they are not, normally at least, spores or cysts, the name heterocyte has been proposed to replace heterocyst. There is no doubt that the normal function of heterocysts is related to the utilisation of molecular nitrogen (N_2-dinitrogen) as a nutrient by a process commonly called nitrogen fixation. A variety of prokaryotes can fix nitrogen, for example the bacteria in the root nodules of peas, beans, lupins and other legumes. Other kinds of algae and plants are dependent for their nitrogen on compounds such as nitrates and ammonium salts.

The enzyme system permitting the use of nitrogen gas is harmed by oxygen, consequently most bacteria can only fix nitrogen when gaseous oxygen is present in minute amounts or absent, that is under anaerobic conditions. In contrast, Cyanophytes possessing heterocysts can fix nitrogen in the presence of oxygen in air or water. Cyanophytes such as the *Anabaena* species shown often are abundant and having gas vesicles can be present in enormous numbers in waterblooms. The photosynthesis of such large populations leads to large amounts of oxygen in the water around the Cyanophytes, indeed the water can become supersaturated with oxygen. Yet nitrogen fixation can take place, provided heterocysts are present. These cells have a system protecting them from oxygen, one

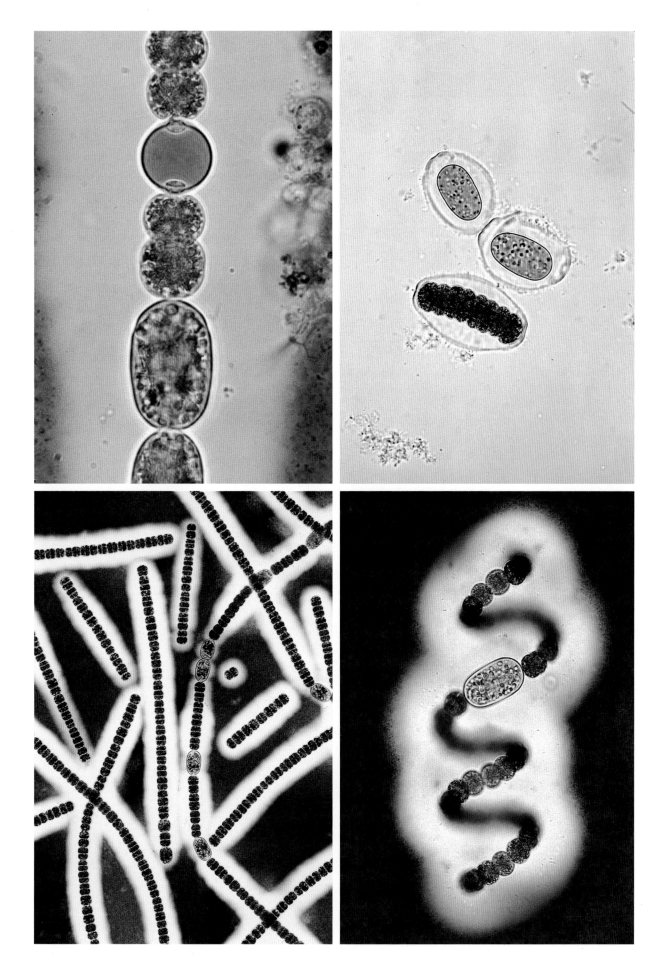

211

feature of which is that they themselves do not release oxygen in the process of photosynthesis. Since the enzyme system of Cyanophytes basically is the same as that of bacteria fixing nitrogen anaerobically, it is not surprising that some of them can fix nitrogen in the absence of heterocysts provided very little or no oxygen is present.

Bacteria in the swellings (nodules) on the roots of certain plants are protected from oxygen and can fix nitrogen. It is this capacity for nitrogen fixation which makes peas, beans and other leguminous plants such valuable agricultural crops. The world's most important crops are grasses such as rice, wheat and corn. If genetic engineering could transfer such bacteria or their genes for nitrogen fixation to such plants the benefit to mankind would be enormous, though a catastrophe to the fertiliser industry. Cyanophytes are an alternative. Some forms already live with certain plants and supply them with much or all the nitrogen they need (see Chapter 11, Symbiosis). Since the Cyanophytes concerned possess heterocysts, protection from oxygen is not necessary. Though such heterocystous Cyanophytes are found in or with certain ferns, fern allies (cycads) and liverworts, they are present in only one genus of flowering plants. In Asia, adding such Cyanophytes to rice paddies and so increasing the nitrogen content of the water and soil is being investigated.

Since water contains much nitrogen gas in solution and the atmosphere above it is a limitless source, heterocystous Cyanophytes have a nutritional advantage over other algae in a well oxygenated environment. For example, in an abundant planktonic population of algae, the available supplies of nitrates and ammonium compounds may be very sparse or exhausted. Then, the growth of all eukaryotic algae is severely limited or ceases but the growth of heterocystous prokaryotic Cyanophytes is not held up because they can use the nitrogen dissolved in the water.

Like *Gomphosphaeria*, *Anabaena* sometimes has other organisms attached to it but in general it is striking how few they are in numbers and species. Indeed, if one considers the great range of algae living on inanimate objects, then one may also ask why so few live on other algae. As mentioned earlier, it is notable that algae with rough cell walls such as *Cladophora* and *Oedogonium* have more algae on them than those with smooth, mucilaginous walls (e.g. *Spirogyra* and allied genera). Are there relatively few algae growing on the mucilage of other algae because it is "slippery", is continuously being produced and its surface lost, or do some algae pass repellant substances into the mucilage?

Vorticellids are unicellular animals (ciliates). They are particularly common on some planktonic species of *Anabaena* (fig. 407). There is a bell-shaped "head" bearing

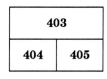

Figs 403-405. *Anabaena*.

Fig. 403. Colonies around floating leaves taken from a lake scum; oak leaf 11 cm long.

Fig. 404 as fig. 403; part of a colony (c. 0.5 mm across) at a higher magnification. A clump of mature spores is present in the lower tangle of cells.

Fig. 405. Filament containing three heterocysts (spherical cells without gas vescicles). The elongate sausage-shaped cell (20 μm l.) below the upper heterocyst (right) is an immature spore.

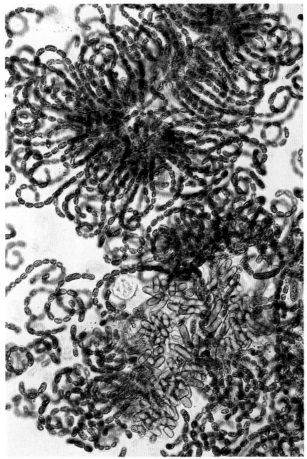
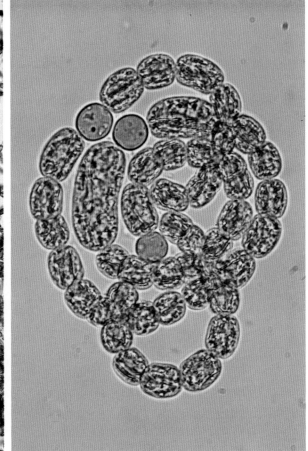

an apical ring of cilia (flagella – see p. 80). The beating of the cilia produces a current of water which enables the ciliate to "capture" small microorganisms such as bacteria and small algae. Vorticellids may also have live algal cells within them (fig. 455). The stalk attaching the vorticellid to the *Anabaena* can contract into a tight spring and then uncoil. Vorticellids on this *Anabaena* (fig. 407) are carried to the surface in quiet weather by the gas in the alga's cells. Experiments have shown that it would take the weight of a much larger number of vorticellids to overcome the buoyancy of a colony of similar size to the one shown.

Fig. 407. *Anabaena*. Tight tangle (234 × 172 μm) of filaments with gas-vacuolate cells to which are attached vorticellid ciliates. N.

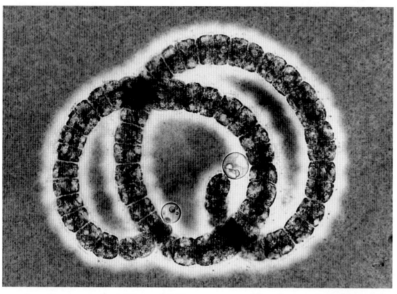

Fig. 406. *Anabaenopsis*. Note heterocyst at each end of the filament (× 640). I.

Anabaenopsis (fig. 406) can easily be mistaken for *Anabaena*. The filaments have similar bead-like cells containing gas vesicles. However, on the average they are shorter than those of *Anabaena*. The characteristic difference is the very common presence of a single heterocyst at each end of a filament in *Anabaenopsis* and its extreme rarity or absence at each end in *Anabaena*. In *Anabaenopsis*, the heterocysts typically arise in pairs and then separate, so producing a filament with a single heterocyst at each end. It is the successive development of paired heterocysts and their separation from one another that produces the relatively short filaments. The filaments of *Anabaena* may break at any point so that the likelihood of having a single heterocyst at each end is very small.

Anabaenopsis is common in the tropics and in continental countries with hot summers. It is much less common or absent in cold climates or countries with cool summers.

Cylindrospermum (fig. 408) is characterised by the terminal cell at one or both ends of a filament becoming a heterocyst. The cell below this terminal heterocyst will develop into a spore. As a result, such filaments are easily recognised as belonging to a species of this genus. Sometimes more than one spore develops below the heterocyst. The cells lack gas vacuoles.

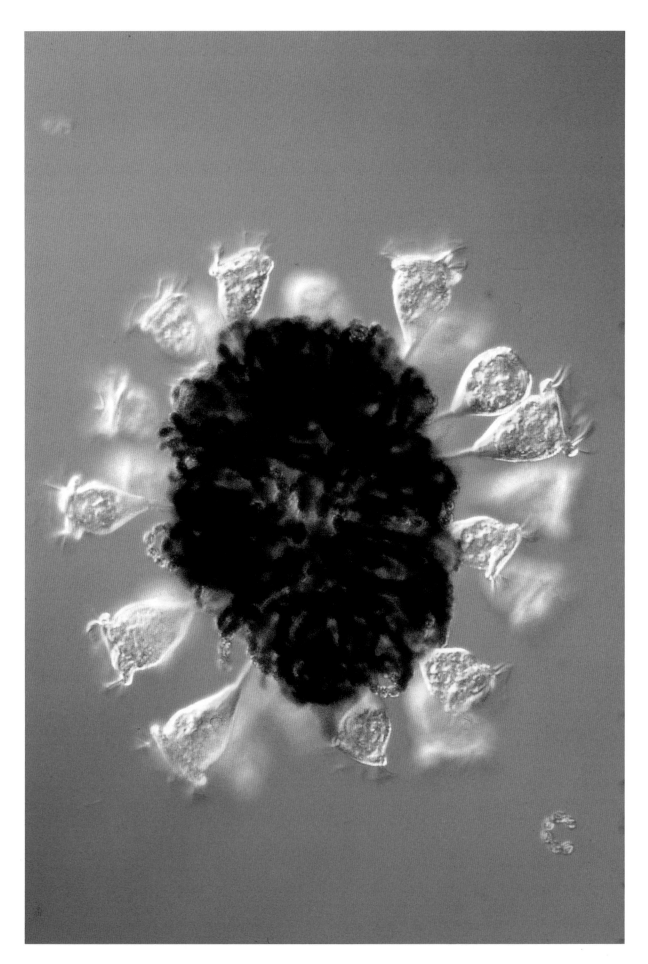

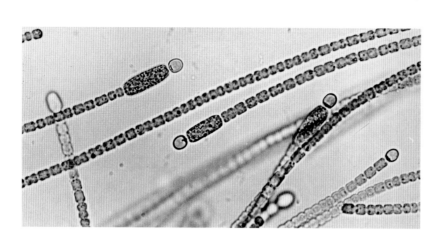

Fig. 408. *Cylindrospermum*. Note terminal heterocyst and spore next to it (not invariably present) (× 640).

Cylindrospermum lives in aquatic and terrestrial habitats. On undisturbed soil or the edges of paths, *Cylindrospermum* may grow so profusely that the surface is smooth, shiny and dark blue-green or black in colour.

Aphanizomenon (figs 409-412) also is filamentous but frequently the filaments are not solitary but form fasciculate bundles (fig. 411). These bundles can be big enough to look like blue-green pencil shavings. Since the cells contain gas vesicles, *Aphanizomenon* can form waterblooms. In fully formed filaments, the terminal cells usually are narrower and longer than the central ones, sometimes much more so; they also contain fewer gas vacuoles and less pigment than the other cells (fig. 412). The terminal cell itself can appear to be virtually devoid of content.

Aphanizomenon can fix nitrogen. The heterocysts and spores (figs 409, 410) are similar to those of *Anabaena*. *Aphanizomenon* is common in waters favourable to the growth of other Cyanophytes producing waterblooms.

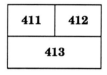

Figs 411, 412. *Aphanizomenon*. Bundles (flakes) of filaments (broadest 437 μm br.).

Fig. 412. Terminal cells (2.7-3.2 μm br.) of filaments, almost colourless and with few gas vacuoles. **P.**

Fig. 413. Cyanophyte waterbloom. All the genera present, *Anabaena, Microcystis, Gomphosphaeria* and *Aphanizomenon*, have been shown in previous pictures (fig. 383 *et seq.*).

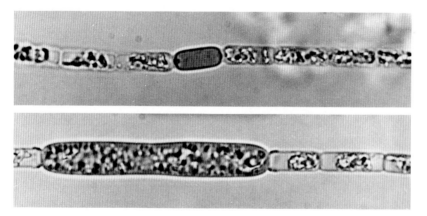

Figs 409, 410. *Aphanizomenon*.

Fig. 409. An elongate heterocyst; and fig. 410, a long spore. Both × 1680.

Certain species of *Anabaena* can form bundles of filaments but the terminal cells are neither devoid or almost devoid of visible contents nor notably narrower than the central ones.

Trichodesmium (figs 414, 415) also forms floating rafts of filaments, it too lacks narrow terminal cells more-or-less devoid of content. It differs both from *Aphanizomenon* and *Anabaena* in the absence of heterocysts and spores.

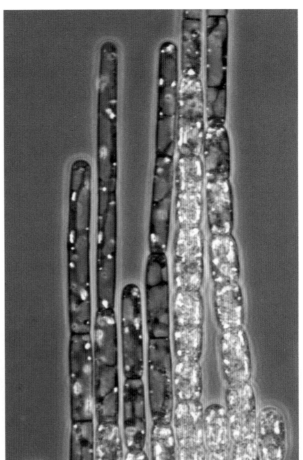
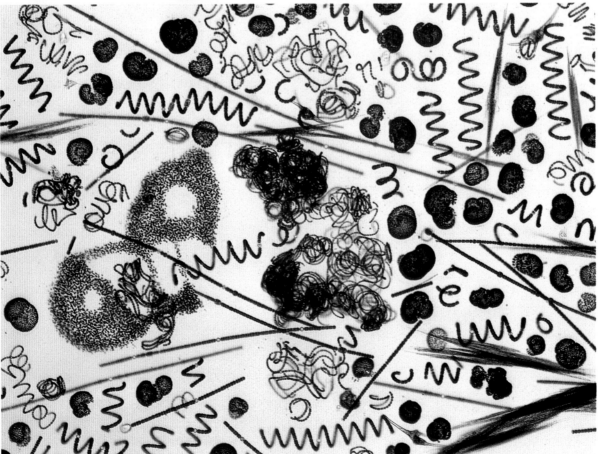

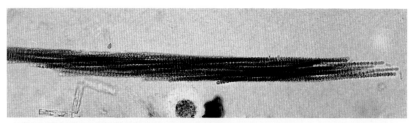

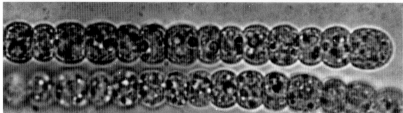

Figs 414, 415. *Trichodesmium*.

Fig. 414. A bundle of filaments (× 160).

Fig. 415. Filaments (× 1600).

Trichodesmium is an uncommon freshwater alga but reddish marine species are familiar objects to those sailing in tropical seas, forming windrows which can be several kilometres long.

Gloeotrichia (figs 416-420) forms globose colonies up to about 2mm in diameter. They consist of radiating filaments, the major parts of which are embedded in mucilage.

In young colonies, all the cells are well pigmented, contain many gas vacuoles and do not narrow markedly from base to apex (fig. 419). As the colonies enlarge so the filaments become more hair-like, narrowing gradually from base to apex. The distal portions project from the mucilage and consist of more-or-less colourless cells with few or no gas vacuoles. There is a heterocyst at the base of the filament (fig. 416), next to which a spore (fig. 417) can form. The spore can develop from a single cell or several which fuse together; as it matures it loses its gas vacuoles.

The species shown is planktonic and can form striking waterblooms. In 1809 it was given the English name of the Little Hedgehog Conferva – *Conferva* being a generic name once given to diverse filamentous algae. There are also non-planktonic species which may be over 2mm in diameter. Their cells lack gas vacuoles.

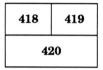

Figs 418-420. *Gloeotrichia*.

Fig. 418. Central colony, 875 μm diam.

Fig. 419. Heterocysts (round cells, c. 7 μm diam.) in centre of a small colony. **D**.

Fig. 420. Inner regions of colony.

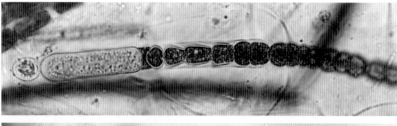

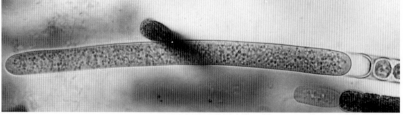

Figs 416, 417. *Gloeotrichia*.

Fig. 416. A short spore and attached heterocyst (× 640).

Fig. 417. A long spore probably arising from the fusion of several cells (× 640).

Waterblooms and eutrophication

At this point a little more should be said about waterblooms and eutrophication. Traditionally and generally today the term waterbloom is applied to surface accumulations of gas-vacuolate Cyanophytes (figs 10, 403, 413). A few other algae produce superficial scums (e.g. *Botryococcus* (fig. 11); *Botrydiopsis* (fig. 204); *Euglena* (fig. 7).

Cyanophyte waterblooms have been noted for centuries and entered folklore. An example of the latter is a Cyanophyte common in Central Europe which produces reddish waterblooms. During the winter of 1825-26 a waterbloom produced by this Cyanophyte turned the water in the Lake of Murten in Switzerland blood red. The local people thought that this sudden change in colour was the rising to the surface of the blood of Burgundian soldiers who drowned in the lake during a battle in 1476. This Cyanophyte is now called the Burgundy Blood Alga. It is a species of the genus *Oscillatoria* (*Planktothrix*).

The appearance of or an increase in the incidence of waterblooms is considered to be a sign of undesirable changes in water quality, that is of excessive eutrophication (p. 12). Enrichment of the water in plant nutrients will support more plankton algae, as well as other algae such as *Cladophora* (p. 72). Like *Cladophora*, waterblooms of Cyanophytes are not unnatural growths, they can be found in pristine lakes. However, in recent times the appearance of waterblooms in lakes which previously were free of them, and the increase in their frequency and massiveness in other lakes, have been shown to be caused by man. As mentioned earlier, the most important sources of enrichment are domestic sewage and agricultural wastes in the form of dung and artificial fertilisers. These sources have increased because the human population has increased but, especially in the more industrialised countries, alterations in sanitation and agricultural practices have exacerbated the eutrophication problem. Human sewage which once passed into the soil (e.g. via septic tanks), now passes into sewage works. The sewage effluent therefrom may be hygienically acceptable but contains large quantities of inorganic plant nutrients, notably phosphorus in the form of phosphates. Plants, and here Cyanophytes may be considered with them, need about 16 chemical elements for growth. In non-tropical countries, phosphorus usually is the major element controlling algal growth in freshwaters. Hence, an increase in the supply of phosphorus is likely to increase algal growth. Human excrement is rich in phosphorus, and another source is the phosphate present in detergents. Most of this phosphorus arrives in lakes and rivers via the effluent of sewage works. The necessity to reduce the phosphorus input into lakes in order to reduce eutrophication has led to the installation of systems which remove phosphates in sewage effluent and to controls on the amount of phosphate permitted in detergents. The former is the most effective method of controlling over-enrichment by phosphorus and in some countries now is standard procedure. There are other less frequently usable methods of controlling eutrophication, for example by diverting the sewage input but the question then arises, to where is the sewage to be diverted? Both the waterblooms shown (figs 10, 421) were caused by enrichment with phosphate, one (fig. 10) from sewage effluent and the other (fig. 421) by adding phosphate fertiliser. In the latter picture, there are two large experimental containers in the lake, both filled with lake water, a waterbloom has formed in the one to which phosphate was added.

The next most common element limiting algal growth is nitrogen in the form of nitrates or ammonium salts. Though human sewage contains much nitrogen, a considerable amount is likely to be lost in its purification in a sewage works. The atmosphere has been enriched with nitric acid derived mainly from the nitrogen oxides produced in the combustion of fossil fuels in automobiles and power stations. The major source of eutrophication by nitrogen is agriculture. In modern agriculture, high yields depend on high use of artificial fertilisers. Most of the phosphorus in them may be held in the soil but highly soluble nitrate or ammonium salts can easily be washed away. Further, intensive farming permits dense populations of animals and so danger from their excreta. If sufficient phosphorus is present, then a shortage of nitrogen often controls all algae except nitrogen fixing Cyanophytes. In tropical countries, lack of nitrogen often is a major factor controlling the growth of algae other than these nitrogen-fixing Cyanophytes. Cyanophytes generally grow best in warm water and are particularly common in tropical countries both in water and on land. In temperate counties, massive growths of Cyanophytes are most likely in summer or early autumn, though occasionally they are abundant in winter, as in the case of the Lake of Murten, and even under ice.

Global warming is likely to increase the abundance of waterblooms in temperate and high latitude lands because the summers will be hotter and longer. When there is thermal stratification (p. 17) in summer, algal growth will be confined to or greater in the warmer upper layer because light is insufficient in the lower depths for active photosynthesis. Oxygen from photosynthesis and from the air above is available to organisms in these upper, more or less well wind-mixed waters; the reverse is true below. When waterblooms are present nearly all the oxygen will be present in the uppermost layers of the water. The Cyanophytes form a coating or "blanket" of actively photosynthesizing algae. The lack of light below this blanket and the products of the death of algae and other organisms which pass into the lower regions will use up the store of oxygen in them. Indeed, in part at least of this lower region all the free oxygen will be used up and anaerobic conditions, fatal to animals and many algae, will prevail.

Those fish which need cool, well-oxygenated water will be caught between the deoxygenated cool water of the lower layers and the too hot upper waters and in danger of their gills being contaminated by the Cyanophytes. Further, as mentioned, a considerable number of these gas-vacuolate Cyanophytes contain toxic substances. The more of these that are tested, the greater becomes the number which have been found to be toxic. In the population composing the waterbloom in fig. 413, all but one of them can be toxic. The cells of these bloom-forming Cyanophytes often die and break up (lyse). A large part of a waterbloom may change into a blueish, milky liquid in a few hours. Such a change has taken place in the part of the waterbloom in fig. 10 near to the boats and shore. Both toxins and pigments will have been released into the water.

Ducks can often be seen swimming in a waterbloom, apparently without harm. Yet deaths of aquatic birds do occur in waters rich in waterblooms and can be related, in part at least, indirectly to the Cyanophytes. The anaerobic conditions in the mud, which also is rich in organic matter, support the growth of many bacteria which do not need a supply of oxygen. Among such "anaerobic" bacteria is a species which produces the powerful botulinus toxin. Birds which root about in the mud can become infected and it seems that most deaths of aquatic birds when waterblooms are present are from botulism.

Apart from their effect on other organisms and on water quality, waterblooms affect amenity. They are aesthetically disliked (except to some who study them!) and, being rich in proteins, can stink when they decompose. Further, warnings have to be given of the potential dangers if they are swallowed. It has been claimed that they can produce skin rashes. As with *Cladophora* (p. 76), a more probable cause is an animal parasite. Nevertheless, it must be remembered peoples' skin can be sensitive to a wide variety of organisms or substances.

Oscillatoria (figs 422-424, 430) has unbranched filaments. As the figures show, certain species look very different from others and recently this genus has been subdivided into several new ones, namely *Planktothrix* (fig. 422), *Tychonema* (fig. 423) and *Limnothrix* (fig. 424). However, the majority of the species originally included in the genus *Oscillatoria* have not been given new generic names. For example the generic name of the species shown in fig. 430 still is *Oscillatoria*.

The species of *Oscillatoria* (*Planktothrix*) shown in fig. 422 was the major component of the artificially induced

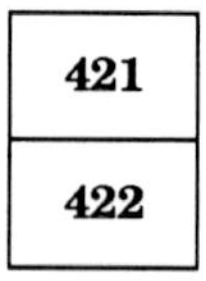

Fig. 421. Enclosures in small lake nearer one with waterbloom which arose after fertilization with phosphate.

Fig. 422. *Oscillatoria* (*Planktothrix*); filaments (circa 6 μm br.). This species was dominant in the waterbloom seen in the previous picture.

waterbloom shown in the previous picture. This species has gas vesicles but that in fig. 423 (now called *Tychonema*) does not, though it has a rather curious cell structure which might be mistaken for gas vesicles. An increase in the latter species can be a sign of undesirable eutrophication in lakes in northern Britain and in Norway. The specimens shown in fig. 423 came from Windermere, England's largest lake and measures are now being taken to remove most of the phosphate from the sewage effluent entering the lake which is in the extremely popular tourist region, the English Lake District. The abundance of the Burgundy Blood Alga (p. 220) also can be a sign of excessive eutrophication and so of the need to control the input of sewage effluents and other sources of enrichment. Lake Washington in the U.S.A. is a famous case of eutrophica-tion by this alga and its control, based on the work of Professor W. E. Edmonson (seen in fig. 258) and colleagues at the University of Washington.

The *Oscillatoria* (*Limnothrix*) in fig. 424 is curious in having all or nearly all the gas vesicles concentrated at the ends of the cell, so producing large gas vacuoles. These appear in fig. 424 as "white" regions and in some cases are seen to be paired. The gap between the pairs of gas vacuoles marks the position of a cross-wall between two cells. It is an alga of highly eutrophic (nutrient rich) waters.

The previous three species are planktonic but the one in fig. 430 is not, nor do its cells contain gas vacuoles. It lives on a surface such as mud on the bottom of a pond or lake. The filaments can move, creeping or gliding over the substratum (cf. *Merismopedia* p. 198). A number of filamentous Cyanophytes can move in this way but it is not yet clear how they do so.

Oscillatoria is extremely common, though its planktonic species are absent in many oligotrophic waters (i.e. waters poor in plant nutrients).

Phormidium (figs 425, 426) is another large genus which is in the process of being divided up into a set of new genera. The filaments commonly are within a sheath (fig. 425). They are unbranched and look like those of *Oscillatoria*. The whole organism, filament plus sheath, cannot move but the filament can move within its sheath. Part of the filament may break off from the rest, glide out of its sheath and move away to form a new sheathed filament. Hence, this use of motility is a form of reproduction. Filaments may remain without sheaths for a long time, when they are indistinguishable from *Oscillatoria*. A characteristic feature of *Phormidium* is that numerous threads can aggregate together to form sheets or colonies of varied shape, size and thickness (fig. 426).

Phormidium is particularly common in wet or damp terrestrial habitats, including soil. In soil the threads often are solitary and without sheaths and so easily mistaken for

423	424
425	426

Fig. 423. *Oscillatoria* (*Tychonema*). The curious appearance of the cells is not caused by gas vacuoles which are absent in this species. Filaments 5-7 μm br.

Fig. 424. *Oscillatoria* (*Limnothrix*). White areas, gas vacuoles. Filaments 1.5 μm br. **P.**

Fig. 425. *Phormidium*. Ends of two filaments (5.5 μm br.) **N.**

Fig. 426. *Phormidium*. Filaments tightly coiled together. Central hole 50 × 48 μm.

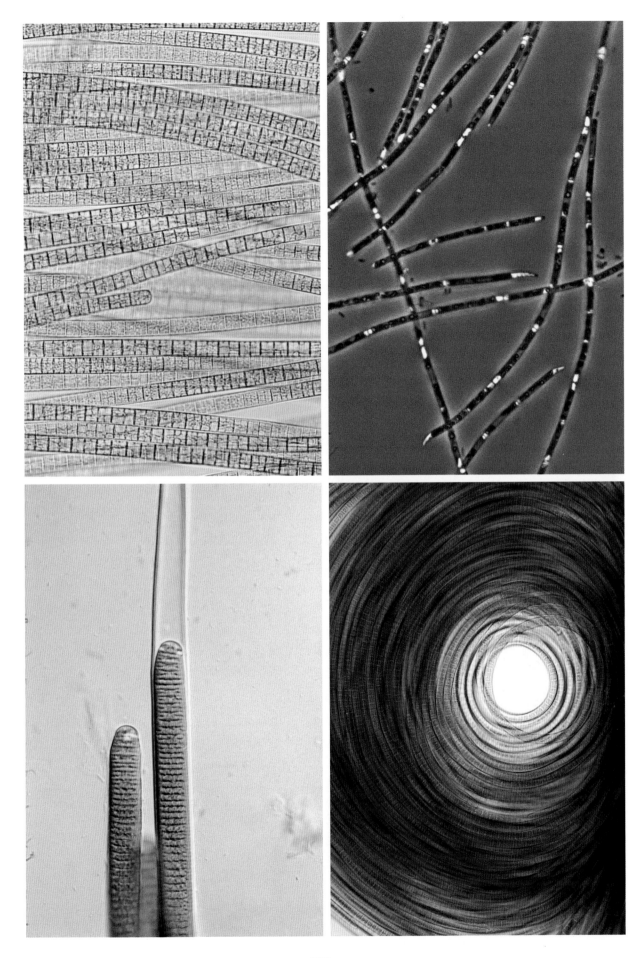

Oscillatoria. Neither *Phormidium* nor *Oscillatoria* produce spores or heterocysts.

Phormidium is an example of the differences in the naming of Cyanophytes by bacteriologists and phycologists (algologists). The species seen in fig. 425 would be placed in the genus *Lyngbya* by bacteriologists. The generic name *Lyngbya* is used also by phycologists but with a different circumscription to that of bacteriologists. The reason why naming Cyanophytes is based on two partially different Codes of Nomenclature, one botanical and the other bacteriological has arisen from the history of our understanding of what Cyanophytes are (see p. 194). Such an unsatisfactory situation is unlikely to last much longer for everyone accepts that harmonisation of the two Codes is desirable, though, for reasons outside the scope of this introduction to algae, there still are some problems needing resolution.

Fig. 430. *Oscillatoria*. Filaments 13 μm br. A species living on muddy substrata.

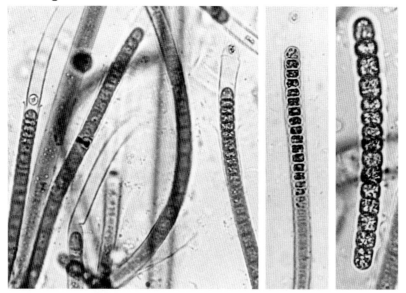

Figs 427-429. Hormogonia of an unidentified Cyanophyte. Note more bead-like cells in hormogonia before (figs 427, 428) and after (fig. 429) separating from the rest of the filament. Hormogonial cells of filaments in figs 428, 429 contain gas vacuoles.

Figs 427, 428 (× 800); fig. 429 (× 1260).

The gliding of part of a filament out of its sheath is the only method of reproduction in *Phormidium* and other sheathed, filamentous Cyanophytes which lack spores. The part that glides out is called a hormogonium. Hormogonia also exist in filamentous Cyanophytes which possess spores. When the filaments are unsheathed, hormogonia may be difficult to distinguish from fragments arising from other causes (e.g. mechanical breakage). However, in many other Cyanophytes the hormogonia are delimited from or distinguishable from other parts of the filament in various ways. A common example of delimitation is the special development of a cell which when dead forms a separation disc between hormogonium and the rest of the filament. Cells of hormogonia can look different from the rest, as in figs 427, 429 where they are more bead-like. In a few Cyanophytes, they are so different that one might call them spore bodies. A further difference can be that the cells of hormogonia contain gas vacuoles whereas the other cells in the filaments do not (fig. 428).

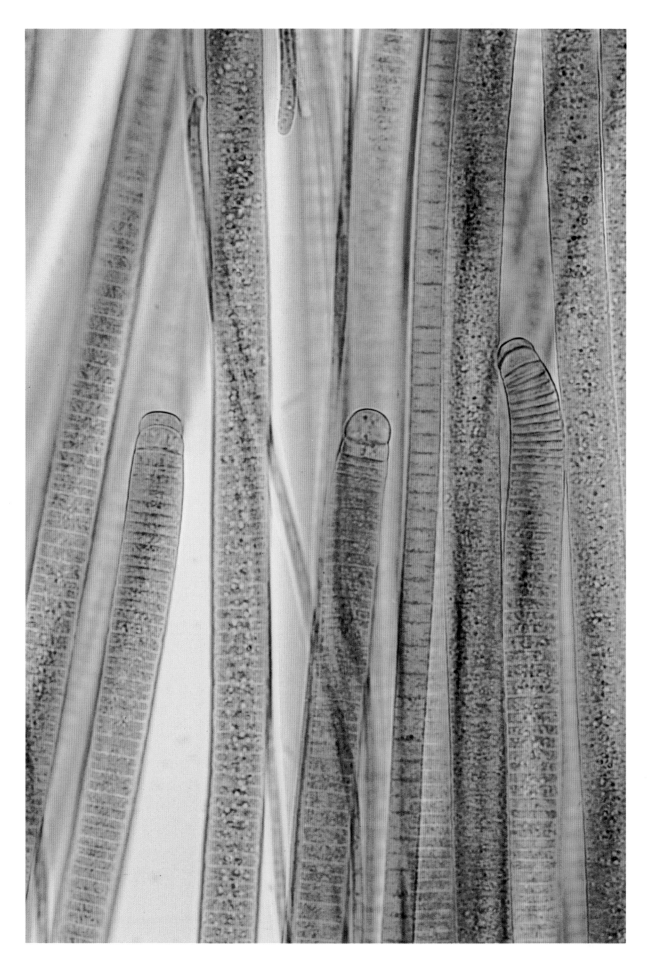

Hapalosiphon (figs 431, 432) is common in boggy, peaty waters. The filaments are uniseriate (one row of cells) and branched. In the species shown, growth is apical. Branches are initiated by cell divisions in the plane of the longitudinal axis of the filament. In the succeeding genus *Tolypothrix* and many other filamentous genera, occasionally even in *Hapalosiphon*, branches are initiated by cells which grow through a break in the filament sheath.

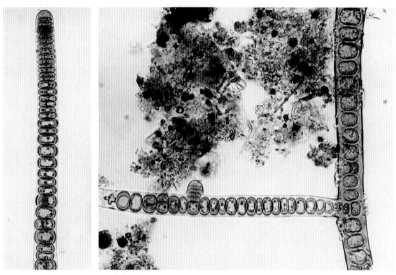

Figs 431, 432. *Hapalosiphon*.

Fig. 431. Terminal region, note short apical cells; this is where cell multiplication is occurring (× 640).

Fig. 432. Note branch at right angles to the main filament and on this branch, a very young branch (5 cells) (× 640).

Tolypothrix (figs 433, 434) has branched filaments, the branches often arising where there is a heterocyst. The cells of an unbranched filament are interrupted in some way at this point and one part grows out from the sheath to form the branch. In *Scytonema* (fig. 435) both halves of the filament usually grow through the ruptured sheath to form paired branches. A further distinction is the rarity of heterocysts at the base of the paired branches in *Scytonema*. There is however some doubt whether *Tolypothrix* and *Scytonema* are separate genera. Whereas most branches in *Tolypothrix* arise separately as in figs 433 and 434 and those of *Scytonema* in pairs as in fig. 435, *Tolypothrix* occasionally produces paired branches and *Scytonema* occasionally single ones.

Species of *Tolypothrix* usually are aquatic whereas those of *Scytonema* are commonest in terrestrial or semi-terrestrial habitats. The sheaths around the filaments of *Scytonema* are more often brown in colour (see Professor G. E. Fogg's painting, fig. 435), a feature commoner in terrestrial than in aquatic habitats and perhaps related to protection from the sun's rays. As can be seen in fig. 435, the sheath usually consists of two or more layers. In moist terrestrial habitats part of the sheath can be much inflated and spirally striated, as brilliantly and beautifully depicted in Professor Fogg's painting (fig. 436). This Cyanophyte has been placed in a separate genus, *Petalonema* but there appear to be all integrades between it and a typical *Scytonema* such as that depicted in fig. 435.

Figs 433, 434. *Tolypothrix*.

Fig. 433. Note heterocysts where filaments (10 μm br.) branch.

Fig. 434. Close-up of branching. Filament 10 μm br.

Figs 435, 436. *Scytonema*.

Fig. 435. Painting by Professor G. E. Fogg.

Fig. 436. *Petalonema* stage. Painting by Professor G. E. Fogg.

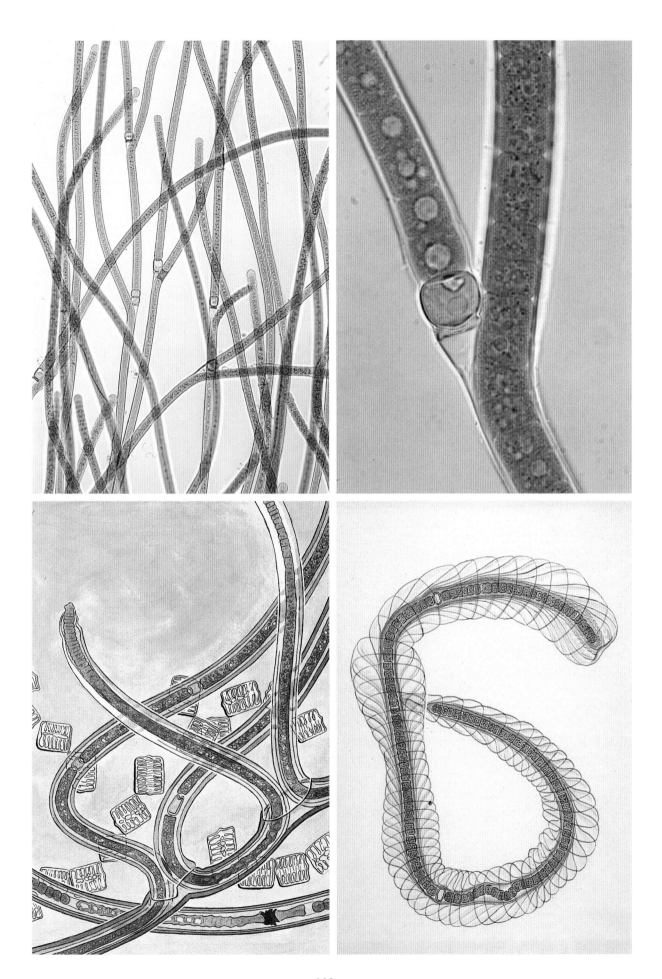

Spirulina (figs 437-444) used to be divided into two genera on the basis of whether cross-walls were present in the filaments. Species of *Spirulina* lacked cross-walls, those of *Arthrospira* possessed them. Most supposed *Spirulina* species have been found to possess cross-walls (fig. 437) by examining them at high magnification or by using stains. In any case, a coiled filament without cross-walls would now be placed neither in *Spirulina* nor in *Arthrospira* but in a different family of Cyanophytes.

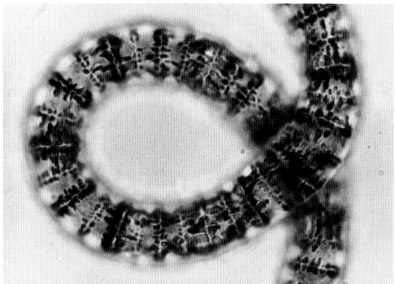

Fig. 437. *Spirulina*. Note cross-walls (× 1600).

Despite *Spirulina* having been restricted to species allegedly lacking cross-walls, the name is retained because it was described properly (i.e. in a manner prescribed by the International Codes of Nomenclature) before the genus *Arthrospira* was created. Further taxonomic trouble has arisen because *Spirulina* virtually is an *Oscillatoria* with wavy or coiled filaments. Hence, it is not surprising that some authorities consider its separation from *Oscillatoria* is unjustified. All this is yet another example of the unsatisfactory state of classification in this group and why new classifications are being proposed.

The cells of some species contain gas vesicles and are planktonic. The species shown in figs 439 and 443 is common in salty tropical, or semi-tropical, lakes (e.g. soda lakes) and its filaments can produce very tight, top-shaped spirals. In the region of Lake Chad in Africa *Spirulina* is collected and eaten and it appears that the Aztecs of Mexico also ate it. In view of the large crops which can be produced in such waters and the rate of growth of the alga at the relatively high temperatures typical of the hot countries where it abounds, it is under investigation as a crop which can be grown on an industrial scale (cf. *Chlorella*, p. 245). At present it is something of a "cult" food, widely advertised in health magazines and the subject of some books. In 1982 the "health food" firms and shops in the USA sold 18 million dollars worth of *Spirulina* powder and tablets (fig. 438).

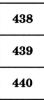

Figs 438, 440. *Spirulina* as health food.

Fig. 438. *Spirulina* powder and, on top of bottle, two tablets of *Spirulina*.

Fig. 439, A coiled filament (156 μm l.).

Fig. 440. A nutritious edible snack.

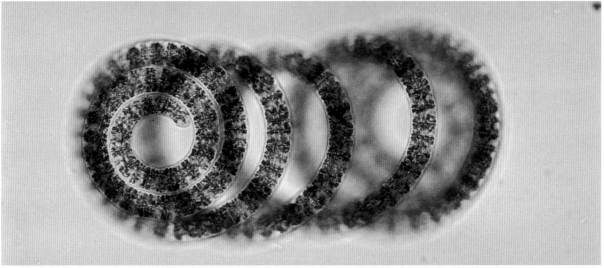

Advice has even been given as to how much you should put in your nuclear bomb shelter. There is no doubt that it is non-poisonous (unlike some Cyanophytes, p. 205), edible and nutritious, its protein content being similar to or better than that of steak. It is less clear that it has any unique or miraculous properties.

Incidentally, it seems that some toxic Cyanophytes (e.g. certain species of *Microcyctis*) are harmless when pasteurized and dried, judging by eating Ukrainian material.

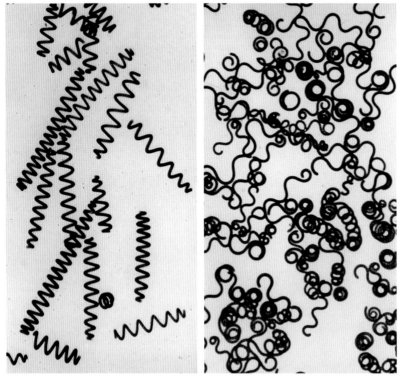

Figs 441, 442. *Spirulina*.

Fig. 441. Filaments in a drop of water on a microscope slide and fig. 442, after a microscope coverglass has been placed on the drop of water. (Both × 64).

The filaments of *Spirulina* in fig. 441 are viewed in a drop of water on a microscope slide. After placing a coverslip on the water, the regular spring-coiled shape of the filament may be temporarily lost (fig. 442). The less the water between coverslip and slide, the more the shape is likely to be altered. Species of planktonic, spirally twisted gas-vacuolate *Anabaena* (figs 397, 413) can present an identical configuration to the *Spirulina* in figure 441.

In addition, filaments of *Anabaena* can be devoid of both heterocysts and spores, a condition which characterizes *Spirulina*. Although at a low magnification it could thus be difficult to differentiate between these two genera, this may not be so if higher power optics are used because the beaded filaments of *Anabaena* are never seen in *Spirulina*.

In temperate regions, species which lack gas vesicles and are not planktonic are common. The best place to look for them is on the muddy bottom of a standing water, provided, of course, that it is not so far below the surface that there will be insufficient light for photosynthesis. Like many other filamentous Cyanophytes they can creep over a substratum and as fig. 444 shows may intertwine with one another.

Figs 443, 444. *Spirulina*.

Fig. 443. A plankton species with gas vacuoles. Top-shaped spiral 106 μm l.

Fig. 444. A benthic species lacking gas vacuoles. Filaments 6 μm br.

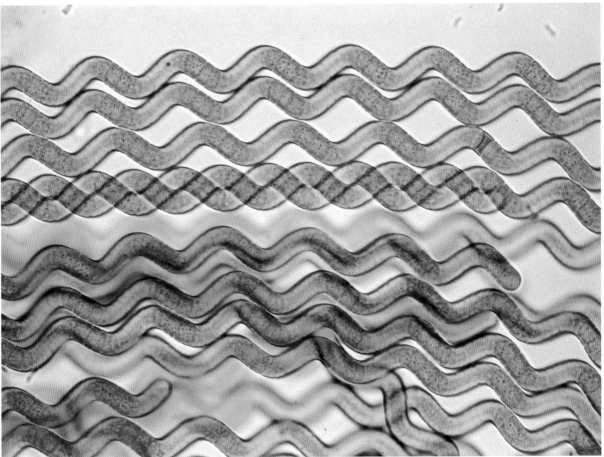

Nostoc (figs 445-447, 449, 452) has *Anabaena*-like filaments intertwined within mucilage which can be soft or so firm that it is almost leathery in texture. The mucilage often is yellow or brown in colour. In the species in fig. 446 the outer mucilage is colourless but that immediately surrounding the filaments is brownish. This species is microscopic but others can reach several centimetres in length, breadth or thickness. Species of *Nostoc* grow in water, on land and in soil. The one shown in figs 445 and 446 is common on wet rocks. All contain heterocysts and so can fix nitrogen.

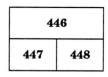

Figs 446, 447. *Nostoc*.

Fig. 446. Colony 150 µm diam. **I**.

Fig. 447. Sold as a culinary delicacy (shi), three-fifths natural size.

Fig. 448. *Azolla*. An aquatic fern containing *Anabaena* as a symbiont. The discoid, paler green plant is duckweed (*Lemna*). Approximately natural size.

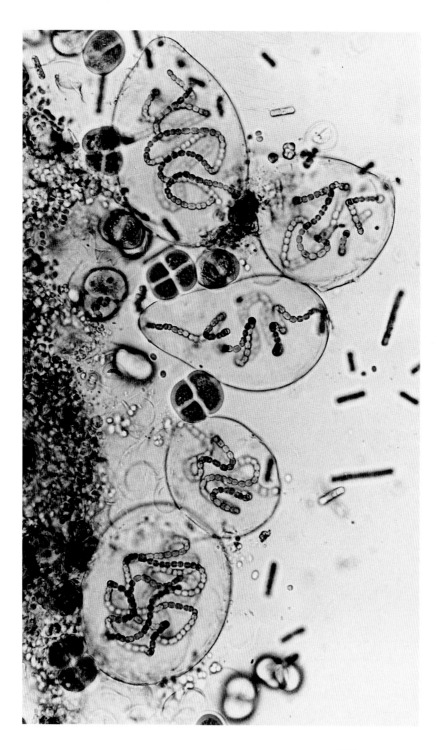

Fig. 445. *Nostoc* from a wet rock face, also present *Chroococcus* and small *Aphanothece*-like cell masses (× 370).

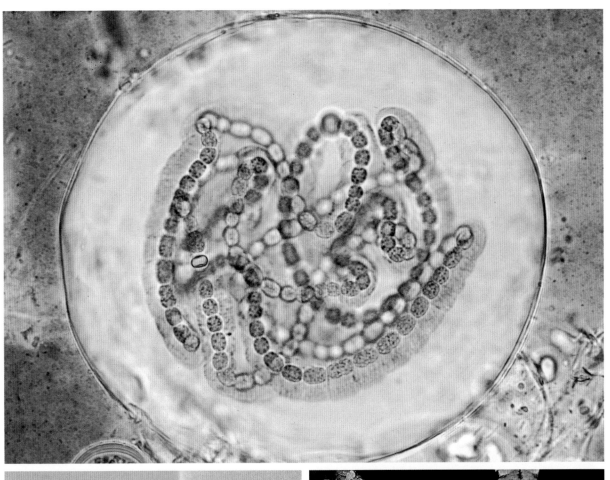

Some of the terrestrial species can live in very harsh climates, for example in parts of Antarctica where the air temperature never rises above freezing point and little snow falls. Such places are called dry valleys or polar deserts. Members of the Commonwealth Transantarctic Expedition (1955-58) found so much *Nostoc* at over 80° South that they ate it to see if it could be a possible source of food. Unlike the Chinese (see below), they found it unpalatable, perhaps because it was uncooked.

Many other Cyanophytes also are major algae in tropical and sub-tropical deserts. The ability of some algae, predominantly certain Cyanophytes and Chlorophytes, to live in stony deserts was mentioned on p. 10. Cyanophytes can be very abundant in parts of the clayey deserts of Asia which are bordered by mountains. In spring there is a short period when such areas are flooded by snow melt from the mountains and by rain. The growth of slimy Cyanophytes can become so great that even camels slip and, for a short time, the traditional caravan routes have to be changed. During the succeeding hot dry weather the Cyanophytes dry up, forming papery crusts and flakes which can be blown about by the wind. In this dried state the Cyanophytes live through the heat of summer and cold of winter. They can remain alive in air-dried soil for many years and *Nostoc commune* (fig. 449) has been revived from a herbarium sheet set up 107 years before.

In Asia, especially in China, certain species of *Nostoc* are considered to be delicacies. Fig. 447 shows part of a highly prized delicacy called shi or facai. The hair-like, filamentous colonies (see the bits sticking out in the picture) do not grow in masses like the block on sale. Considerable time and effort must be needed to collect such a large mass of *Nostoc* and free it from extraneous matter. This *Nostoc* is abundant in parts of some Asian, clayey deserts (e.g. the Gobi Desert). In summer, the temperature can rise above 30°C and in winter, below −20°C. Further, even in summer there can be large diurnal fluctuations in temperature. The *Nostoc* threads usually are narrow (less than 1 mm) and often long (over 10 cm long) even when not aggregated into larger clumps. When dry, the threads adhere firmly to the clayey soil or are blown far and wide.

Nostoc commune (fig. 449) may not be as common as its name suggests but when present can be distributed over considerable areas and easily seen. It colonizes a wide variety of substrata, for example in Britain, coarsely sandy or gravelly ground, footpaths or in flat valleys between sand dunes (slacks). Sometimes it is a pest on lawns but then generally when part of a lichen (Chapter 11). It has a world-wide distribution. The *Nostoc* from 80°S probably was *N. commune* but the sample received was not suitable for certain identification.

In the dry state, it is easily overlooked, forming a thin coating resembling a small sheet of dried-up gelatine.

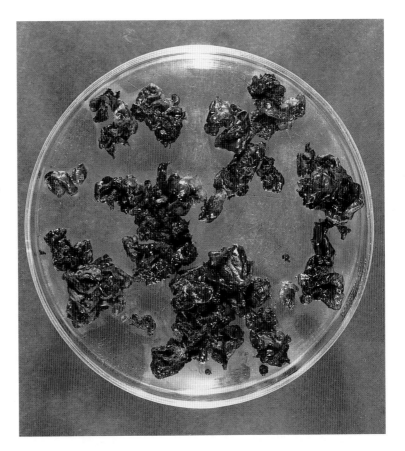

Fig. 449. *Nostoc commune*, an alga famous in folklore. Three-fifths natural size.

When wetted it soon swells up to form a soft but firm leathery gelatinous mass (fig. 449) of irregular outline and with a plain or wrinkled surface. Its colour is variable, olive-green, blue-green, almost black to brown or yellowish-brown.

Nostoc and certain other jelly-like substances found on the ground are often referred to in folklore. It was not until late last century that what some people already knew became generally recognized, namely that *Nostoc* was different from other "jellies" and should be accepted as a "plant". It is probable that its sudden appearance after rain led to the belief that it fell from the heavens. The commonest belief was that it came from shooting (falling) stars or the moon. It interested alchemists and was even thought to be the Philosopher's Stone. A variety of European names were given to it, for example, starshot, star jelly, witches' or fairies' butter; beurre magique, crachat d'un diable, crachat de lune, purgations des étoiles; Sternschnuppe, Kuckucksspeichel and Meteorgallerte – the last name less from folklore than from naturalists.

Chapter Eleven

Symbiosis

Cyanophytes and plants

It is convenient to start the section on algae living in association with other organisms by continuing with the last mentioned genus of Cyanophytes, *Nostoc* (fig. 452) even though the species concerned, *N. punctiforme*, very recently has been transferred to the genus *Trichcormus* (see p. 241). In the literature on its symbiosis with *Gunnera* and in textbooks the generic name found will be *Nostoc*.

Gunnera (figs 450, 451) is a genus of perennial, herbaceous flowering plants containing both very small and very large species. The gigantic herb shown here is commonly grown in wet places in botanic and other large gardens. It needs considerable space because its leaves can be over two metres high and one metre across.

If the stem is cut below the soil surface in the area between where the base of the leaf-stalk fuses with the stem and where roots first appear, bright blue-green patches will be seen (fig. 451). Examination under the microscope (fig. 452) shows that these blue-green patches are composed of filaments of *Nostoc*. Despite living in the tissues of the *Gunnera* and in darkness, the *Nostoc* is as blue-green as any species living freely in the light. If filaments are removed from the *Gunnera*, they can be cultured and will photosynthesize. Since there is no *Nostoc* in the seeds of *Gunnera* the young plants must be infected by *Nostoc* living in the soil. *Nostoc* moves in the hormogonial stage (p. 226) from the soil into glands on the stem of *Gunnera* via copious mucilage release by these glands. Finally it is incorporated within living *Gunnera* cells.

When *Nostoc* is living in *Gunnera*, most of the cells are like heterocysts which suggests that it can actively fix nitrogen and tests have shown that it does. Certainly, small species of *Gunnera* obtain almost all their nitrogen from the *Nostoc*. It is hardly likely that this is so in the large species shown here. Clearly, the *Nostoc* is of benefit to the *Gunnera* and the *Nostoc* may obtain carbon from the *Gunnera*. Since the *Nostoc* is in the dark it cannot manufacture essential carbon compounds from carbon dioxide by means of photosynthesis.

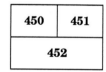

Figs 450, 451. *Gunnera*.

Fig. 450. Leaf. Fig. 451. Piece of tissue containing patches of *Nostoc*.

Fig. 452. *Nostoc* from *Gunnera*. Largest cells 8 × 7 μm diam.

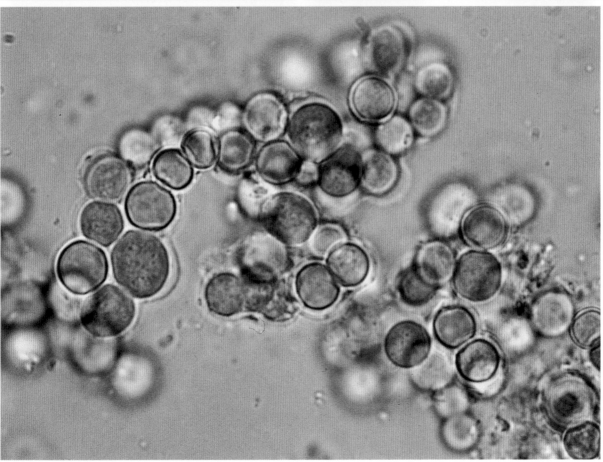

Gunnera is a genus of plants living in the southern hemisphere, most of the species are South American but others live in New Zealand, parts of Asia and, in the Hawaiian islands, far from any other landmass. In addition, some species, such as the giant *G. manicata* (circa 2 m high) or the dwarf (circa 10 cm) ground-cover plant *G. magellanica* from S. America are widely grown in gardens in the northern hemisphere. Whether growing in their natural (and varied!) habitat or gardens, all the species contain *Nostoc* as a symbiont. The species concerned, *N. punctiforme*, is the same everywhere which means that this *Nostoc*, unlike species of *Gunnera*, has a world-wide distribution in soil. However, the criteria applied by botanists to determining what is a species may not be the same as those of bacteriologists; often they certainly are not. It is known that not all strains of *N. punctiforme* will infect all species of *Gunnera*. Hence, the best we can say at present is that there does not seem to be a clear justification for claiming that these strains of *Nostoc* are so different from one another that they should be placed in separate species.

The living together of two (occasionally more than two) organisms which have certain interrelations with one another is called symbiosis. In symbiosis the organisms obtain benefit from one another. The benefits obtained by each partner can be unequal, sometimes extremely so. Sometimes it is not clear which partner obtains the greater benefit. For example, is the *Nostoc* obtaining more benefit from the *Gunnera* than the *Gunnera* does from it? A variety of classifications has been suggested to cover various kinds of symbiosis.

In some cases, one partner may obtain so much more benefit than the other, that the symbiotic relationship comes close to parasitism. Indeed, there is no clear boundary between symbiosis and parasitism. Here, by parasitism we mean that one organism does nothing but harm to another, which is what people usually mean when they say some organism is a parasite. One organism – we shall meet examples – may virtually "farm" another and in the process "cull" the population of the organism it farms if it has become too large and threatens the biological balance between them. This still is symbiosis by our terminology because though one partner is dominant, the other does obtain some benefit from being "farmed". The situation is analogous to human farming because the population of farm animals is maintained by the farmer within certain limits.

Gunnera is the only flowering plant with a symbiotic Cyanophyte though they live in this way with several other kinds of plants such as liverworts, cycads and fungi.

A species of *Anabaena* lives symbiotically with the small aquatic fern *Azolla* (fig. 448). *Azolla* grows on the surface of water and is particularly abundant in hot countries. The filaments of *Anabaena* live in a special cavity in the fern's leaf. Like the *Nostoc* in *Gunnera*, the *Anabaena* in *Azolla* can fix nitrogen and is rich in heterocysts. The fern can obtain most or all of the nitrogen it needs from the *Anabaena* and can be used as a green manure rich in nitrogen. In China and Vietnam it has been used as a fertilizer in the cultivation of rice for a long time. Much work is being done to extend its use in other countries. Fertilization of rice with *Azolla* is labour intensive and so most suitable for countries where wages are low and the cost of buying artificial fertilisers (e.g. nitrates and ammonium salts) too high. Nevertheless, a source of additional phosphate may be necessary to ensure active fixation of nitrogen by the *Anabaena* and maximal harvest of *Azolla*. This phosphate can be obtained from dung. Free living, nitrogen-fixing Cyanophytes also are used as fertilisers for rice, notably in India. Rice paddies contain many algae, including Cyanophytes. To ensure effective nitrogen fixation in these algal populations, selected species of Cyanophytes are cultivated and added to the paddy fields. It is not yet clear that this is an economically viable form of nitrogenous fertilisation.

In the past, doubt has sometimes been expressed about the symbionts of *Gunnera* and *Azolla* belonging to different genera. There is no reason why the symbiont of *Gunnera* should be placed in *Nostoc*. It does not produce the *Nostoc* type of colony and is more like the *Anabaena* of *Azolla*. The latest classification places both in the genus *Trichormus*.

In addition, most of the non-planktonic species of *Anabaena* without gas vacuoles are now included in *Trichormus* (e.g. the species seen in figs 453, 454). All the planktonic species with gas vacuoles shown earlier remain in *Anabaena*. The difference between the two genera, *Trichormus* and *Anabaena*, as at present understood, depends on the developmental relationships between the heterocysts and spores. Such a matter is beyond the scope of this book and in any case not always easy to determine.

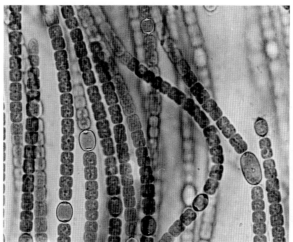

Figs 453, 454. *Trichormus* (× 640). Note rows of spores in fig. 454.

Chlorella and invertebrates

Chlorella (figs 455-459) is a common unicellular Chlorophyte and the commonest genus living symbiotically with freshwater invertebrates. Facts about its free-living (non-symbiotic) existence are given later.

The species or forms of *Chlorella* which are found in invertebrates have been placed in a separate genus *Zoochlorella* and often are referred to as zoochlorellae. There does not seem to be justification for placing the symbiotic forms in a separate genus, though not all strains of a given species are equally acceptable to an animal host.

Several ciliates contain symbiotic chlorellae (figs 455-457). An example is the vorticellid ciliate in fig. 455. Most vorticellids, including the one on *Anabaena* shown earlier in fig. 407 do not have *Chlorella* as a symbiont.

The other two ciliates shown here (*Euplotes* fig. 456 and *Frontonia* fig. 459) are free-swimming species. A feature which immediately distinguishes these two is the presence of some especially thick cilia (flagella) in the *Euplotes* in fig. 456. These actually are fused groups of cilia. Such compound cilia may be used as "walking legs" or for food gathering.

The presence of cells of *Chlorella* or other minute Chlorophytes in a ciliate is not in itself evidence of symbiosis. They may be being used as food. When a ciliate regularly contains live cells of *Chlorella*, it is reasonable to suspect that they are symbionts, though experimental evidence may be necessary to confirm this assumption. When the *Chlorella* cells in the ciliates shown are observed under higher magnification (fig. 458), it can be seen that they are green and have undamaged chloroplasts; in short they look just like healthy cells of a free-living *Chlorella*. It is virtually certain that they are living symbiotically in these ciliates and not simply aggregations of food particles. However, it should be pointed out, for reasons which will become clear, that if every part of such a ciliate was examined, a few cells in the process of being digested might well be found. The *Chlorella* can continue to multiply in the ciliate. Clearly there is only room for a limited number of *Chlorella* cells and even before this limit is reached the number of *Chlorella* cells could be so large that it interferes with the functioning of the ciliate's cell. Hence, the animal must have some method for removing excess *Chlorella* or controlling its rate of growth. Cells could be ejected from the ciliate or killed and digested. However, it seems that the main control is by restricting the growth-rate of *Chlorella*. Since the *Chlorella* concerned can be grown outside these animals, it can be shown that under suitable conditions of light and temperature it grows much faster outside than inside the animals.

It can be argued that such animals are the dominant partners in this symbiosis. They capture the *Chlorella*,

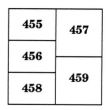

Figs 455-457, 459. *Chlorella*, symbiotic in: fig. 455, a vorticellid ciliate (body 63 μm l.); fig. 456, the ciliate *Euplotes* (87 μm l.), note the thick compound cilia; fig. 457 N., the ciliate *Frontonia* (120 μm l.) and fig. 459, the tentacle of the freshwater polyp *Hydra*.

Fig. 458. Cells (to 8 μm diam.) of *Chlorella* more highly magnified, occurring in *Frontonia*.

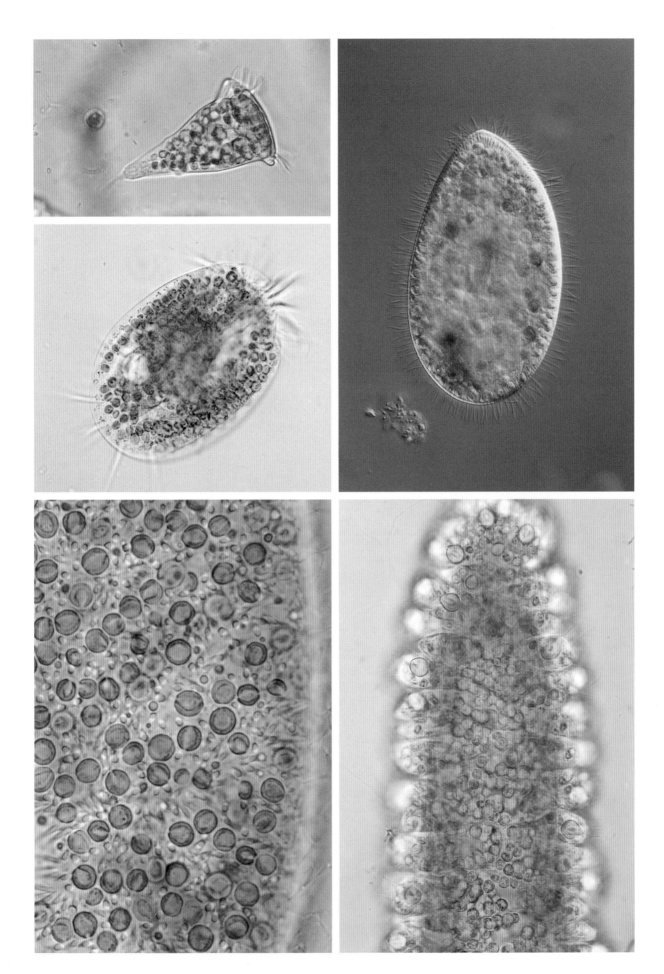

243

utilise products of its photosynthesis and control its numbers. In a few cases, ciliates containing *Chlorella* do not need other food. To return to our earlier analogy, they farm *Chlorella*. Nevertheless, there are advantages in being so "farmed". The host animal obtains carbon, mainly in the form of the sugar maltose from the alga's photosynthesis. The *Chlorella* obtains nitrogen from the animal and possibly some carbon dioxide from the animal's respiration. Inside a ciliate, *Chlorella* is protected from the outside world, in particular from being ingested and digested by some other animal, may be even by another kind of ciliate and from infection by a parasite.

Symbiotic *Chlorella* can be removed from a ciliate and grown in culture media such as those suitable for a variety of algae. Under normal culture conditions, the massive release of sugar which takes place when it is in a ciliate ceases. However, if *Chlorella* is grown in exceptionally acid media (e.g. at pH 4) release of sugar from its cells does take place. In the animal, the cells of *Chlorella* are maintained within vacuoles. It could be that the fluid within the vacuole containing *Chlorella* is very acid but other hypotheses have been advanced to explain how the cells of *Chlorella* are maintained in a state favourable to its ciliate host.

Figs 460, 461. *Hydra*, in retracted and expanded (4.5 mm l.) states.

Most of what has been said about the symbiosis between *Chlorella* and ciliates is applicable to that between it and the common freshwater polyp *Hydra* (figs 460, 461). Like ciliates, only some of the species of *Hydra* contain symbiotic chlorellae. In *Hydra*, the cells of *Chlorella* are restricted to certain regions of its body and tentacles (fig. 459).

Chlorella has only one method of reproduction – by the formation of replicas of itself which are released when the cell wall breaks up. It is extremely common as a free-living alga but in small or even moderate abundance can easily be overlooked. Its cells usually are 5 μm or less in diameter under natural conditions. The cells of terrestrial species are abundant in soil. It can resist being air-dried. Consequently, its cells are blown far and wide as dust and have been found high up in the air. They neither grow nor reproduce in the air.

Chlorella is very easy to grow and so was one of the first "laboratory algae". It was used in some classic work on photosynthesis. Previously, leaves had been used but their bulk and structural complexity limited their usefulness. *Chlorella* had the great practical advantage that the whole plant is a single, small cell. From *Chlorella* work spread to *Euglena* and to *Chlamydomonas* (p. 82) which, like *Euglena*, can grow in both light and in darkness and has the advantage over both *Chlorella* and *Euglena* that it reproduces both asexually and sexually. Sexual reproduction permits work on the genetical basis of photosynthesis. *Chlorella* was also one of the first algae grown as a possible food additive (cf. *Spirulina*, p. 230) and still is grown for this purpose to some extent, notably in Japan and Taiwan.

A further use of *Chlorella*, *Spirulina* and other small algae being studied is as food and as part of bioregulatory systems for space travel. For short journeys in space or living in space stations, food is taken up in the spacecraft or sent up to the space station. For longer journeys or life in space, food will have to be produced, oxygen supplied and, if possible, waste products recycled. Clearly, apart from producing food, photosynthetic organisms are a source of oxygen and potential utilisers of the carbon dioxide respired by animals. Algae such as *Chlorella* do not need a dark period (night), so in space they can photosynthesize and grow continuously. The oxygen arising from that photosynthesis can be breathed by the space travellers and the carbon dioxide exhaled by them used in the photosynthesis of such algae. Further, a certain amount of the food eaten is returned as excrement which, as in sewage treatment plants, could yield vital nutrients for the algae, such as ammonia, nitrates and phosphates. Hence, it is possible that algae could form a valuable part of a recycling system for life support in space.

Some animals do not have algal cells as symbionts but only their chloroplasts. It does not seem that these chloroplasts can multiply, therefore they will have a limited lifetime inside an animal. This, as we have seen, may be the situation (or near to it) in the case of the Cryptophyte inside the Dinophyte (p. 180). If the chloroplasts could multiply as well as the animal host and especially if they multiplied at the same time as the host, the situation would be more akin to the formation of a new organism than to symbiosis. This chloroplast-containing organism would be a photosynthesizing animal. Perhaps we should say it would be a plant, since there is no fundamental difference between plants and animals at the level of algae (see p. 1) unless one erects photosynthesis as the separating feature. Examples of an animal and an alga containing symbiotic photosynthetic organelles are shown later (cyanelles, p. 252). Further, such an organism as an animal with symbiotic chloroplasts may be an example of how certain stages in the evolution of plants began.

Lichens

Lichens (figs 462-464) are familiar objects on trees, rocks, walls and tombstones. They are dual organisms composed of an alga (sometimes more than one alga) and a fungus (rarely more than one).

Lichens are unique in that their shapes and forms (morphology) are quite different from those of their constituent algal and fungal components. It can be said that the appearance of a lichen depends more on the fungus than the alga. The same algal species with different fungal partners can be part of lichens of very different form and appearance.

Almost all the algal partners are Chlorophytes or Cyanophytes and so green or blue-green in colour. The algae usually are located within a mat of fungal filaments (hyphae) and so in the lichen their colour is likely to be masked to some extent by that of the fungus (e.g. see the yellow lichens on the wall, fig. 464 and the light grey ones on the tombstone, fig. 463). In about 90% of the 15,000-20,000 known lichens the algal partner is a Chlorophyte. In some 500 lichens the algal partner is a Cyanophyte or there are two algal partners, a Chlorophyte and a Cyanophyte.

The chlorophyte genus *Trebouxia* is the commonest algal partner in lichens. Some people split this genus into two (*Trebouxia* and *Pseudotrebouxia*) but, considering it as a single genus, then it is the algal component of about 50-70% of all lichens. Superficially *Trebouxia* looks very like *Chlorococcum* (fig. 13). It differs in that its chloroplast does not line the outer part of the cell but fills most of it and is somewhat lobed. The single pyrenoid is in the central region of the chloroplast.

Trentepohlia is the second commonest algal component in lichens. It is a branched filamentous Chlorophyte (figs 465, 466) which also is common as a free-living plant on rocks, boulders, trees and, especially in the tropics, on leaves. *Trentepohlia* is cosmopolitan but more abundant in tropical than in temperate regions, both free-living and in lichens. It favours shady places where it often forms visible yellowish, orange or brownish red patches, carotenoid pigments masking the green of its chlorophyll. Fig. 466 shows a pear-shaped sporangium sessile on a filament. Sporangia may be sessile or stalked. They produce flagellate spores or gametes.

A closely related genus, *Cephaleuros*, is parasitic. The plant enters the leaves or bark of a wide variety of tropical and sub-tropical plants including a number of commercial value. It is the cause of red rust in tea plants, which can be a serious disease.

Stichococcus (fig. 467) is the algal component of a number of lichens. This Chlorophyte can be unicellular or the cylindrical cells are joined to form short or, less

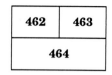

Figs 462-464. Lichens on a tree-trunk, tombstone and stone wall.

commonly, long unbranched filaments. Its species vary in their predisposition to be unicellular or filamentous. Environmental conditions also influence the balance between the unicellular and filamentous states. In lichens, *Stichococcus* nearly always is unicellular. How far this state is controlled by inherent (i.e. genetic) factors, the physico-chemical environment or the direct influence of the fungal partner is unknown. *Stichococcus* is very common in soil and other terrestrial habitats but can also grow in water. Like *Chlorella*, its only method of reproduction is by cell division, in this case producing a single copy of itself. However, whereas in *Chlorella* the daughter cells are produced inside the mother cell, in *Stichococcus* the mother cell simply divides by the formation of a cross-wall into two daughter cells. Some cells which have divided recently and are still attached to or close to one another can be seen in fig. 467.

It is not only that the partners in a lichen unite to form an organism which looks so different from each of them separately, there also are changes in their microscopic structure, in physiology and ecology. Algal cells inside a lichen often look somewhat different from their appearance when living outside. Indeed, in a number of cases the correct identification of the algal partner has only been possible after cells have been extracted from the lichen and grown in the laboratory. The fungus also has a major effect on the algae's physiology (cf. ciliates and *Hydra*). On present evidence, the fungus receives far more nutrients from the alga than the alga does from the fungus. Most of the carbon, perhaps at times even 80% or more, produced by the photosynthesis of the alga is rapidly released from its cells and utilised by the fungus. In the case of Cyanophyte partners which fix nitrogen, this nitrogen too largely passes to the fungus. Clearly, the fungus operates a mechanism which makes such nutrient transfers possible because if, the algal symbionts are removed from a lichen, the effluxes of carbon or nitrogen from their cells rapidly decrease to the lower levels characteristic of non-symbiotic forms. Various kinds of carbon compounds (carbohydrates) are released by algal partners of lichens. Some at least of the Cyanophytes release glucose and, if they fix nitrogen, ammonia. It is not clear whether the fungal partner directly or indirectly supplies nutrients to the alga. Since growth of the alga in a lichen usually is slow, it may well obtain all the nutrients it needs from the atmosphere and rain. In addition, dust and soil particles fall on lichens and supply nutrients.

Like the algal-invertebrate symbioses, there are many aspects of the algal-fungal association in lichens which are not understood. A major mystery is what are the conditions which make the formation of lichens apparently so easy in nature, judging by their abundance from one end of the world to the other. Lichens often flourish in harsh climates

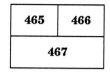

Figs 465, 466. *Trentepohlia*, the second commonest algal partner in lichens. Cells 6.2-12.5 μm br.; lateral sporangium in fig. 466. 20 × 17 μm.

Fig. 467. *Stichococcus*, the algal partner in some lichens. Cells up to 10 μm l. P.

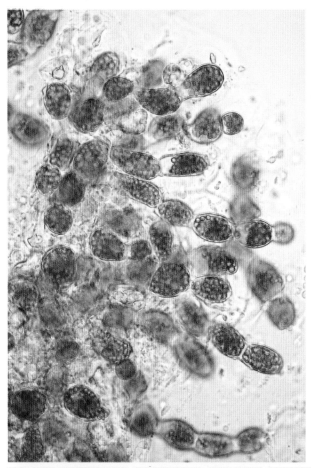
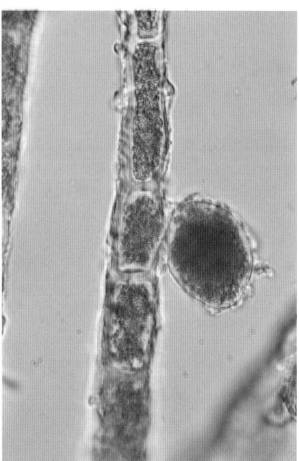
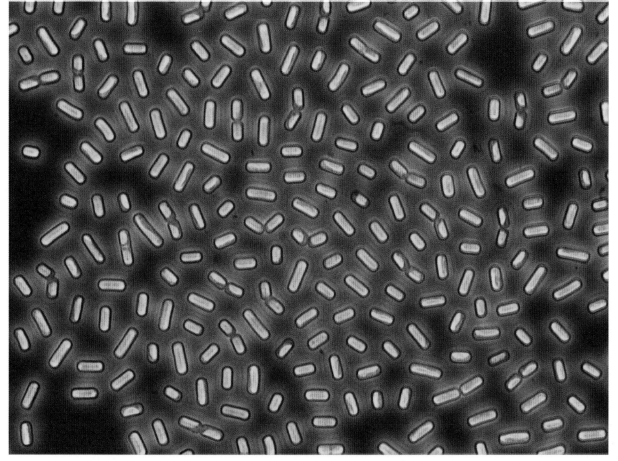

249

or in apparently unfavourable habitats in more equable climates (e.g. the tombstone, fig. 463). Yet in the laboratory it is very difficult to create lichens from their component organisms. Even today few fully developed lichens such as one sees in nature have been synthesized artificially. One of the problems in work on lichens is the slow growth of so many of them. A crustose lichen such as that on the tombstone may only increase in diameter by about one millimetre per year. An investigator needs plenty of time and patience.

Many of the lichens on old tombstones are themselves old. In temperate climates, lichens can be hundreds of years old and it has been calculated that some Antarctic lichens are thousands of years old. Though the fungus can be considered as the dominant partner in a lichen, to be captured by it is not without benefit to the alga. Lichens often exist in habitats which would make algal growth on their own impossible or very restricted. Most of the algal partners are terrestrial species and so able to withstand greater ranges of heat, cold and drought than aquatic species. Nevertheless, in a lichen a terrestrial alga can continue to live and multiply, albeit very slowly, under conditions it could not do so outside the lichen. Consequently, they are present in places where they are very rare or virtually absent as free-living organisms. The tombstone (fig. 463) is an example, though the churchyard is in a relatively mild, oceanic climate. Tombstones in much more severe climates have lichens on them. Free, the alga would be exposed to severe and often rapid climatic change. Further, this tombstone is a slab of smooth, acid slate to which it would be difficult to remain attached and from which little or no nutrients can be obtained.

The fungi on their own are in an even poorer nutritional position than the algae. They cannot photosynthesize and depend for life on preformed organic matter. There is no organic matter in slate or rock in general yet, despite these disadvantages, on millions of tombstones fungi and algae meet and form lichens.

Many species of algae in lichens also are common free-living organisms. Nevertheless, it may be that only certain races or varieties are acceptable as partners of fungi. A few extremely common terrestrial algae are uncommon or unknown as components of lichens. *Chlorella* is an example, despite being the commonest component of invertebrate symbioses. Two other examples are the Chlorophytes, *Apatococcus* (fig. 468) and *Desmococcus* which are the commonest terrestrial algae covering the bark of trees, woodwork, walls (figs 1, 2) and other surfaces with a rather pale and powdery green coating. They exist as small groups or packets of cells, often in pairs or quartets and as single cells. Both are so alike that originally they were considered to belong to a single genus

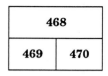

Fig. 468. *Apatococcus* cells and cell groups with fungal filaments (colourless threads) growing among them; sample taken from palings seen in Fig. 1. *Apatococcus* cells c. 6-13 μm in longest dimension.

Fig. 469. *Glaucocystis*. Colony (58 × 46 μm) of two cells containing cyanelles.

Fig. 470. *Paulinella*, an amoeboid protozoan living in a case (30 μm l) and containing two cyanelles.

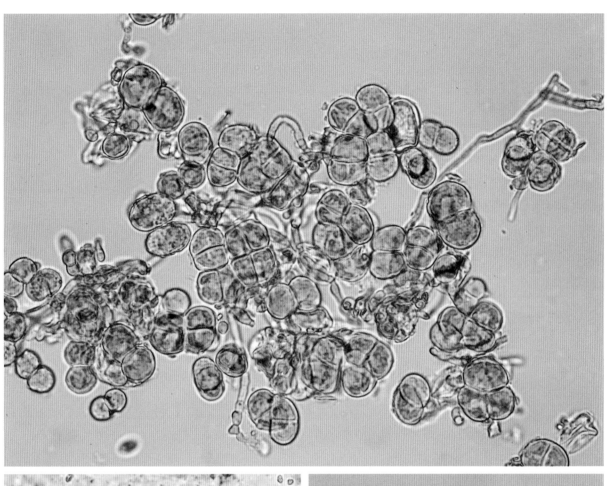
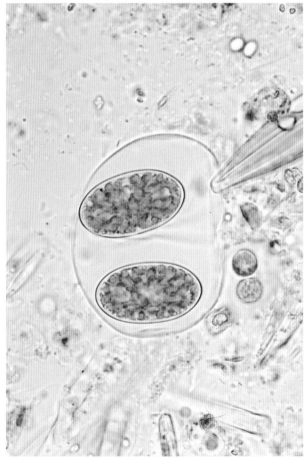
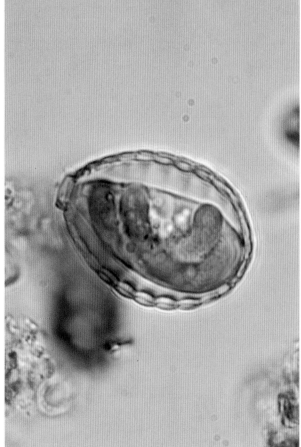

called *Pleurococcus*. Such algae are still generally referred to as pleurococcoid algae.

Since *Apatococcus* is so common in habitats rich in lichens it is remarkable that it does not appear to be the algal partner of any lichen. Further, it is not uncommon to find fungal filaments among the cells and cell groups of *Apatococcus* as in fig. 468 (the colourless threads). What the relationship, if any, between the alga and fungus may be is unknown. Apart from the fact that it is not an early stage in lichen formation, it is not clear whether the fungus is parasitizing the alga, utilizing organic matter released by live algal cells or that of dead cells.

The bark of trees in towns often is coloured green by pleurococcoid algae but few, if any, lichens are found on them. Pleurococcoid algae are much more resistance to pollution by cars and industry and so also to acid rain than are lichens. The lichen flora becomes poorer and poorer as one passes from the relatively pure air of the countryside to that of towns or industrial areas. Consequently, lichens are useful indicators of increasing atmospheric pollution.

Cyanelles

With figs 469, 470 we return to Cyanophyte or Cyanophyte-like symbionts. These symbionts are called cyanelles.

Glaucocystis (fig. 469) is an alga of uncertain taxonomic position. The cells are solitary or in groups. Each cell contains two stellate groups of cyanelles and each cyanelle looks like a curved *Cyanothece* (p. 195). *Glaucocystis* reproduces by producing internally two to eight copies of itself. These daughter cells may remain within the enlarged mother cell wall for some time so that temporary colonies arise. Starch is produced and during reproduction rudimentary flagella and contractile vacuoles are formed. These features suggest that *Glaucocystis* is a Chlorophyte with cyanelles in place of chloroplasts. Against this view is a statement that the starch present is not the kind found in Chlorophytes but is the kind typical of Rhodophytes. However, no Rhodophyte has contractile vacuoles or flagella even of rudimentary construction.

Paulinella (fig. 470) is a unicellular amoeboid protozoan living in a case or shell. Inside the animal are two cyanelles once again looking like curved *Cyanothece* cells and possessing a similar method of division. The amoeba always contains two cyanelles except during and shortly after it divides into two daughter cells. Each daughter cell receives one cyanelle which then divides into two. Some other animals and algae contain blue-green symbionts called cyanelles but it is uncertain to what extent these cyanelles are related to one another. The cyanelles differ from the symbiotic Cyanophytes seen previously in that they cannot be grown outside the host organism. They can be considered to be symbionts which in the course of time have become an integral part of the host cell, that is organelles just as chloroplasts are organelles. *Paulinella* can then be looked upon as a photosynthetic animal.

Glaucocystis is found in muddy or peaty places, not uncommonly in bogs among sphagnum moss. *Paulinella* occurs in similar habitats but less commonly in Sphagnum.

Even though only a few of the many kinds of symbiosis are seen here, they do give an indication of the variety of symbiotic relationships. In *Azolla*, the Cyanophyte symbiont lives in a leaf cavity; in *Gunnera* in the host's tissue; in *Hydra* and ciliates, the Chlorophyte *Chlorella* lives within the animals' cells, and, as mentioned, some animals incorporate only the chloroplasts of an alga (see also the case of the Cryptophyte taken in by the Dinophyte *Gymnodinium*). Finally, in the cases of *Paulinella* and *Glaucocystis* there is some doubt whether one should consider the Cyanophyte cyanelle as a symbiont or a cell organelle. The suggestion that the chloroplasts of plants arose via a symbiotic pathway is one of the pillars of the symbiotic theory of the evolution of eukaryotes. Cyanophytes are prokaryotes and therefore do not belong to the plant or animal kingdoms but to the bacterial one. Since Cyanophytes are both photosynthetic and among the oldest known group of organisms (p. 195), they are prime candidates for the symbiotic theory of how the chloroplasts of plants arose. Other organelles of eukaryotic organisms also may have originated by a symbiotic pathway. Nor should it be thought that yet other kinds of symbiotic transfers have not taken place.

The symbiotic evolutionary pathway may be from the union of a prokaryote with a eukaryote or one eukaryote with another. A present day organism could have resulted from the incorporation of more than one organism into another. For example, if we accept that the blue-green species of *Gymnodinium* arose from the incorporation of a blue-green Cryptophyte or its chloroplast into a colourless *Gymnodinium*, then a further possibility arises. If the Cryptophytes arose from the entry of one cell into another (p. 180) with eventual coalescence into a new unitary organism and if one of the original partners has a common ancestry with Rhodophytes, then the presence of a Rhodophyte kind of blue photosynthetic pigment is not surprising. In this case, a blue-green *Gymnodinium* has evolved, at least partially, by a symbiotic pathway from three otherwise unrelated ancestors.

The symbiosis theory is based on a variety of facts and assumptions far beyond the scope of the commentaries on the pictures included in this book. The theory is mentioned because of the series of symbiotic systems encountered above and because it is a fascinating possibility. This symbiotic evolutionary pathway may only be a theory but it is one for which there is strong factual support.

Chapter Twelve

Animals feeding on algae

Algae are "eaten" by many animals. Biologists usually avoid the verb to eat because it does not cover the diversity of methods used by animals and, notably, microorganisms to obtain and process food. The verbs usually used are to ingest, to graze upon or to feed on.

Freshwater and terrestrial algae are important food for many animals, most of which are invertebrates, though there are herbivorous fish and *Spirulina* and other algae are a major food source for flamingoes in certain saline African lakes. Nearly all the invertebrates shown here are microscopic or only just visible to the naked eye. However, the great importance of larger invertebrates in controlling algal numbers should not be forgotten. Examples are snails, caddis flies and stoneflies.

Protozoans

Protozoa, like algae, are a diverse group of organisms. All are eukaryotes. As mentioned before, there is no absolute distinction between animals and plants. At the level of protozoa some of the organisms are classified in both the animal and plant kingdoms. Ciliates have no counterparts in the plant kingdom. Some of the amoeboid or flagellate protozoans pictured later do have somewhat or markedly similar photosynthetic and so plant-like counterparts.

Ciliates

Ciliates have been encountered previously (pp. 212, 242-244). The *Lembadion* in fig. 471 has ingested about 20 cells of a single species of *Cryptomonas* (p. 170) a cell of which can be seen outside the ciliate. The wide oval area on the right of the mass of cryptomonads is its "mouth".

Frontonia feeds on a variety of algae and can ingest whole filaments of the Cyanophyte *Oscillatoria*, which when first taken in become coiled around its body (fig. 472). Note that this species of *Frontonia* does not contain cells of the symbiont *Chlorella* (cf. the *Frontonia* shown in fig. 457). The ciliate in fig. 473 has been feeding on non-planktonic pennate diatoms, the cells of which are in various stages of digestion. Three live cells are present outside the ciliate. The ribbed funnel-shaped organelle in the ciliate can be expanded so that the diatom cells can

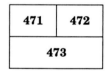

Fig. 471. *Lembadion* (92 × 72 µm), a ciliate which has ingested many cells of *Cryptomonas*. Outside is a single, living *Cryptomonas* cell (19 × 9 µm).

Fig. 472. *Frontonia* (420 × 234 µm). The ciliate has ingested several filaments of *Oscillatoria* (*Planktothrix*) which are coiled round inside it.

Fig. 473. A ciliate (125 × 75 µm) with (right) its ribbed funnel-like feeding organ (cytopharyngeal basket), through which several diatoms have entered its body and now are in various stages of digestion. **P.**

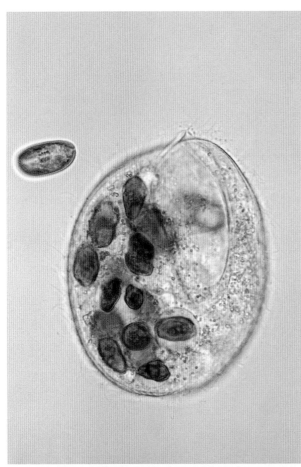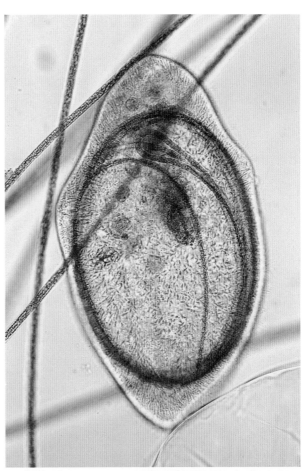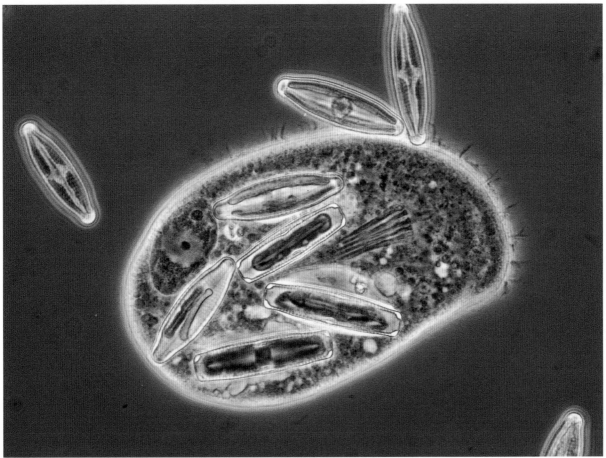

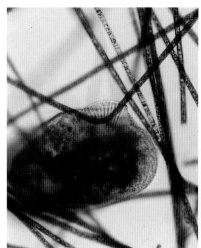
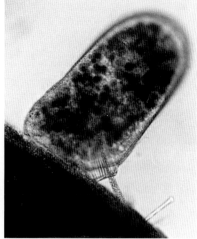
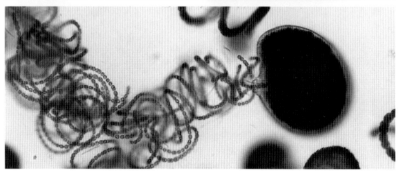

Figs 474-476. *Nassula* ingesting filamentous Cyanophytes via its cytopharyngeal basket.

Fig. 474. A filament of *Oscillatoria* (*Planktothrix*) is just beginning to be taken in (× 256).

Fig. 475. Two filaments have been detached from a raft of *Aphanizomenon* (large black area) (× 286).

Fig. 476. Feeding on *Anabaena* cells (× 205).

pass down into the ciliates body. It is called a cytopharyngeal basket.

Some species of *Nassula* feed on a variety of Cyanophytes, others on different kinds of algae. The cells or filaments are ingested via a cytopharyngeal basket the tubular mouth of which can be seen in figs 474-476.

The *Nassula* in figs 477, 478 is feeding on the Cyanophyte *Aphanizomenon* (p. 216). In fig. 477 there is a single oval-oblong, black body among the filamentous colonies of *Aphanizomenon*. The filaments of *Aphanizomenon* are black because their cells contain gas vesicles (see p. 202) and the oval-oblong body, a cell of *Nassula*, is packed with *Aphanizomenon* and so with gas vesicles. Fig. 478 shows the plankton of the same lake a week later. The *Aphanizomenon* population now is reduced to very small colonies or single filaments. Large numbers of *Nassula* are present. At the end of such an outburst as this, and as food becomes short, *Nassula* forms resting cysts (fig. 481) which then sink into the bottom sediments. Within the cyst, *Nassula* retains the framework of its cytopharyngeal basket which can be seen in fig. 481 as a central circle of rodlets.

As mentioned earlier (pp. 220-222) large populations of planktonic, gas-containing Cyanophytes have undesirable effects on water quality and other aquatic animals. Further, *Aphanizomenon* is one of the toxic Cyanophytes implicated in the deaths of mammals such as dogs, cattle and sheep. *Nassula* is one of the few organisms ingesting

Fig. 477. A waterbloom of *Aphanizomenon* (colonies to 187 × 125 μm). Single oval-oblong body, *Nassula*, which is feeding on the Cyanophyte.

Fig. 478. The same waterbloom a week later. Note many *Nassula* and remnants of the *Aphanizomenon*.

Aphanizomenon and other planktonic Cyanophytes such as *Oscillatoria* and *Anabaena*. Most aquatic animals avoid them. Therefore, *Nassula* is a possible organism to use for biological control of toxic or otherwise undesirable Cyanophytes.

Whether this is a practical possibility remains to be seen. Large numbers of the ciliate would need to be kept under cultivation or its cysts collected and stored ready to "seed" a lake at the appropriate moment. The idea of using *Nassula* in this way is not new. It was suggested over 50 years ago by a Swiss scientist Dr E. A. Thomas in relation to his study on blooms of the Burgundy Blood Alga *Oscillatoria (Planktothrix) rubescens* (p. 220) which were threatening the water quality in lower Zürichsee. Biological control is preferable to adding toxic chemicals, which may affect other aquatic organisms and are not acceptable in drinking water. Further, whereas chemicals simply kill the organisms concerned, *Nassula* both removes the Cyanophytes and is itself a source of food for other organisms such as fish. Removal of the Cyanophytes may also reduce deoxygenation of the water.

The *Nassula* in fig. 482 is in the process of ingesting a filament of a quite different and non-gas-vacuolate *Oscillatoria*. The filament is passing in via the cytopharyngeal basket. Once inside the *Nassula* the *Oscillatoria* filaments soon become broken down at the cross-walls into short pieces prior to their ultimate digestion.

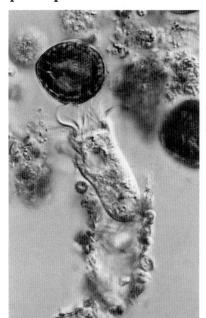
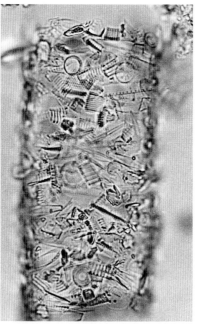

Fig. 481. A resting cyst of *Nassula* (81 µm diam.). Below the cyst are spores of the *Anabaena* species upon the vegetative cells of which (see figs 404, 405) the ciliate had earlier fed. N.

Fig. 482. *Nassula* (100 × 64 µm). The ciliate is in the process of ingesting a filament of *Oscillatoria* via its cytopharyngeal basket (seen at 12 o'clock). Discoid remains of other filaments are visible in the cell.

Figs 483, 484. Tintinnid ciliates.

Fig. 483. The case (146 × 66 µm) is coated with centric diatoms. P.

Fig. 484. A case (119 × 56 µm) with kephyrioid Chrysophytes.

Figs 479, 480. Tintinnids.

Fig. 479. A cell partially outside its case (× 640). N.

Fig. 480. An empty case covered mainly by broken pieces of walls of diatoms (× 800).

Tintinnids are ciliates which live in a tubular or sac-like case (fig. 479). The case commonly has the remains of small algae attached to it. In general only fragments of long diatoms are present (fig. 480). In fig. 483 the vast majority are small centric diatoms. Resistant remains of other algae can be attached to the tintinnid's case, for example the cases in which kephyrioid Chrysophytes (fig.

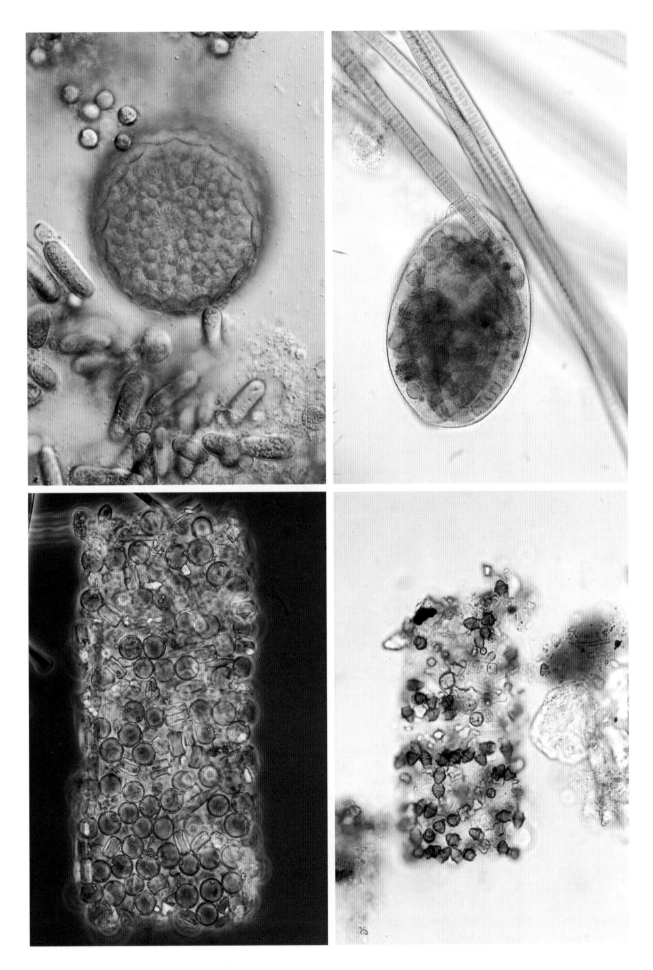

259

484 and p. 158) live. The case of the species shown is colourless when first formed but soon becomes brown. More needs to be known about the source and development of these coatings of algal remains. In their absence, fine silt particles may stick to the tintinnid.

Other protozoans

The flagellate protozoan in fig. 485 has taken in a cell of the Euglenophyte, *Phacus* (p. 102) which is more than half its own size. Just below the insertion of the apical flagella of the protozoan is its prominent nucleus and nucleolus (dark-blue sphere).

The name of this protozoan is uncertain. It possibly is a species of a colourless (non-photosynthetic) biflagellate (parts of both flagella are visible in the figure) Chrysophyte called *Spumella* but the proof of this depends on finding a chrysophycean cyst.

The genus *Monas* consists of similar protozoans which apparently do not form chrysophycean cysts. We say "apparently" because such cysts may one day be found in one or more of these *Monas* species, when they will be transferred to *Spumella*. A common name for flagellate unicellular protozoans or algae is monad which means the same thing as *Monas* (Greek: *monos* – solitary).

Amoebae come in many shapes and sizes. In common with other amoebae, the specimen shown (fig. 486) moves by means of extensions of its protoplasm, a number of which can be seen. These extensions are called pseudopodia. In some other amoeboid organisms the pseudopodia look somewhat or very different from those of this amoeba and are given different names, for example very fine pseudopodia are called filopodia. Here, all such protoplasmic extensions are called pseudopodia. If a suitable source of food is encountered, the amoeba's pseudopodia flow round it. In fig. 486 several small centric diatoms have been captured and are now located in food vacuoles in various parts of the amoeba's body. When digestion is complete, the amoeba will discard the siliceous remains.

In some amoeboid organisms, digestion of food is only completed after the animal has ceased to move, rounded off and secreted a firm wall around itself. Such a structure is called a digestion cyst, examples of which are shown in figs 487, 488.

Two digestion cysts belonging to the amoeboid organism, *Asterocaelum*, which also has been feeding on centric diatoms, are present in fig. 487. Unlike the previous amoeba, it possesses long fine pseudopodia in its vegetative feeding stage (fig. 494). The cysts are ornamented with very delicate spines which are often difficult to see. When first formed the spines are filled with protoplasm (fig. 489). However, this protoplasm soon

| 485 | 486 |
| 487 | 488 |

Fig. 485. A protozoan (monad), probably a species of *Spumella* (52 × 37 μm) with an ingested cell of *Phacus*. **P.**

Fig. 486. An amoeba (112 × 62 μm) containing cells of a centric diatom. **N.**

Figs 487, 488. *Asterocaelum.*

Fig. 487. A large (45 × 50 μm) and a small (19 μm) spiny digestion cyst containing centric diatoms and surrounded by mucilage. The small cyst contains only one diatom. **I.**

Fig. 488. Resting spore (19 × 22 μm) developed within a digestion cyst. Note remains of centric diatoms. **I.**

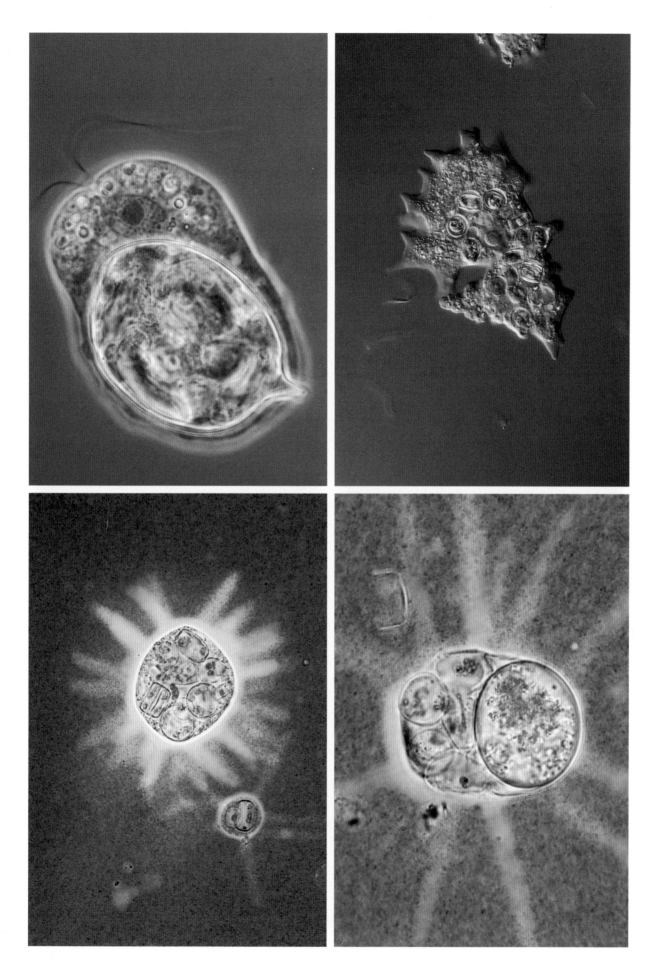

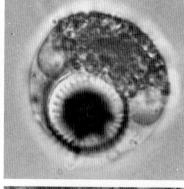

Figs 489–493. *Asterocaelum*.

Figs 489, 490. A spine on a digestion cyst containing protoplasm and one now devoid of protoplasm. Note mucilage sheathing each spine (× 1600). I, P.

Figs 491 (× 1600), 492 (× 1008) I. Early stages in resting spore development, before and after condensation of the protoplasm into a solid lump; note single digested cell of *Stephanodiscus* in fig. 491.

Fig. 493. Mature resting spore (the large winged body) surrounded by diatoms digested by *Asterocaelum* cells (× 640). I.

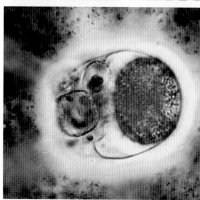
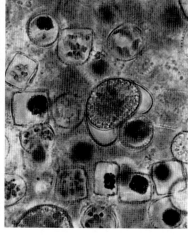

passes from the spine into the developing cyst, so leaving the spine empty and transparent (fig. 490). A layer of mucilage surrounds both the cyst and the spines. The Latin words *aster*, a star and *caelum*, the sky, reflect the stellate appearance of the cysts especially well seen in Indian ink (fig. 487).

Eventually, one or more amoebae (figs 495–498) will emerge from the cyst leaving it empty. These daughter amoebae may be small (fig. 498) or large (fig. 497). The fusion of one amoeba with another can take place as they leave a cyst or while wandering free in the water. The giant feeding individual in fig. 494 probably has arisen in this manner.

Towards the end of a period of abundance and as food becomes scarce, the content of some cysts, instead of developing into new amoebae, become transformed into a resting spore (fig. 488). First, the animal rids itself of undigested algal material and then becomes surrounded by a firm wall (fig. 491). Next the protoplasm shrinks away from this wall and condenses into an oval or spherical body (fig. 492). This is the young resting spore, around which later a thick wall will develop (fig. 488).

The *Asterocaelum* shown here occurs in the plankton of lakes and rivers. Although it feeds predominantly on centric diatoms, small green algae can also be ingested. Another *Asterocaelum* is known to devastate planktonic populations of *Anabaena*.

Figs 494–498. *Asterocaelum*. Amoebae.

Fig. 494. Giant amoeba (140 × 50 µm).

Fig. 495. Four amoebae (14–35 µm diam.) issuing from a cyst.

Fig. 496. A single large amoeba (37 µm diam.) coming out of a cyst.

Fig. 497. The same amoeba (now 47 × 34 µm) free in the water and with fully formed pseudopodia.

Fig. 498. Four small amoebae (19 µm diam.) and the empty cyst from which they have emerged.

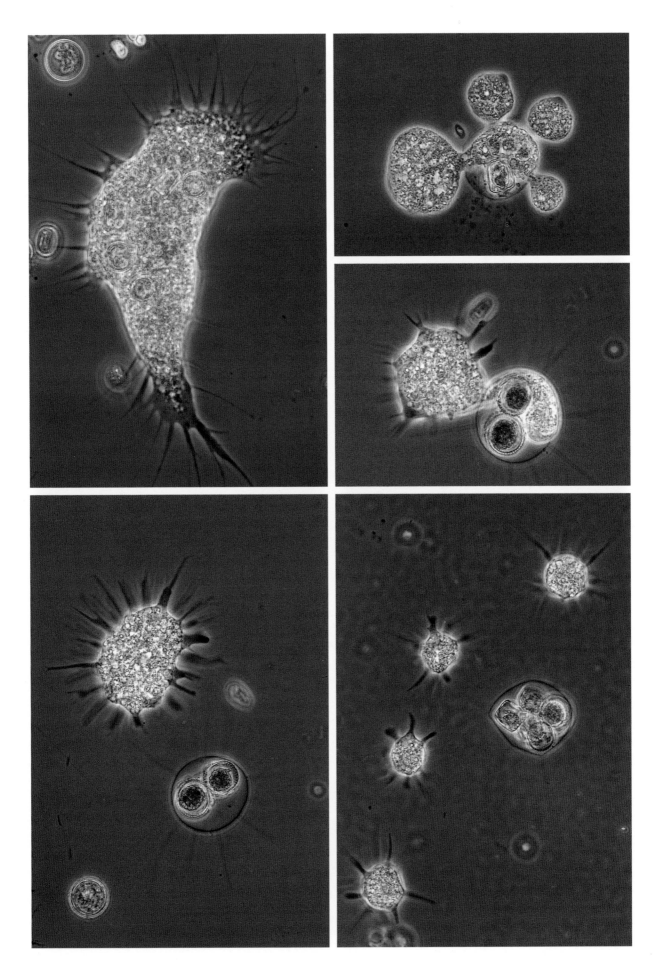

Another amoeboid organism has formed the cyst in fig. 502. Many needle-shaped pennate diatoms have been ingested and are now packed parallel to one another in the cyst. In due course new amoeboid individuals will issue forth leaving the dead diatom frustules behind.

Vampyrella (figs 499-501, 503-519) does not engulf the whole of its prey and eventually become a digestion cyst like *Asterocaelum* (fig. 487). Instead, it extracts the contents ("sucks out" – hence the name) of algal cells (predominantly Chlorophytes) and then is transformed into a digestion cyst. In fig. 503, two vampyrellas are creeping among filaments of *Oedogonium* (p. 62). The uppermost one has made contact with a cell by means of fine pseudopodia, some of which are in focus. As seen here, the protoplasm of a *Vampyrella* often possesses an overall orange or reddish colouration.

In figs 499 and 500, the contents of a cell of the green alga *Geminella* (p. 66) are in the process of being extracted by a *Vampyrella*.

502	503
504	505

Fig. 502. Digestion cyst. (94 × 37 μm) of an amoeboid organism containing cells of a pennate diatom.

Fig. 503. *Vampyrella*. Upper cell (31 × 25 μm) making contact with *Oedogonium*, the content of which it will eventually extract. **N**.

Figs 504, 505. *Vampyrella*. **I**.

Fig. 504. A young digestion cyst on an empty cell of *Geminella* (7 μm br.).

Fig. 505. Two digestion cysts; the upper more mature than that in fig. 504; the lower containing only residues because its content has issued as one or more amoebae.

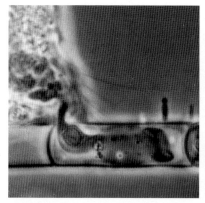
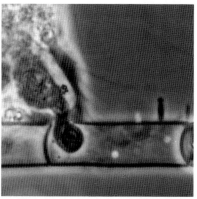

Digestion cysts of the *Vampyrella* are shown in figs 504 and 505. In the cyst seen in fig. 504 the ingested algal material is still green indicating that it has been recently taken in and that little if any digestion has as yet taken place. The upper cyst in fig. 505 represents a more advanced stage in the digestion process. The algal content has now become a shrunken brown lump and the orange coloured globules in the *Vampyrella* protoplasm more prominent. The lower cyst in fig. 505 contains only the undigested remains of the *Geminella* cells, the *Vampyrella* itself having emerged from the cyst in the form of one or more new amoebae (fig. 501).

499	500

Figs 499, 500. *Vampyrella*, main contents of a cell of *Geminella* passing into it. The black rod on top of the algal cell is a bacterium (× 1800). **P**.

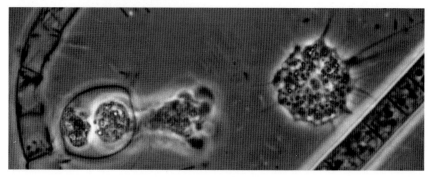

501

Fig. 501. *Vampyrella*. On the left, an amoeba issuing from a digestion cyst; on the right, the amoeba which had previously come out of the cyst (× 1475). **P**.

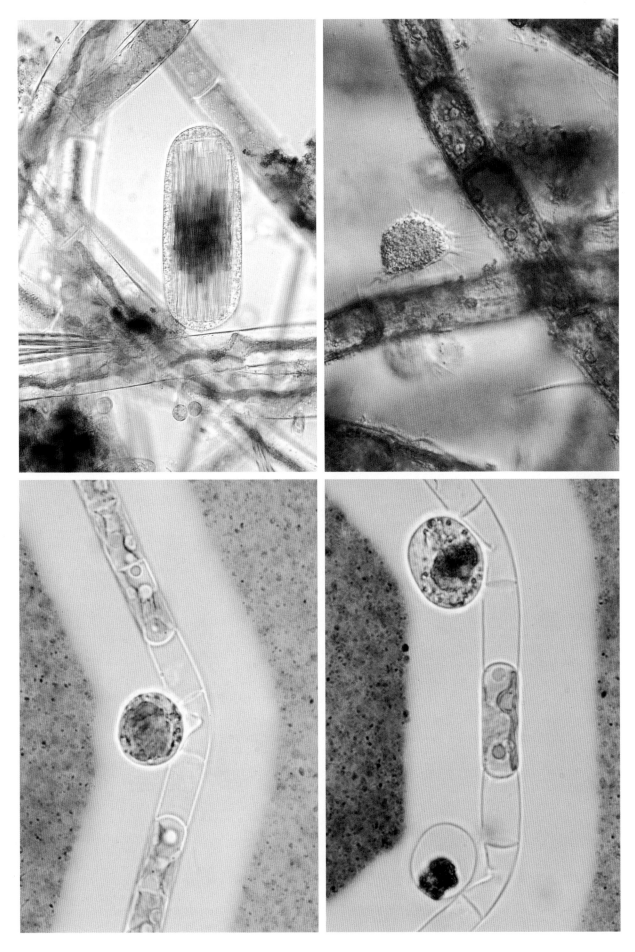

265

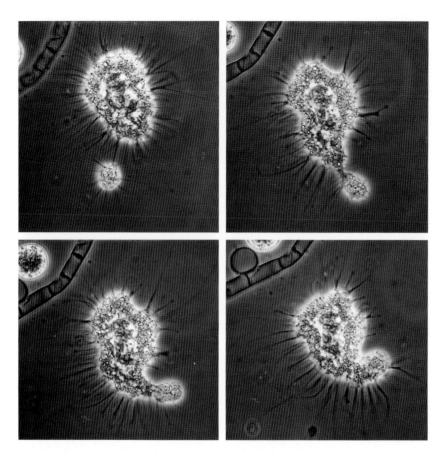

Figs 506-509. *Vampyrella*. Stages in the fusion of a small and large individual (× 585). **P.**

In *Vampyrella* amoebae can fuse with one another and the participating individuals be somewhat similar or very dissimilar in size (figs 506-509). Amoebae of *Asterocaelum* also exhibit the same behaviour.

Further examples of the infestation of *Geminella* are shown in figs 510 and 511. Along the bottom filament in fig. 510 two *Vampyrella* cells can be seen making contact with cells of *Geminella*. Straight pseudopodia penetrating the mucilage envelope (here not visible) join as yet unharmed algal cells to the amoebae. Later these pseudopodia will shorten and so bring the main body of the *Vampyrella* into direct contact with the algal cell wall. To the left of these "vampyrellids" is one which is still creeping around freely in the water. On the filament above this specimen (left) is a *Vampyrella* which has penetrated a cell and the alga's contents have just passed into it and are still green. Once a hole is made in the alga's cell wall most or all of its contents flow rapidly into the *Vampyrella*. This may cause its body temporarily to become greatly extended. Other examples of this stage are present in the top and bottom filaments in fig. 511. The *Vampyrella* usually creeps from one *Geminella* cell to another extracting their contents successively. Sometimes it may remove the contents of two cells simultaneously. Both cysts undergoing digestion and those which have liberated amoebae are evident on the filaments. Ultimately such an attack as this can result in the death of 90% or more of the algal population.

Figs 510, 511. *Vampyrella*. Severe infestation (epidemic) on *Geminella* (cells 6 μm br.). Note various stages of development, e.g. a "free" amoeba (fig. 510), amoebae contacting algal cells, extracting their content, young and old cysts. **P.**

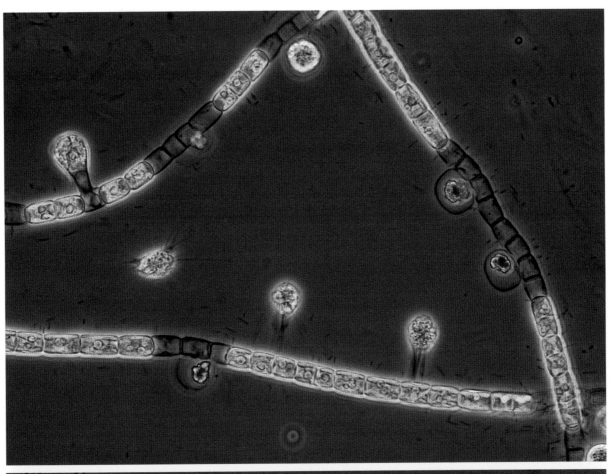
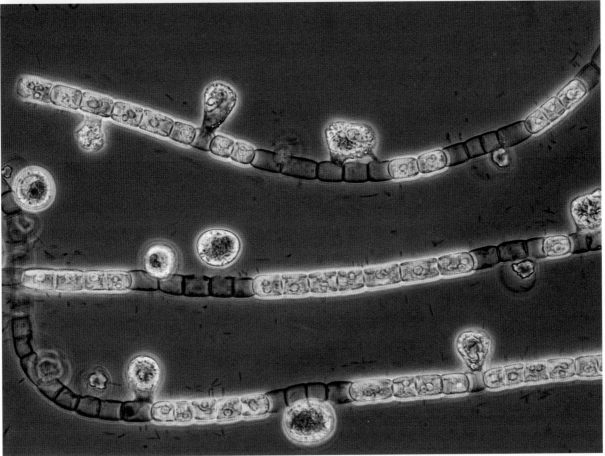

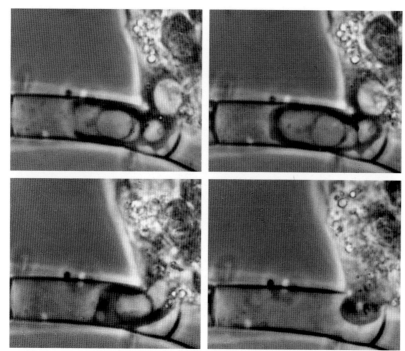

Fig. 512-515. *Vampyrella*. Stages in the extraction by a pseudopodium of the last remnants of a *Geminella* cell (× 1800). P.

One interesting facet in the feeding process of *Vampyrella* is the passage of a pseudopodium into an algal cell after the main mass of its chloroplast has been extracted. Such a pseudopodium, as it were, licks up any of the cell content that has remained and then takes it back into the *Vampyrella* (figs 512-515).

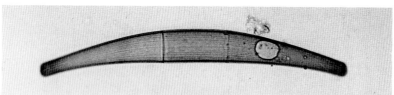

Fig. 516. *Closterium*, with hole in its wall made by a *Vampyrella* (× 320).

In boggy places, where cells of the desmid *Closterium* (p. 40) are common, empty specimens with a hole in the wall (fig. 516) may be found. Such algal cells will have fallen prey to a *Vampyrella*.

In fig. 517, a cell of *Closterium* has a hole in its yellow striated cell wall, near the top of the picture. On the lower part, in the region where the brown (parent semicell) and colourless (daughter semicell) parts of the wall meet, is a large reddish-brown body which is the digestion cyst of a *Vampyrella*. Just above the cyst is the piece of the wall which has come out of the *Closterium* (note its striations and yellow colour). The *Vampyrella* in its amoeboid stage has "cut out" this piece of wall and so gained entry into the *Closterium*. Then it has removed every scrap of protoplasm from the desmid. Since *Vampyrella* has no cutting organ, the piece of *Closterium* wall must have been excised by an enzymatic process. However, exactly how the *Vampyrella* does this is not known. In the case of *Geminella* it appears that a piece of wall is not "cut out" but that the hole is made by complete dissolution of the algal wall at the point of contact with the *Vampyrella*.

Fig. 517. *Vampyrella*. Digestion cyst (63 × 58 μm) of a species feeding on *Closterium*. The cyst lies to the right of and below a hole made in the wall of the desmid by the amoeboid stage. The striated piece of wall it "cut out" is immediately above the cyst.

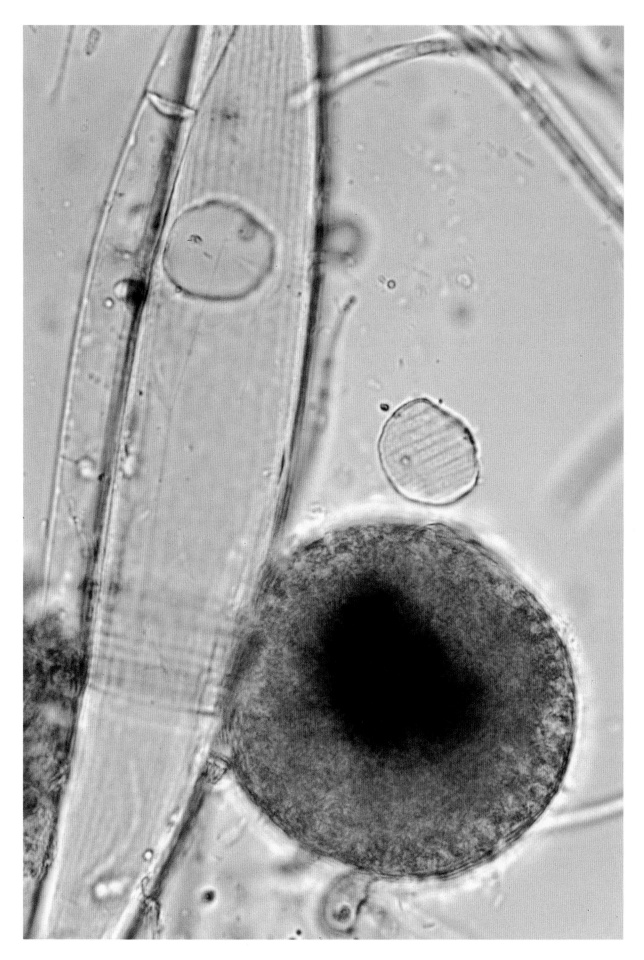

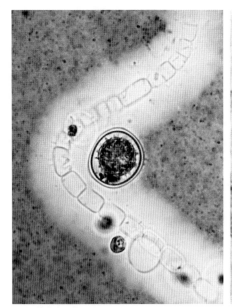 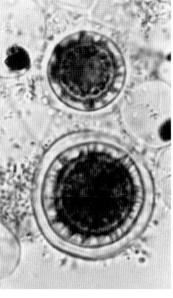

Figs 518, 519. *Vampyrella*.

Fig. 518. Cyst containing a resting spore attached to empty *Geminella* cells (× 820). **I.**

Fig. 519. Spiny-walled mature spores (× 1600).

Just as in *Asterocaelum*, the cysts of *Vampyrella* formed after feeding, may not liberate new amoebae but produce a resting spore instead. Such a cyst with its developing spore is attached to the completely grazed filament of *Geminella* seen in fig. 518. When mature, the spores of this species are covered with short spines (fig. 519).

Vampyrellids usually appear rather specific in their choice of algal food, attacking only species of a single genus or those of closely related genera.

Although many of the amoeboid-type of creatures which graze on algae have been known for a very long time (*Vampyrella* 129 years) they have remained a neglected area of study. Indeed, some species have never been recorded since they were originally described, others only once, and it is evident that some still await discovery.

The next two organisms shown provide examples of these two latter categories.

The genus *Enteromyxa* was created in 1885 for a long worm-like animal up to 0.5-1 mm long which devoured filaments of the Cyanophyte *Oscillatoria*.

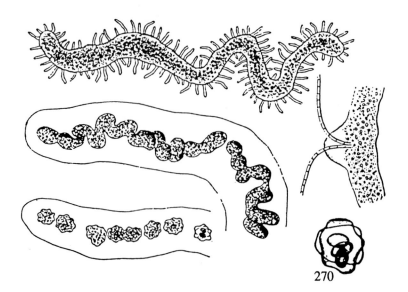

Figs 520-524. *Enteromyxa*. Original drawings by L. Cienkowski.

Fig. 520. The organism with its array of pseudopodia.

Fig. 521. Ingesting *Oscillatoria*.

Figs 522, 523. Segmentation leading to spore formation.

Fig. 524. A segment with three spores.

It had been found in a pond in Germany by one of the great naturalists of the time, L. Cienkowski. His drawings illustrating the life history of the animal are reproduced in figs 520-524. The body of *Enteromyxa* bears long, rather stiff, pseudopodia (fig. 520) which capture the filaments of *Oscillatoria*. At the point of entry of the alga into the animal a dome-shaped area of hyaline protoplasm is produced (fig. 521). The initial stage in the development of resting spores involves the withdrawal of pseudopodia and segmentation of the *Enteromyxa* body (fig. 522). The segments may be rather unequal in size or form a chain of many much smaller units. Within these segments one or more resting bodies will eventually be formed (figs 523, righthand segment; 524).

It was not until 78 years after Cienkowski's description that *Enteromyxa* was again recorded and this time from a pond in our garden nor does it seem to have been seen since. Photographs of our material confirm the original description. They are part of a large specimen (fig. 525), the ingestion of *Oscillatoria* via a hyaline dome (fig. 526), and segmentation of the body leading to the production of spores (figs 527-530).

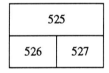

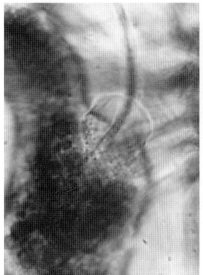
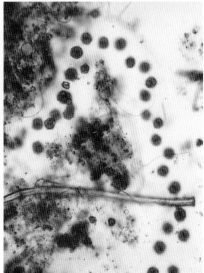

Figs 525-527. *Enteromyxa*.

Fig. 525. Part of the vermiform-shaped animal (× 600).

Fig. 526. A filament of *Oscillatoria* entering via a dome-shaped pseudopodium (× 768).

Fig. 527. Part of a cell with contents transformed into a central row of segments (× 120).

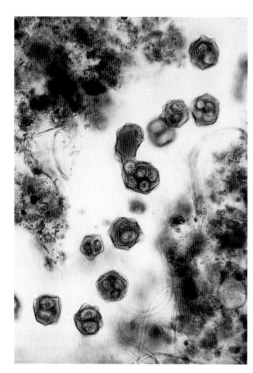
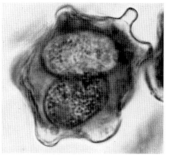
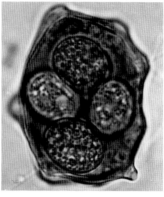

Figs 528–530. *Enteromyxa*.

Fig. 528 (× 300). Segmentation into angular bodies containing spores which are seen at higher magnifications (× 1600) in figs 529 and 530.

On germination, these spores each produce a small amoeboid individual which consumes *Oscillatoria* (figs 531, 532) and grows into the vermiform adult stage.

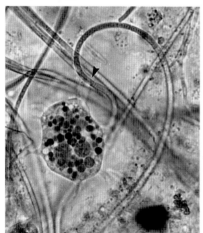
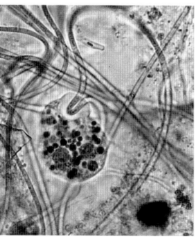

Figs 531, 532. *Enteromyxa*. A young amoeba which has captured a filament of *Oscillatoria* (fig. 531, arrowed) and is taking it in. Both × 615.

The organism in figs 533, 534 is so little known that it can neither be named nor put into a definite category. It appears to be a protozoan which forms an amoeboid network and is reminiscent of members of a group called slime moulds. There is much uncertainty to what extent slime moulds should be considered as protozoans or fungi. The present organism may well belong to a new genus but much more knowledge of its life history is needed. Protoplasmic networks of this kind which can migrate from place to place are called plasmodia. This plasmodial organism invades *Volvox* (p. 88). It consumes the individual algal cells and grows throughout the entire colony. Finally, when food has become more or less exhausted, the plasmodium spreads out extending far beyond the *Volvox* colony (fig. 534). In this manner further colonies are caught.

Figs 533, 534. An undescribed plasmodial protozoan in *Volvox*.

Fig. 533. Early stage in its growth within *Volvox*.

Fig. 534. Colony of *Volvox* completely destroyed and plasmodium now extending outwards from it. Width of the part of the plasmodium shown (560 µm). N.

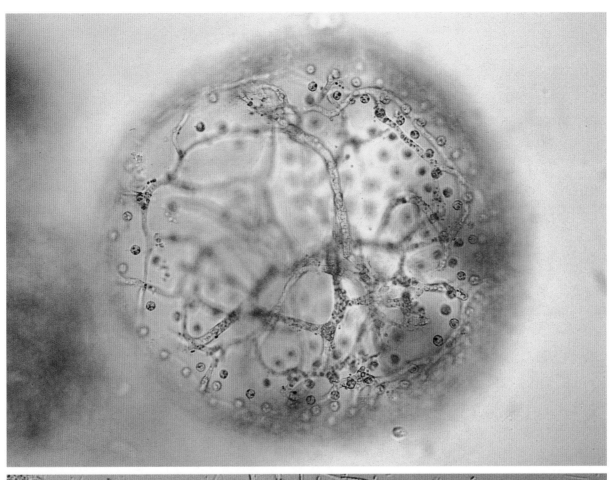
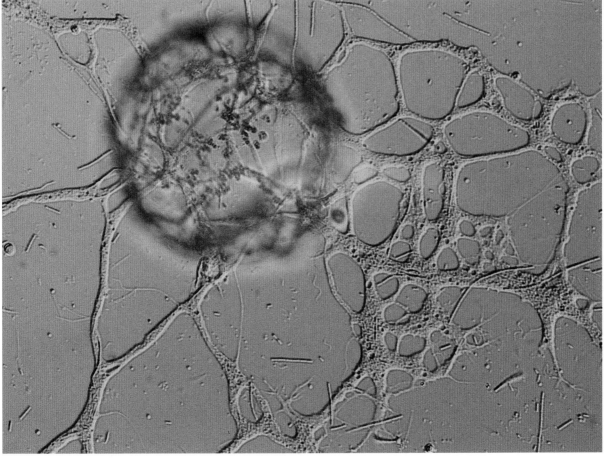

Some non-protozoan animals

Bosmina (fig. 535) and the Water Flea, *Daphnia* (fig. 536), are crustaceans which filter particles such as small algae out of the water. The filtration system involves complex procedures for collecting and processing the particles so that they can be ingested. The Water Flea can break up certain kinds of algae which cannot be filtered and be ingested whole. Examples are the planktonic diatoms *Tabellaria*, *Asterionella* and *Aulacoseira* (figs 239, 240, 255). Other colonial algae and a variety of filamentous forms cannot be broken up into pieces of suitable size. In fig. 535, the colonies of the Cyanophyte (*Gomphosphaeria*) cannot be ingested by the *Bosmina*, though the cells themselves could be. However, the cells lie within a covering of mucilage (figs 391, 393) and so cannot be dislodged by the movements of the crustacean's "limbs". On the other hand the cells of *Mallomonas* (the oval, yellowish-brown cells; compare with fig. 269) could be ingested. The colonies of *Tabellaria* (far left, at 7 o'clock and bottom of the picture) could not have been caught in the animal's collecting system, but a cell might be broken off and could then be ingested.

The brown colour of the gut contents of the *Bosmina* and the *Daphnia* are not necessarily a guide to the kinds of algae ingested. Diatoms, *Mallomonas* and other brown algae will give a brownish gut content but so can green algae because chlorophylls turn brown at an early stage of digestion. Sometimes, green algae do remain green in the gut, especially if they are so numerous that they are not quickly decomposed or if they are resistant to digestion. Some Chlorophytes, for example *Sphaerocystis* (p. 22) can pass through the gut without being killed. The extruded animal faeces (fig. 537) contains a live colony of this alga.

Aquatic worms ingest detritus. The part of such a worm shown in fig. 538 contains several algae. In the centre of the picture is a cell of *Euastrum*, above it is another desmid, *Tetmemorus* (for desmids see p. 38 et seq.). To the right of the *Tetmemorus* is a two celled colony of *Chroococcus* (p. 196). Some boat-shaped pennate diatoms (? *Navicula* spp) are also present. Midge larvae can graze on similar sorts of algae. Worms and midges often are present in large numbers in mud, the adult stage of the latter producing clouds of flies.

Rotifers

Rotifers (figs 539-546) are a group of small animals (nearly all less than 1 mm long) which are very common in freshwater and graze to a large extent on algae. They have a variety of methods of ingesting algae; collection through the currents produced by the beating of a corona of cilia; seizing and swallowing or piercing and sucking.

535	536
537	538

Fig. 535. *Bosmina* (700 × 450 μm), a crustacean feeding on microorganisms. The surrounding colonies of *Gomphosphaeria* are too large to ingest.

Fig. 536. *Daphnia* (1190 × 625 μm), the Water Flea, a "filter-feeding" crustacean.

Fig. 537. An animal faecal pellet containing a group (28 μm diam.) of *Sphaerocystis* cells.

Fig. 538. Part of an aquatic worm (*oligochaete*) which has ingested desmids (*Euastrum* is in the centre of the picture), diatoms, *Chroococcus*, etc. The worm is 300 μm broad.

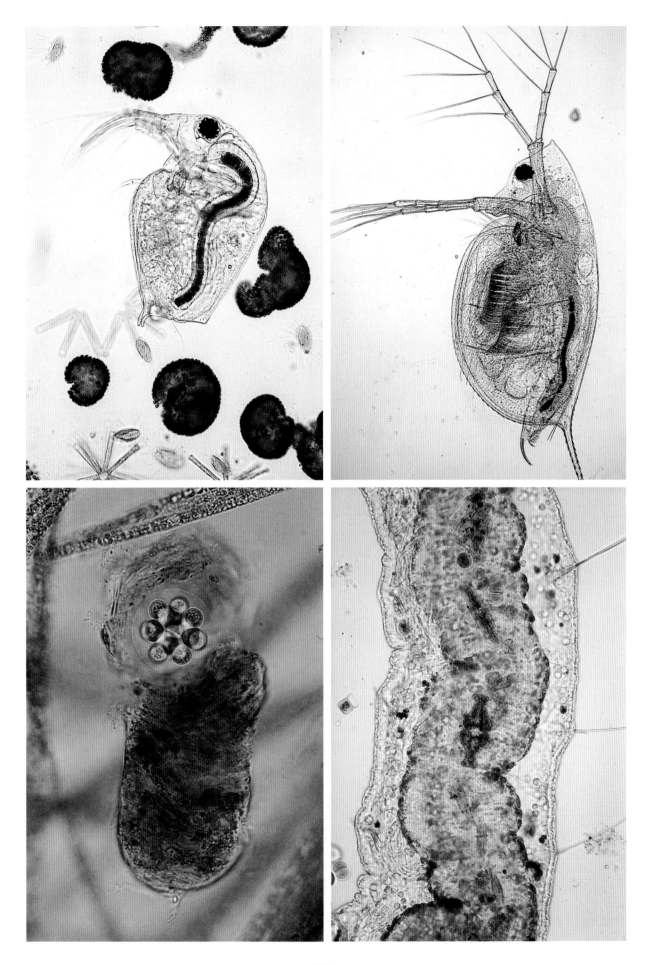

The rotifer in fig. 540 has been feeding among weeds or on the bottom of a waterbody and ingested a variety of unicellular algae, the majority of which are diatoms. Diatoms are both very abundant freshwater algae and a very important source of food for crustaceans, rotifers, protozoa and other invertebrates. This may seem surprising since they live in an inedible siliceous case. However, just as diatoms can only live in a siliceous "box" if it has holes in it, so digestive juices of animals can pass through these holes. Also the overlap of the two halves of the siliceous case offers a source of entry for digestive juices or these halves may be separated. The silica is not digested and so examination of an animal's gut or excrement will show which diatoms have been ingested. The cell contents of all diatoms seem to be easily digested. In contrast, some Chlorophytes are resistant to digestion and whereas all diatoms seem to be acceptable as food some Cyanophytes are avoided as food.

Asplanchna (fig. 541) is a planktonic rotifer which can ingest larger algae than can *Bosmina* or *Daphnia*. Here it has recently ingested four colonies of *Eudorina* (p. 88) and some cells of *Asterionella* are also present. By some unknown means *Asplanchna* can select its algal food. It is not uncommon to find that the relative abundances of certain algae within it are quite unlike their relative abundances in the plankton. The rotifer *Keratella* shown in fig. 169 can also be ingested by *Asplanchna*.

540	541
542	543

Fig. 540. A rotifer (170 × 110 µm) which has ingested pennate diatoms and other algae.

Fig. 541. *Asplanchna* (440 × 290 µm), a rotifer containing four colonies of *Eudorina* and fragments of *Asterionella* cells. I.

Fig. 542. *Ascomorpha* (137 × 94 µm) making contact with *Ceratium*. Note the rotifer's corona of cilia.

Fig. 543. *Cephalodella* (119 × 62.5 µm). This rotifer has entered a colony of *Uroglena* and the brown mass within it consists of the algal cells it has ingested.

The species of the rotifer *Ascomorpha* shown in fig. 542 has an unusual method of feeding on the planktonic Dinophyte *Ceratium* (p. 172). Having sought out a cell of *Ceratium* (fig. 542), the rotifer establishes a firm grip upon it. Then its jaws are applied to the algal wall (fig. 539) which is pierced and the cell contents are removed. These, now empty, skeletons of *Ceratium* frequently become colonized by a saprophytic fungus (p. 298).

Certain other rotifers are known which enter hollow ball-like colonial algae and ingest their cells from within. The species of *Cephalodella* shown in figs 543–546 enters colonies of the Chrysophyte *Uroglena* (p. 158).

539

Fig. 539. The rotifer *Ascomorpha* with its jaws applied to a cell of *Ceratium*, the contents of which it will later remove (× 640).

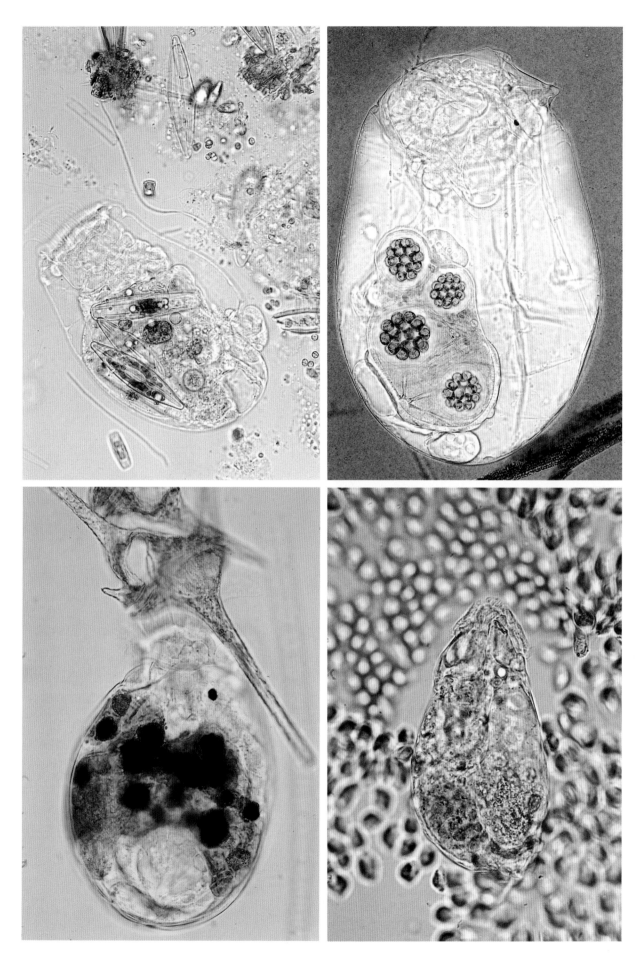

Esconced within the protection of the colony the *Cephalodella* can consume many cells with ease. Eventually whole areas of cells disappear creating "holes" (fig. 545) among the normally very regular pattern of cells lining the outer edge of a *Uroglena* colony (fig. 546).

Another rotifer does the same thing in colonies of the Chlorophyte, *Volvox*.

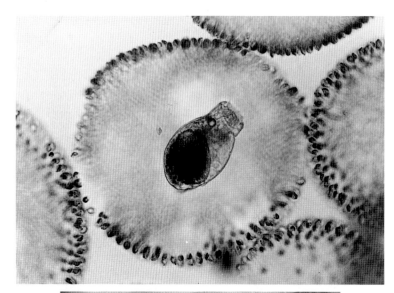

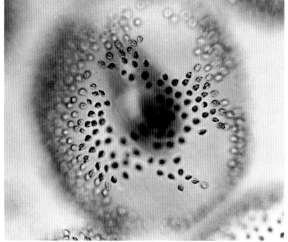

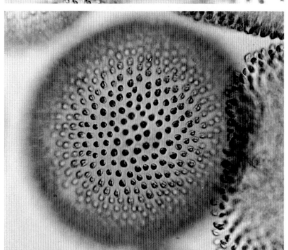

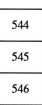

Figs 544-546. *Cephalodella*.

Fig. 544. Animal in a colony of *Uroglena*.

Fig. 545. Holes in a colony of *Uroglena* indicating where cells have been removed by the rotifer inside (out of focus).

Fig. 546. A complete colony of *Uroglena*, no cells removed by grazing. All × 256.

Chapter Thirteen

Fungi living on and in algae

Parasites and saprophytes

The fungi found on or in algae may be harming or more usually killing them, that is, they are parasites. If they are decomposing already dead cells, then they are saprophytes. The types of growth and reproduction of parasites and saprophytes often are alike. Indeed, a given genus may contain both parasitic and saprophytic species.

The difference between parasitism by fungi and grazing by animals may seem obvious if we equate grazing with eating. However, we have explained why we avoid the word eating since the various methods by which animals ingest algae are so diverse. In fact, just as some interactions in symbiosis come close to one organism parasitizing another, so some of the protozoa grazing on algae can be called parasites.

Vampyrella does not engulf its prey, it extracts its cell contents. Is this so different from a tick, louse or flea which extracts blood and are they not called parasites?

Fitting Nature into the clear-cut compartments beloved of Man often does not work because of the almost infinite diversity of the living world. For the purpose of compartmentalism in this book but not necessarily to ensure universal acceptance, we make the following distinction between grazing and parasitism.

Organisms grazing on algae (for short, grazers) digest the algae within themselves. Thus, *Asterocaelum* engulfs its prey before digestion and *Vampyrella* incorporates the protoplasm of its prey into its own cell before digesting it. Parasites, in contrast, do not digest their "prey" within their own cells but externally. They release enzymes which break down all or part of the host cell's contents into substances which they can then utilize. For example, starches may be broken down into sugars or proteins into amino acids. These sugars and amino acids can then pass into the parasite and form the building blocks of its protoplasm. This is just what a rust or mildew does when parasitizing a wheat plant or a rose. Indeed, toadstools extract their nutriment from rotting wood and non-parasitic moulds from jam in the same way.

It should be mentioned that there also are bacterial parasites, which likewise do not ingest algae, and viral ones.

The majority of the fungi infesting algae are contained within two very different groups, one informally referred to by us as chytrids, the other as biflagellate fungi. An almost universal method of asexual reproduction among these fungi is by motile spores (zoospores) produced in a spore bearing body, the sporangium.

In chytrids, the zoospore possesses a single, long or short, flagellum (figs 547, 548) which usually is attached to its posterior end. When swimming, the flagellum is directed backwards thus driving the body of the zoospore along in front of it. In the biflagellate fungi the zoospore has two flagella, commonly with one projected forwards and the other backwards during motion (figs 549, 550). These two flagella also differ in their fine structure, like those of certain algae (e.g. Chrysophytes).

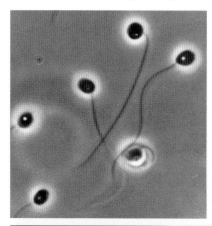 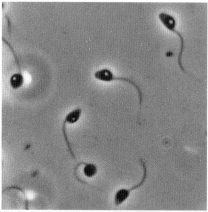
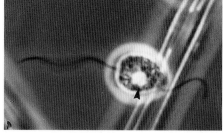 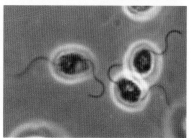

547	548
549	550

Figs 547, 548. Zoospores of chytrids with long and relatively short flagella respectively; one per cell. Fig. 547 ($\times 1600$); fig. 548 ($\times 2000$). **P.**

Figs 549, 550. Zoospores of biflagellate fungi. Both $\times 1600$. **P.**

Arrow indicates contractile vacuole.

As a result of electron microscopy and other modern techniques, the study and classification of species within these groups of fungi is presently undergoing much change. By continuing to use the general terms chytrid and biflagellate fungi we are following what is known as the earlier "classical system" of classification. Some examples of the fungi contained in these two groups are illustrated as well as the basic stages which occur in their life-histories. Although the examples shown only touch upon the remarkable diversity of form exhibited by these fungi, it is hoped that they will enable the reader to recognize these organisms, which frequently are present in collections of algae but which so often are overlooked. A few other fungi living on algae belong to groups other than those just mentioned. We have included two, *Blastocladiella* and *Anisolpidium*.

As has been done in the sections on symbiosis and grazing, when an alga not shown previously is encountered, a few words are written about it.

Chytrids

Stages in the life history of algal parasites and saprophytes

The life history of a chytrid begins with one of the motile zoospores finding and securely attaching itself to or near to an algal cell. A firm wall then develops around the settled zoospore and it is said to have encysted. Besides a nucleus and other organelles which are usually only detectable by electron-microscopy, most chytrids possess zoospores with a characteristic single, large or small, refractive lipid globule. This, together with the single posteriorly attached flagellum, make them readily recognisable at the moment of liberation from a sporangium (figs 571-573), free in the water (figs 547, 548, 551) or when encysted on an algal cell (figs 554, 601).

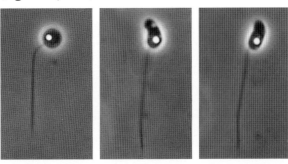

Most chytrid zoospores are less than 4 μm long or broad. They usually (figs 547, 551) but not always (fig. 548) are spherical to globose in shape when swimming. However, they are not rigid bodies and their shape may vary when they are moving over or through something, for example, mucilage (figs 552, 553). Minute contractile vacuoles can be present (fig. 552) but probably have often been overlooked.

When algae are surrounded by mucilage, zoospores may either traverse this mucilage until a cell is reached or encyst on its margin or within it. In the last two cases, the zoospore makes contact with an algal cell by the production of a thin thread.

The growth and development of chytrid sporangia takes place in a number of different ways. One of the commonest methods is for a fine germ tube produced by the encysted zoospore to penetrate the algal wall and so reach the cell content. There the fungus establishes a nutrient gathering system, like the roots of a plant. It is called the rhizoidal system and can be very varied in its form, as is seen later (figs 563-568, 570). Enzymes produced by these rhizoids break down the alga's content which is then transferred via the rhizoids back into the encysted spore. As a result,

551	552	553
554			

Fig. 551. A chytrid zoospore free in the water (× 1600). **P**.

Figs 552, 553. A zoospore becoming elongate as it traverses the mucilage surrounding an alga. In Fig. 552, near the anterior end of the zoospore is a minute white sphere, which is an open contractile vacuole. In Fig. 553 it has closed and so no longer is visible (× 1600). **P**.

Fig. 554. An encysted zoospore on a cell of *Closterium* (× 928). **P**.

Note single large lipid globule in each zoospore.

growth of the latter takes place and a sporangium develops. The protoplasm in the fully grown sporangium eventually divides up into further zoospores. Finally, these are liberated when the sporangium wall opens by one means or another, a stage called dehiscence. If a zoospore swims off and infects another algal cell the process can be repeated.

Sexual reproduction of various kinds can take place and the zygote so formed can become a resting spore. Similar types of resting spore can also arise asexually.

Sporangia

The colourless round bodies (Fig. 555) attached to the colony envelope surrounding the cells of *Eudorina* (p. 88) are sporangia of a chytrid *Rhizophydium*. The sporangia are connected to an algal cell by a fine thread, the germ tube of an encysted zoospore, which branches within the cell to form rhizoids. In this fungus, the rhizoids consist of a few minute branches reminiscent of those shown in fig. 563. Normally they are impossible to see or may only be found with difficulty after squashing or breaking a host cell. The minute round body (fig. 555 at 4 o'clock) is a zoospore which has germinated so recently that as yet it has barely increased in size.

In this host-parasite relationship, the fungus itself parasitizes a single algal cell. Other examples will be shown later where one fungus can parasitize and kill many algal cells. An interesting feature of the fungus on *Eudorina* is that apparently it does so little harm to the cell it parasitizes. This is in marked contrast to most of the parasites to be seen later, where a single infection will kill the host cell. In *Eudorina*, whereas one infection alone appears insufficient to kill a cell, multiple infections eventually will do so.

In the species of *Zygorhizidium* parasitizing the Chlorophyte *Chlamydocapsa* (fig. 556), the zoospore passes through the mucilage in which the algal cells are embedded and attaches itself directly to a host cell. The fungal sporangium is saddle-shaped and the remains of the chloroplasts of the three infected algal cells are brown in colour. *Chlamydocapsa* is like *Sphaerocystis* (p. 22), in that globose cells are arranged more or less randomly in mucilage and can divide into further such cells. However, one of the differences from *Sphaerocystis* is that the cells can reproduce by zoospores or themselves turn into flagellate cells. *Chlamydocapsa* is not uncommon in the plankton. It used to be called *Gloeocystis*.

In fig. 557, a young spherical sporangium of an unidentified fungus is visible on a cell of a lateral branch of the Rhodophyte, *Batrachospermum* (p. 184).

Another species of *Rhizophydium* (fig. 558) does not infect *Eudorina* or any other Chlorophyte but the

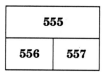

Fig. 555. *Rhizophydium*. Sporangia (to 13 μm diam.) on *Eudorina*. I, N.

Fig. 556. *Zygorhizidium*. Sporangia (to 26 × 17.5 μm) on *Chlamydocapsa*. I.

Fig. 557. Sporangium (14 μm diam.) of unknown chytrid on *Batrachospermum*.

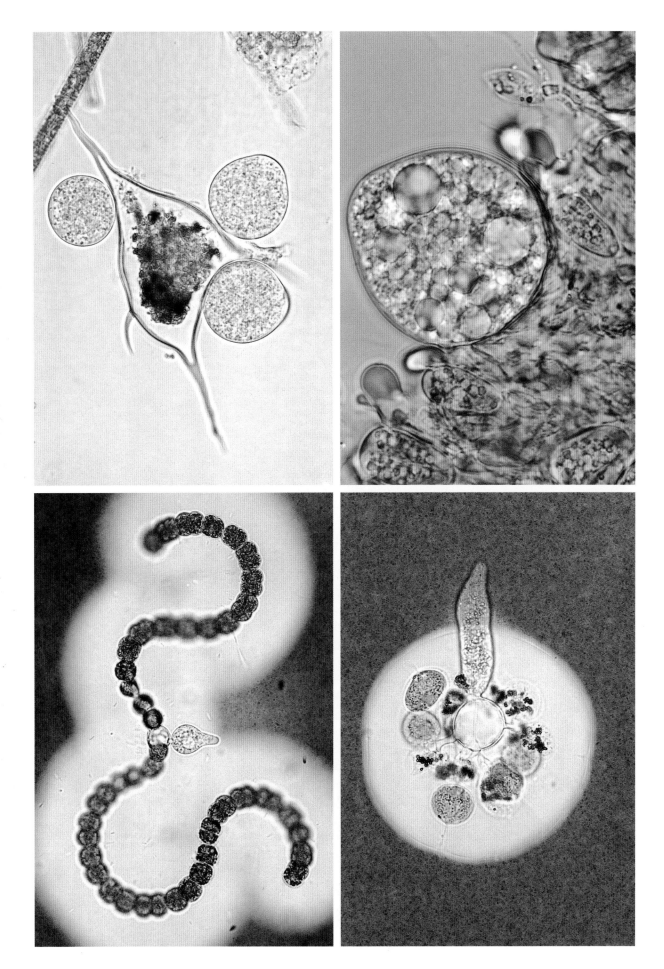

Rhizoids

Figures 562 and 563 show two cells of the Chlorophyte *Kirchneriella* (p. 30) which have been parasitized by a species of *Zygorhizidium*. In fig. 562, a sporangium sits on the algal cell and the material inside the alga is made up of dead algal residues and rhizoids of the fungus. In another specimen (fig. 563), the rhizoidal system has been made visible by squashing a host cell. It consists of a small bunch of short, branched finger-like rhizoids arising from a single thread which was connected with the sporangium.

Yet another species of *Rhizophydium* is seen in fig. 564. Four infections which are just beginning to enlarge into sporangia are situated on the apex of a filament of the Cyanophyte *Oscillatoria* (*Planktothrix*, p. 222). From them arise long straight rhizoids which pass through many cells of the *Oscillatoria*. It is a peculiarity of this parasite that the zoospores singly or often in groups settle on and infect the Cyanophyte via its end cells. It could be that these cells release some substance to which the fungal spores are attracted. If so, another fungal parasite of *Oscillatoria* is not so attracted because it encysts on the lateral surfaces of the filaments and infects cells via their longitudinal walls. The rhizoids which are similar to those of the apically placed parasite can then extend in both directions along the filament to cause its destruction. Quantitatively, this fungus reduces populations of *Oscillatoria* much more rapidly than the one which exhibits restricted apical infection.

The large reddish-brown body on the left hand side of fig. 565 is the sporangium of a species of *Rhizidium* parasitizing the Chlorophyte *Pseudosphaerocystis*. This colonial alga has numerous cells embedded in mucilage, which here appear to be joined to one another by threads, rather in the manner of *Dictyosphaerium* (p. 24). In fact, the threads are rhizoids of the parasite. By the time that the fungal sporangium is mature, virtually all the algal cells will have been killed.

There are many colonial Chlorophytes in the plankton in which round green cells are embedded in mucilage. Examples already seen are *Sphaerocystis* (p. 22) and *Chlamydocapsa* (p. 282). Such algae can be difficult to distinguish from one another. The name *Pseudo-sphaerocystis* refers to the fact that it could be mistaken for *Sphaerocystis* and it probably has been, though it has features which are very different from those of *Sphaerocystis*. For example, the cells may bear flagella or pseudocilia (see Glossary) at the same places on the cell but not at the same time. Cells with flagella can swim out of the colony. The cells usually are in pairs, facing one another, sometimes forming quartets. *Pseudosphaerocystis* is a common planktonic alga. It used to be called *Gemellicystis*.

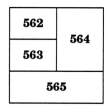

Figs 562, 563. *Zygorhizidium*. Fig. 562. Sporangium (11 × 7.5 μm) and fig. 563, rhizoids, of a species parasitic on *Kirchneriella*. P.

Fig. 564. *Rhizophydium*, a species parasitic on *Oscillatoria* (*Planktothrix*). Germinated zoospores on apex with rhizoids (white lines) penetrating 60 μm down the filament. P.

Fig. 565. *Rhizidium*. On the left, a sporangium whose branched rhizoidal system (to 106 μm l.) has infected many cells of *Pseudosphaerocystis*. Stained with iodine-potassium iodide solution. P.

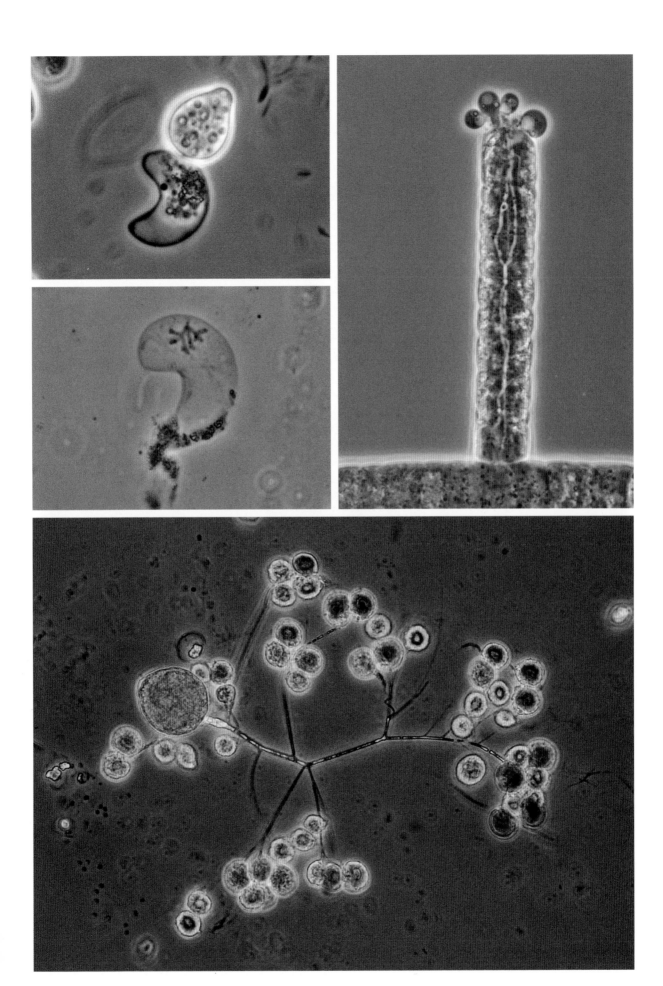

In fig. 566 a cell of the Chlorophyte *Characium* is attached by a stalk and a brown cushion-like base to a filament of the Xanthophyte *Tribonema* (p. 110). On the upper surface of the *Characium* is the developing sporangium of a species of *Chytridium*, the genus which gave this group of fungi the common name of chytrids. This parasite has quite a different type of rhizoidal system from those shown previously. Instead of a bunch of short threads or long threads it is just a round swelling. It can be seen as a colourless body filled with globules inside the *Characium*.

Characium is a unicellular Chlorophyte living attached to a substratum. It reproduces by zoospores. The attaching base or "foot" of *Characium* and a variety of other algae not uncommonly are coloured brown by oxides of iron and manganese. Why this is so is not known. There is a reasonable explanation of where the iron and manganese may have come from. What is not clear is why oxides of these metals should be accumulated in the basal disc of the *Characium* and not on any other part of its cell, or the *Tribonema* or on the *Chytridium*.

Just as some chytrid parasites infest more than one cell of an algal filament by growing through the filament (fig. 564) or reach spatially distant cells by means of branched rhizoids (fig. 565), others do so by running along the outside of a filament. An example is a species of *Scherffeliomyces* (fig. 567) parasitizing *Anabaena*. The large body on the upper side of the Cyanophyte looks like a sporangium but is in fact a young resting spore (see p. 294 for a section on resting spores). From each side of this spore a broad external rhizoid runs along the *Anabaena*. One *Anabaena* cell after another can become infected as minute lateral branches arise from the rhizoid, penetrate the cell wall and immediately enlarge to form a small globose rhizoidal sac. A few of these structures are just visible in fig. 567 (i.e. from apex of filament downwards, cell 5 and the cell directly beneath the spore (two sacs) as well as its adjoining cell.

Mature sporangia

That the zoospores of chytrids characteristically possess a single refractive globule has been referred to earlier (p. 281). During the early stages of growth of a sporangium the protoplasm frequently contains numerous refractive globules of varying size. However, by the time it is fully grown and its content has become divided up into zoospores (i.e. it is mature), a number of lipid globules of almost equal size are seen. Each globule indicates the presence of a zoospore. Such mature sporangia are shown in figs 568-570).

Fig. 568 is a species of *Rhizophydium* on *Asterionella* (p. 132). The individual zoospore globules are clearly demarcated. The contents of the diatom, apart from two

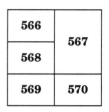

Fig. 566. *Chytridium* parasitic on a cell of *Characium* (22.5 × 7 µm) which is attached to a filament of *Tribonema*.

Fig. 567. *Scherffeliomyces*. Young resting spore (18 × 15 µm) and external rhizoid on *Anabaena*; round haustoria inside some cells (see text).

Fig. 568. *Rhizophydium*, a species parasitic on *Asterionella*. Note rhizoidal thread in the diatom cell. Stained by iodine-potassium iodide solution. Sporangium 7.5 µm diam.; rhizoid 25 µm l.

Fig. 569. *Podochytrium* parasitic on *Pinnularia* (80 × 25 µm). Each globule in the sporangium (27.5 µm l) indicates the presence of a zoospore.

Fig. 570. Unnamed chytrid parasite of *Chlamydomonas* (29 µm diam.). Note bunches of rhizoids attached to the sporangia which are up to 17.5 × 15 µm.

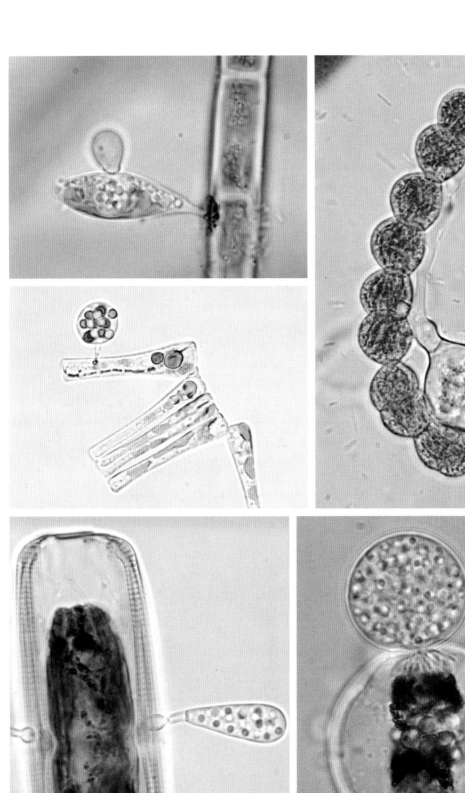
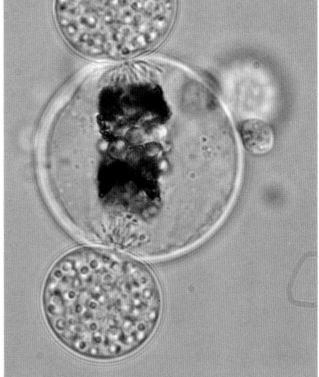

289

globular residues, have been almost wholly broken down into compounds which have passed from the long straight internal rhizoid, through the external germ thread and so into the sporangium.

Fig. 569 shows a sporangium of *Podochytrium* parasitizing the diatom *Pinnularia* (figs 207, 208). The sporangium here is club-shaped and has become divided by a cross-wall (septum) into a sterile basal and an upper fertile region. By focussing the microscope up and down it should be possible to count the number of lipid globules and so determine the number of zoospores which will be released from the sporangium of this and the preceding fungus. However, this is not so for those sporangia of an unidentified chytrid parasitizing a now unrecognisable species of *Chlamydomonas* (p. 80) shown in fig. 570. There are far too many zoospores for an accurate count to be made. Bunches of rhizoids inside the alga can be seen arising from the two sporangia on opposite sides of the Chlorophyte.

Liberation of zoospores

The stage when a mature sporangium opens and its zoospores are set free is called dehiscence. In many chytrids the zoospores are liberated rapidly and this stage may take no more than a minute or so to complete. In others it lasts a little longer if, as in figure 571, the zoospores emerge slowly in one's or two's, or the zoospores first flow out into a thin-walled vesicle produced from the sporangium (fig. 622, p. 314). Being thus constrained, the zoospores form a ball-like mass until the vesicle bursts and sets them free.

Sporangia open in many different ways. One or more pores large or small may develop by dissolution of the sporangium wall, very rarely the entire wall dissolves away. In some species, a distinct cap or lid of wall material is cast off.

In fig. 571 zoospores are just beginning to emerge from a sporangium. This parasite is living on a species of the desmid (not visible) *Staurastrum* (p. 44) which is surrounded by mucilage. Inside the sporangium the spores are globular but as they move out through the two holes which have developed at the apex of the sporangium they become elongate. This shape is maintained as a zoospore with its trailing flagellum works its way through the mucilage. When a zoospore reaches the surface of the mucilage and escapes into the water it starts to round off. A free-swimming wholly spherical spore with its undulating flagellum can be seen at the top right corner of the picture. Note its single large lipid globule. The moment of dehiscence of a different chytrid is seen in figs 572, 573. Here a wide apical opening has developed for the release of the zoospores.

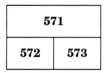

Fig. 571. Zoospores (c. 3 μm br.) issuing one after the other from a sporangium and passing through colourless mucilage. **P**.

Fig. 572. A mass of zoospores liberated from a sporangium (20 × 17.5 μm). Note the large lipid globule (white) in each zoospore. **P**.

Fig. 573. The same but at a different focal plane to show the flagella (up to c. 30 μm l.) **P**.

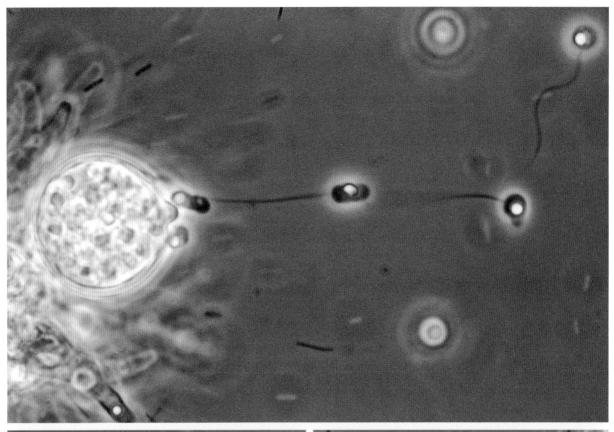
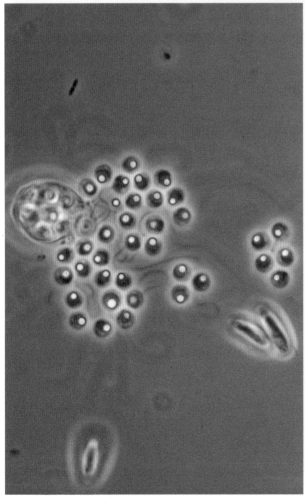
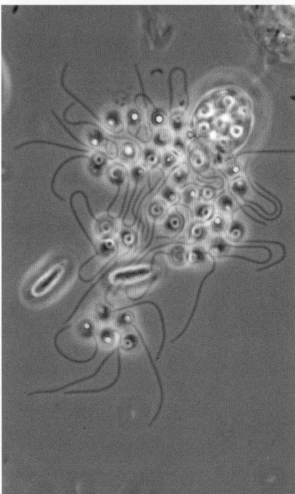

Empty sporangia

After dehiscence all that remains of a sporangium, if it has not wholly dissolved away, is its empty shell (figs 574-578). The usually colourless and thin walls of chytrid sporangia are not always easy to see but they can in some cases be made more obvious by staining with a dye called brilliant cresyl blue (figs 574, 575, 577).

Empty sporangia may give clues to the method of dehiscence, for instance, the sporangium of the unnamed chytrid (fig. 574) has opened by three large holes. The parasitized alga cannot be seen clearly. It is a species of *Chlorogonium*, a biflagellate unicellular Chlorophyte similar to *Chlamydomonas*. Its cells typically are much longer than wide. The lid of the chytrid, cast off at the moment of dehiscence, sometimes remains attached to the apex of a sporangium. This has happened in the *Chytridium* sporangia (figs 575, 576) growing on the diatom *Tabellaria* (p. 132) and the desmid *Staurodesmus* (p. 44) respectively. The parasite of *Staurodesmus* always infects and develops into a sporangium in the region of the isthmus between the two halves of the cell.

Like the previous two chytrids the flask-shaped sporangia of a *Chytridium* (fig. 577) parasitizing the Cyanophyte *Microcystis* (p. 202) have opened by the detachment of a lid although it is no longer present. It is clear that this fungus has almost entirely destroyed a large part of the *Microcystis* colony occupying the bottom half of the picture. Chytrids can be extremely specific in the host plants they parasitize. To the right of the *Microcystis* is the unharmed colony of the related Cyanophyte *Gomphosphaeria* (p. 206). Although *Microcystis* and *Gomphosphaeria* frequently co-exist together in the plankton, this chytrid which parasitizes *Microcystis* does not infect *Gomphosphaeria*. Moreover, as perusal of this chapter shows, a chytrid may only parasitize a certain stage in the life-history of an alga. Examples in the Cyanophytes are chytrids which specifically infect either the heterocysts or the spores.

In the empty sporangia so far depicted, their formation has been as a direct result of the growth of an encysted zoospore. That this type of development does not always take place was described earlier (p. 284) for the *Rhizosiphon* which infects *Anabaena* (fig. 560). This fungus, with its now dehisced and empty sporangium is seen in fig. 578. To the left of and below the empty sporangium, are the remains of the encysted zoospore and its long germ thread which had to pass through the algal mucilage in order to contact an *Anabaena* cell. Once the fungus has established itself inside the alga, the encysted zoospore and its germ tube just become a superfluous appendage. Also clearly visible in fig. 578 is the swollen portion of the parasite which develops in the first algal cell invaded by the fungus.

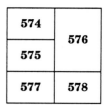

Fig. 574. Unnamed chytrid parasitic on *Chlorogonium*; empty sporangium (14.5 μm diam.) with pores through which the zoospores were liberated.

Fig. 575. *Chytridium*, parasitic on *Tabellaria*. Empty sporangium (9 μm br.), with lid by which it opened still attached to its apex.

Fig. 576. *Chytridium*, parasitic on *Staurodesmus*. Sporangium (19 μm l.) with attached lid. N, I.

Fig. 577. *Chytridium*. Empty sporangia (to 34 × 17.5 μm) of a species parasitic on *Microcystis*. Also present, unparasitized colony of *Gomphosphaeria*.

Fig. 578. *Rhizosiphon*, parasitic on *Anabaena* (see also fig. 560). Empty sporangium (25 μm l.), note swollen *Anabaena* cell beneath it, and knob-like empty encysted zoospore with its infection thread. N. Sporangia in figs 574, 575, 577 stained with brilliant cresyl blue.

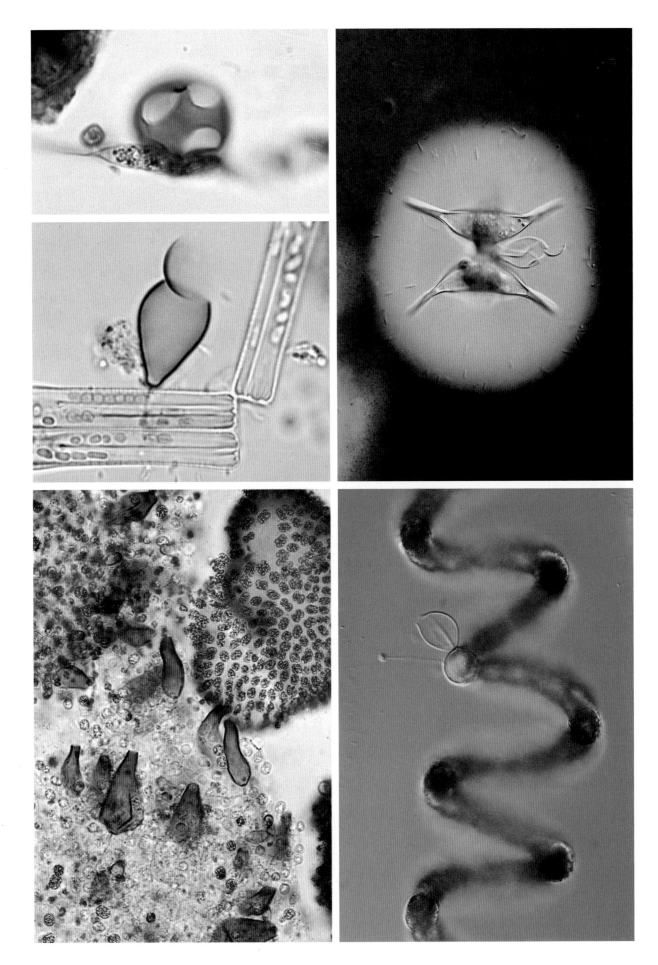

293

Asexual and sexually formed resting spores

A parasite of a cyst (resting spore) of the Dinophyte *Ceratium* was seen in fig. 555. This cyst was infected and collected while still suspended in the water of a lake. Eventually, cysts of *Ceratium* will sink to the bottom and be surrounded by or buried in the silty or muddy bottom deposits. Such a cyst is seen in fig. 579. This cyst is parasitized by a different fungus (as yet unnamed) to that seen in fig. 555. At the base of the *Ceratium* cyst in fig. 579 is a top-shaped sporangium of the fungus and further up an asexually formed resting spore of the same fungus. It is ornamented apically by extensions to its wall which resemble antlers. As mentioned above, the parasite in fig. 555 (*Rhizophydium*) can infect the *Ceratium* cyst before it reaches the bottom of the lake. Cysts of *Ceratium* are in suspension for a short time because they sink very fast. In the lake concerned, they can reach the bottom in a matter of days, yet in this short period a fungus which does not parasitize the *Ceratium* once it reaches the bottom appears as it were from nowhere. Obviously it does not appear from nowhere but where it does come from and how it does so during the period of a few weeks per year when cysts are produced remains a mystery. Another curious fact is that the second chytrid parasite of *Ceratium* cysts (fig. 579) has never been seen in the open water but only on the lake bottom. Therefore, it seems that it only infects cysts after they have sunk on or into the deposits.

In fig. 580, is seen the same species of *Rhizophydium* on the same species of *Eudorina* as that shown in fig. 555. The similarity of the sporangium of the parasite at four o'clock to those depicted in fig. 555 is obvious. However, higher up on the right hand side of the *Eudorina* colony is a sexually formed resting spore. A single cell of *Eudorina* has been infected by two fungal cells attached to the surface of the mucilage surrounding the algal colony. Both fungal cells have produced a germ thread to the *Eudorina* cell and their bodies are united to one another by what is termed a conjugation tube. The small body is the male cell which produced this tube and the large body the female cell. The male cell is empty because its content has passed through the narrow conjugation tube and into the female. After fertilization the zygote has formed a relatively thick wall and become a resting spore.

The chytrid *Pseudopileum* (fig. 581) infects the internally formed spores of the Chrysophyte *Mallomonas* (p. 156) but not the vegetative cell. The larger body on the outside of the *Mallomonas* spore is the female cell which has been fertilized by the smaller empty male cell beside it. The fertilized female cell becomes invested with a thick wall and functions as a resting spore.

Unlike the porose silica wall of a diatom with its overlapping system of girdle bands, there is only one entry

579	580	
581	582	583

Fig. 579. A chytrid parasitic on a *Ceratium* cyst from lake sediment. On the right (5 o'clock) a mature sporangium (44 × 22 µm); on the left (9 o'clock) another, partly hidden sporangium and above (right), a resting spore with antler-like ornamentation.

Fig. 580. *Rhizophydium*, parasitic on *Eudorina* (see also fig. 555). Lower right, a mature sporangium; upper right, large (10 × 6.3 µm) fertilized female cell (zygote) joined by a fine thread (conjugation tube) to a small, empty male cell.

Fig. 581. *Pseudopileum*, sexually formed resting spore on a cyst of *Mallomonas*; (right) the empty male cell (7.5 µm diam.).

Fig. 582. The two dark bodies (33 µm long) in the *Closterium* belong to the chytrid *Micromyces*.

Fig. 583. A spiny body (19 µm diam.) of *Micromyces*, parasitizing *Spirogyra*, is present in the righthand filament.

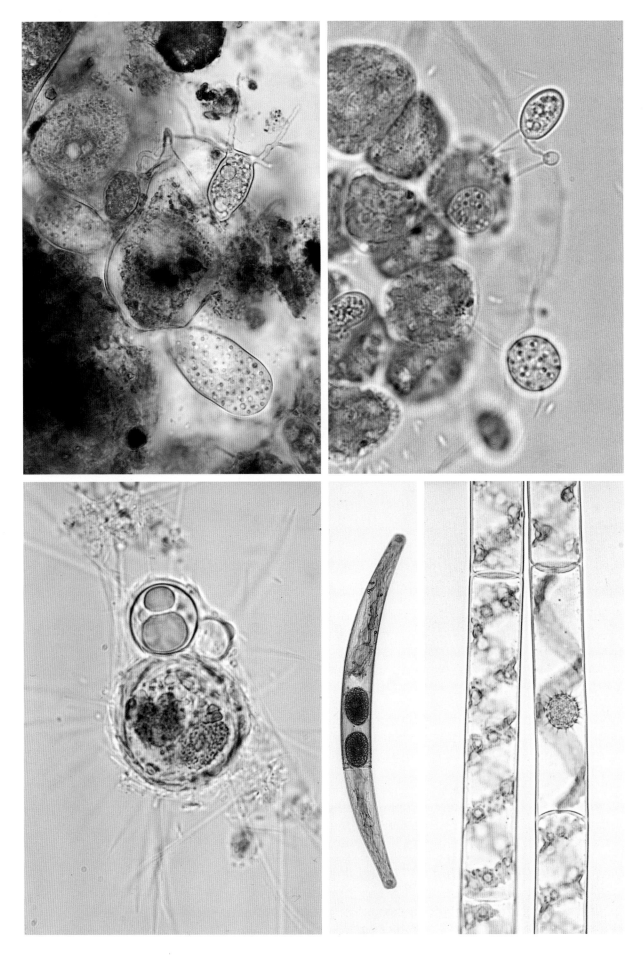

295

point for the germ tube of a parasite of this siliceous *Mallomonas* or, for that matter, any other chrysophyte spore. This entry is via the single hole plugged by non-siliceous matter. Obviously the most direct route of infection is for a zoospore to settle on the plug itself. This occurs in a *Chytridium* which parasitizes cysts of *Ochromonas* (p. 163) and where the dome-shaped plug is contained within a very distinct collar (fig. 584).

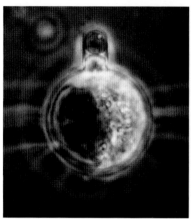
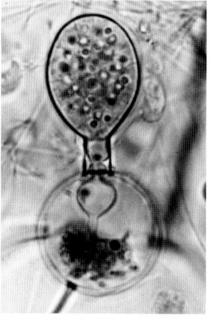
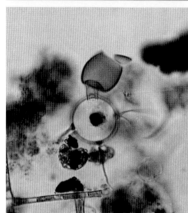
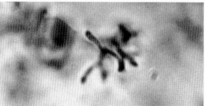

584	585
586	587

Figs 584-587. *Chytridium* parasitizing cysts of an *Ochromonas*.

Fig. 584. Zoospore of fungus lodged in collar of a cyst; bulge below the zoospore is the organic plug of the *Ochromonas* cyst (× 1600). **P.**

Fig. 585. Sporangium with zoospore globules. Inside the cyst there is a rhizoidal sac with a tube entering the remains of the *Ochromonas* chloroplast (× 1600).

Fig. 586. An empty sporangium with its cast-off lid (× 640).

Fig. 587. Part of a short-branched rhizoidal system which develops as a continuation from the tube-like portion visible in the cyst in fig. 585 (× 1600).

The zoospore grows into a sporangium which extends from the collar like a balloon (fig. 585), and later opens by means of a lid (fig. 586) to release its zoospores. Branched rhizoids are situated deep inside the cyst and connect with the sporangium via a sac-like swelling (fig. 587).

In the case of *Pseudopileum*, the zoospore settles on the envelope of scales surrounding the *Mallomonas* spore, some distance away from the cyst pore. A germ thread then grows through the envelope and over the cyst wall until the pore is reached.

Another ingenious strategy to assure the parasitism of a Chrysophyte spore is to start the process at an early stage of development, before the siliceous wall and the plug are formed. That this does happen in one known case illustrates the extraordinary selectivity of some parasites. Such remarkable selectivity is not confined to the siliceous spores of Chrysophytes. A fungus has been seen which likewise only parasitizes the auxospores of the diatom *Stephanodiscus* while they are still surrounded by a thin, unsilicified wall. It may seem unlikely that a fungus can make entry into a Chrysophyte cyst by dissolving part of the silica wall. However, Nature often produces surprises. The penetration of the walls of diatoms and so dissolution of their silica has been recorded in the past. Confirmation of this early work by means of electron microscopy is now needed. The fungi which live on Chrysophyte spores are not known to infect the flagellate vegetative cells. These have their own parasites as is illustrated by the sporangium of a *Rhizophydium* growing on a *Dinobryon* cell (fig. 282) surrounded by a case.

An unusual group of Chytrids commonly found in Conjugate algae

The stellate body in fig. 583 in a cell of the Chlorophyte *Spirogyra* (p. 56) was at one time thought to be part of the alga's life cycle and called an "asterosphere" but this is not so. It belongs to a fungus of the genus *Micromyces* which parasitizes many desmid and other conjugate algae living in boggy pools. Though it is a fungus which is often present in such habitats, when seen, it may still puzzle someone unfamiliar with these organisms. Although reproducing by posteriorly uniflagellate zoospores it has a very different life-cycle from the chytrid fungi already described. After a zoospore has encysted on an alga, its contents pass into a cell, taking the form of a naked, somewhat amoeboid mass (fig. 588). Such bodies become walled, enlarge and frequently appear as prominent "black" spherical or oval areas within the alga, as is seen in the *Closterium* cell (fig. 582). The wall may remain smooth or become spiny (fig. 583). In the next stage of the fungus's life cycle the cell contents of this body, or "asterosphere" (if spiny), are released and form a globose ball (fig. 589) either inside or

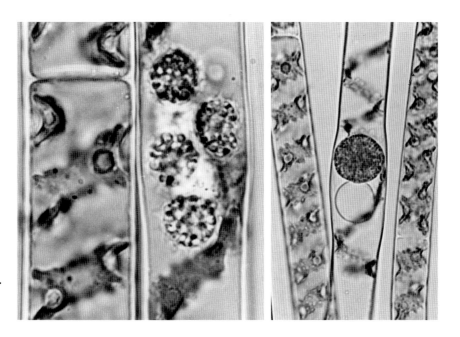

Figs 588, 589. *Micromyces* parasitizing *Spirogyra*.

Fig. 588. Four infections in the naked amoeboid stage. Cell of lefthand filament of *Spirogyra* uninfected (× 1600).

Fig. 589. Later stage (empty, walled body) has "budded out" a globose ball which will become segmented into sporangia (× 640).

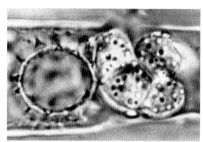
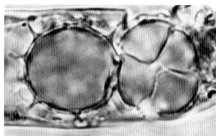

Figs 590, 591. *Micromyces* formation of sporangia (× 1600).

Fig. 590. On the right, a segmented body, each segment of which acts as a sporangium and produces zoospores.

Fig. 591. Dehisced empty sporangia; their outlines are now more clearly visible.

outside the algal host; in the latter case at the end of a tube. The content of this second body segments, each segment acts as a sporangium (figs 590, 591) and produces zoospores. Unlike the previous chytrids depicted, this fungus is devoid of a rhizoidal system. Cells of filamentous conjugates infected by *Micromyces* often exhibit swelling and considerable elongation.

Saprophytes

Saprophytic chytrids live on a variety of plant and animal remains, indeed pollen grains and cast skins of both invertebrate and vertebrate animals have long been used as "baits" to "catch" them. One common on the planktonic, free-swimming stage of *Ceratium* is *Amphicypellus* (figs 592, 615). Its round sporangia with a small knob beneath each are found on dead and frequently empty cells. A much branched system of fine rhizoids arises from the small sporangial knob and extends into the surrounding water. This fungus also occurs on *Peridinium*.

The cell wall of *Ceratium*, like that of *Peridinium* is composed of a set of plates. In cases involving the extraction of the cell content by the rotifer *Ascomorpha* (p. 276) and parasitism by the biflagellate fungus *Aphanomycopsis* (p. 312) the wall of a dead *Ceratium* cell is left virtually intact. When the cells die from other causes, the wall usually breaks up into bits so that these dead cells look quite different from the unbroken ones seen in figs 592 and 615.

It is impossible from fig. 593 to know whether the chytrid infecting the desmid *Cosmarium* (p. 40) is a parasite or a saprophyte. The *Cosmarium* cell clearly is dead but was its death caused by the fungus? Careful and prolonged observation has shown that no sporangial stages of the fungus ever occur on live *Cosmarium* cells or on any of the other algae it lives on. Most frequently it is found when an alga has succumbed to infestation by a chytrid or biflagellate parasite. An early stage in the development of this *Rhizophydium* is present on the *Anabaena* filament (fig. 620) which has been killed by a biflagellate fungus.

If, like certain chytrid parasites of *Asterionella* both host and fungus can be cultured, then experiments can be performed to determine whether a fungus is a parasite or saprophyte.

Fig. 592. *Amphicypellus*, a saprophyte of *Ceratium*: sporangia (to 12.5 × 11.5 μm) subtended by a small swelling; many fine rhizoids radiate out into the water. **P.**

Fig. 593. *Rhizophydium*. A species saprophytic on *Cosmarium*. Large sporangium 22.5 μm diam. **P.**

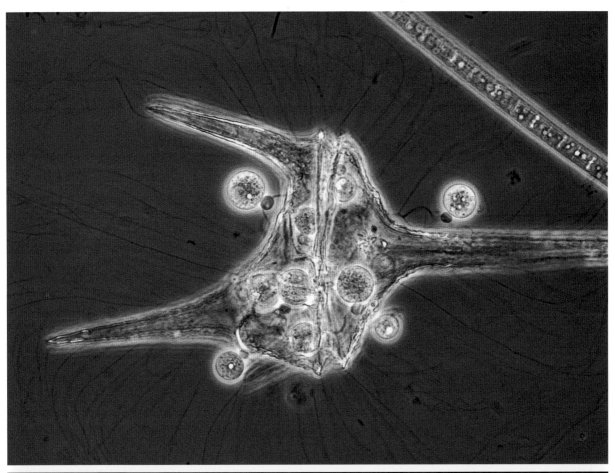

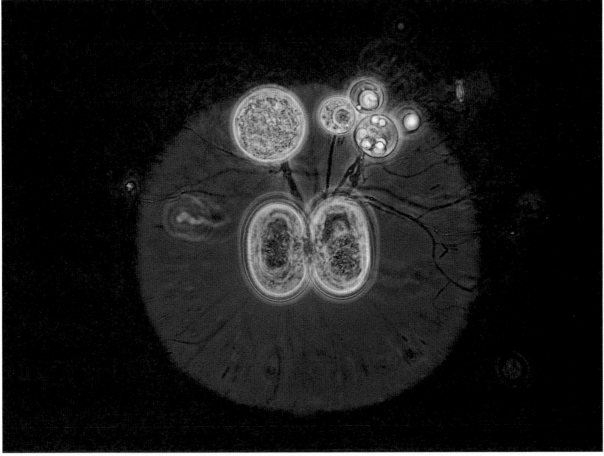

Some aspects of the ecology of fungal parasitism

Epidemics

Like the protozoan infestations of algae, those of fungal parasites can reach epidemic proportions, destroying much or almost the whole of an algal population. Figs 596-598 are of algae taken from populations severely parasitized by chytrids. The cells of the colony of the Chlorophyte *Sphaerocystis* bear numerous developing sporangia of a species of *Zygorhizidium* (fig. 596). The large body in the tangle of *Anabaena* filaments (fig. 597) is a sporangium of a species of *Blastocladiella*. It possesses an extensive, much branched rhizoidal system which has already killed most of the cells in the *Anabaena* colony. Those cells which still remain alive are coloured black since they contain gas vesicles. The *Asterionella* colony (fig. 598) is heavily parasitized by a species of *Zygorhizidium*. Many empty sporangia are present and very little content is left in the *Asterionella* cells. Looking at these pictures, as well as the *Pseudosphaerocystis* (fig. 565) and *Microcystis* (fig. 577), it is not surprising to learn that over 90% of an algal population can be killed in a single epidemic. Parasitism of the desmid *Staurastrum* by *Myzocytium*, a biflagellate fungus (see pp. 310-319), illustrates this very clearly. Fig. 594 shows a healthy algal population; fig. 595 the same population 10 days later.

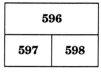

Fig. 596. *Zygorhizidium* severely parasitizing colonies of *Sphaerocystis* (*Eutetramorus*). Chytrid sporangia up to 12 μm long. P.

Fig. 597. *Blastocladiella*. A large tangle of *Anabaena* filaments in the process of being destroyed. Fungal sporangium (85 × 63 μm) with broad rhizoidal axes.

Fig. 598. *Zygorhizidium* on *Asterionella*, a late stage in a fungal epidemic when many empty (dehisced) sporangia are present. *Asterionella* cells 50 μm l. P.

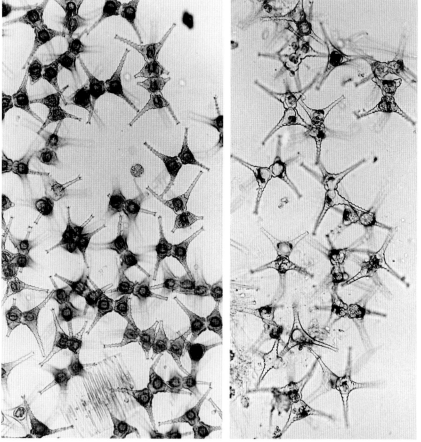

Figs 594, 595. A severe infestation (epidemic) of cells of *Staurastrum* by a biflagellate fungus.

Fig. 595 taken ten days after fig. 594. Both × 256.

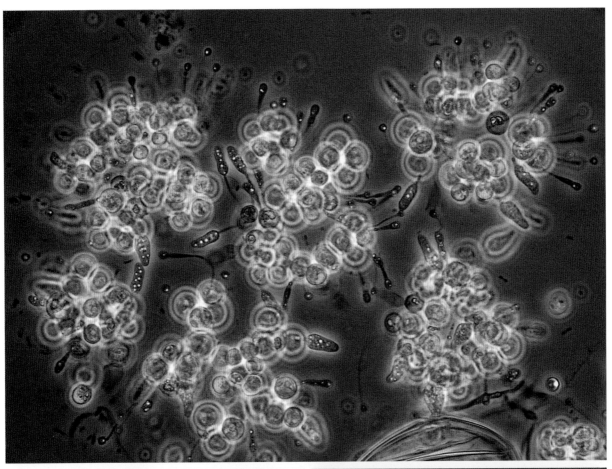
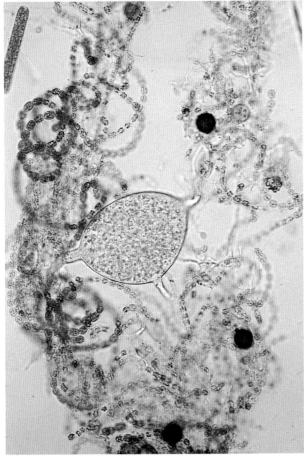
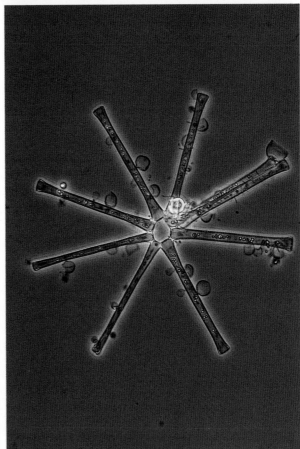

Small flagellated protozoans

A variety of organisms can live attached to algal cells, most of which are unlikely to be confused with fungal parasites. An exception however, are the small colourless bodies seen on *Asterionella* (figs 600, 602) and *Tabellaria* (figs 599, 603). At a cursory glance or to the uninitiated, these bodies could and often have been mistaken for developing chytrid sporangia. Moreover, they can be present in such large numbers that they give the impression of a fungal epidemic. These organisms are in fact very delicate flagellate protozoans, some of which are included in a group known as the choanoflagellates. They feed on bacteria and do not harm the algae to which they are attached. That they are not parasites is easily seen by examining the cells upon which they sit. If such large numbers of fungal parasites were present, all the *Asterionella* cells crowded with the flagellate *Bicosoeca* (fig. 602) would be dead. Indeed a single parasite will kill an *Asterionella* cell.

In figs 599 and 603 the cells of *Tabellaria* infested by the choanoflagellate *Codosiga* look healthy and no different from those without them. At this magnification it can also be seen that the cells of the choanoflagellate are quite different from those of fungal parasites. The apex of each cell is extended to form a short collar (fig. 603). This collar surrounds the basal portion of a long flagellum (fig. 599) which arises from the middle of the anterior end of the animal's body. The flagellum and collar can be seen even more clearly in another choanoflagellate *Salpingoeca* sitting on *Asterionella* (fig. 600).

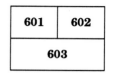

Fig. 601. Cells of *Asterionella* recently infected by chytrid zoospores (2-3 µm diam.). P.

Fig. 602. A mass development of the flagellate protozoan *Bicosoeca* (cells to 8 µm l.) on *Asterionella*, which can be mistaken for a fungal epidemic. N.

Fig. 603. *Tabellaria* with cells of the choanoflagellate *Codosiga* (up to 12 µm l.) attached to it. P.

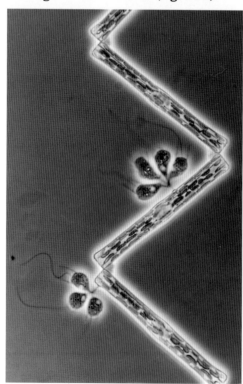
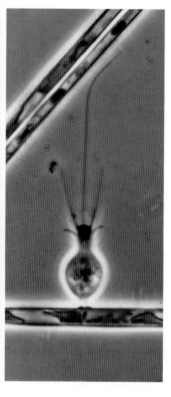

Fig. 599. Three and four-celled colonies of *Codosiga* (note flagella) on *Tabellaria* (× 640). P.

Fig. 600. A cell of *Salpingoeca* (note collar and long flagellum) attached to *Asterionella* by a short stalk (× 1600). P.

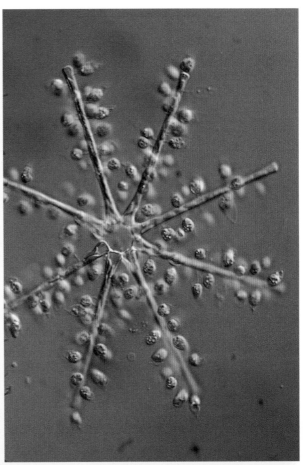
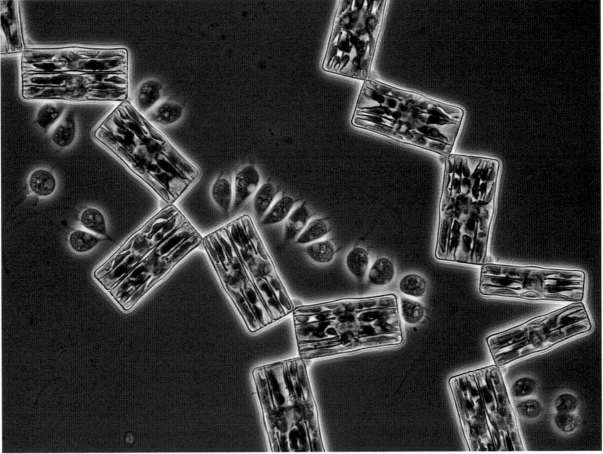

303

Observation and experimentation

Both the course and severity of fungal parasitism are determined by a complex web of interrelationships between each of them and the environment. Only a few facts and possibilities can be touched on here to illustrate this statement. Observations in nature, especially those over long periods and laboratory experiments have revealed what some of these interrelationships are.

Observation has the advantage of telling us what actually happens in nature. This in turn enables us to set up laboratory experiments which are related to the real world outside the laboratory flask. Observation alone has the disadvantage that it is difficult or impossible to disentangle the separate factors determining the observed course of events. Laboratory experiments permit such separation of the influences of various factors, because the environmental factors within the experimental chamber or flask can be controlled. There is, of course, the disadvantage that the world of a laboratory flask is not the same as the natural world. However, it can be said that experimentation is essential in order to unravel the mysteries of nature. These remarks apply to virtually any aspect of algal ecology. Since some of the facts to be mentioned are based on experimental work, something needs to be said about the production of suitable populations for this experimental work. Once again, the principles involved apply to nearly all experimental work, be it on algae, fungi or other organisms.

In order to carry out experiments on host-parasite interactions, the populations of alga and fungus used must be derived from clones. A clonal population is one derived from a single ancestor. In the case of the parasitism of *Asterionella formosa* by *Rhizophydium planktonicum*, which is the main basis of the following discussion, the clones are derived from a colony of the diatom (e.g. fig. 247) and a sporangium of the fungus (e.g. fig. 568). It is reasonable to assume that a population arising from a clone is genetically homogeneous. If so, all the cells will behave in the same way. If populations derived from clones are not used, then they may well be genetically heterogeneous and some cells react in a certain way to a given environmental factor and others in another way. Further, if other organisms are present in an experimental culture, how do we know what their effect has been on the results obtained? Certainly no other alga or fungus or other organism likely to affect the experimental results can be allowed to be present and if possible no other organism of any kind (e.g. bacteria) should be present. Clonal cultures free of all other organisms are said to be axenic or pure. Those used in the investigations considered here contained some bacteria. If it is reasonable to suppose that a clonal population originally is genetically homogeneous, it is also unlikely to remain so indefinitely. The genetical constitution can alter by mutation. However, from what is known about rates of mutation in microorganisms like algae and fungi, it is again reasonable to assume that the genetic constitution of host and parasite will not change significantly, if at all, during the period of a series of experiments which at the most may cover a very few years. If organisms are grown for long periods, then the danger of their becoming genetically heterogeneous will increase the longer these periods are and the more reproduction takes place.

Cases are known where visible changes have taken place in cultured algae, notably in bacterial forms such as the Cyanophytes. For example, Cyanophytes which can produce heterocysts have lost this capacity during cultivation, which in turn affects their ability to fix nitrogen (see p. 210). Further, some taxonomic distinctions are based on the presence or absence of heterocysts, so they can then be misidentified. Another example of changes arising during long cultivation is that of a species of *Micractinium* (fig. 18), a colonial Chlorophyte with spiny cells. The clone concerned lost both its colonial form and its spines. The now solitary spineless cells are indistinguishable from those of *Chlorella* (fig. 458). Where possible, organisms are protected from genetic change between periods of utilisation by keeping them in suspended animation. By using freeze-drying or cryopreservation, they are preserved at temperatures below zero, often far below it (e.g. in liquid nitrogen). As yet such preservation has not been attained with either the *Asterionella* or *Rhizophydium* mentioned here.

Some factors affecting parasitism

Clearly, if a fungal parasite is to produce an epidemic on an alga, the growth-rate of its population must be faster than that of the algal host. The growth of a diatom such as *Asterionella* is simple, a cell divides into two cells. The growth of a fungal parasite such as *Rhizophydium* is more complicated. Zoospores have to contact and infect an *Asterionella* cell. The fungus then grows until a mature sporangium is produced from which a further generation of zoospores is released. The larger the *Asterionella* population, the nearer the colonies are to one another and so the higher the probability that a fungal zoospore will reach and infect an *Asterionella* cell before it, the parasite, dies. The life-span of a zoospore and its ability to infect a diatom cell will depend on various environmental factors, for example the temperature of the water and the number of those animals present which ingest such small cells as fungal zoospores (e.g. filter-feeding crustacea and rotifers, ciliates (fig. 608), heliozoa (figs 604-607) and flagellates, some of which are algae (figs 609, 610).

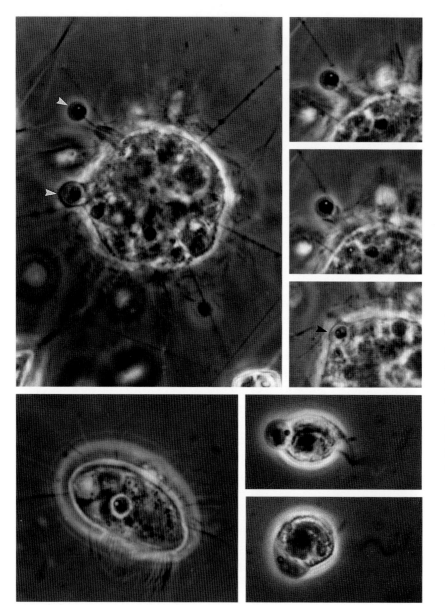

Figs 604-607 (× 1600). Stages in the ingestion of chytrid zoospores (arrowed) by a heliozoan (see text, p. 306). P.

Fig. 608. A chytrid zoospore in the ciliate *Cyclidium* (× 1600). P.

Figs 609, 610 (× 2000). A chytrid zoospore caught by and being incorporated into the base of a cell of the Chrysophyte *Ochromonas*. P.

Heliozoa are unicellular protozoans. Their cells almost always are spherical. They may be free-floating or attached to a substratum by a stalk. Fine threads radiate from the cell's surface giving them the name of sun animalcules. Microorganisms, including chytrid zoospores, can be caught by these threads and conveyed to the surface of the cell and ingested. In fig. 604, a heliozoan has caught two chytrid zoospores which are at different stages of ingestion. The flagellum of the most recently caught zoospore (at 11 o'clock) has been partially engulfed in a tubular pseudopodium produced by the heliozoan. In fig. 605 this same zoospore has been drawn nearer to the "mouth" of the pseudopodium which takes on a cup-shaped form and begins to spread around the zoospore's body (fig. 606). The zoospore becomes wholly enveloped by the pseudopodium (a stage already reached by the zoospore which the helizoan had caught previously; see fig. 604 at 9 o'clock) and finally is taken into the heliozoan cell (fig. 607).

Fig. 611. A race of *Asterionella* exhibiting hypersensitivity (× 805). **P.**

Cells of the colony (left) have been rapidly killed as a result of infection by zoospores of a chytrid parasite; note that the parasite was then unable to grow and the encysted zoospores died. The minute rod-shaped bodies on the dead diatom cells are bacteria. Healthy colony (right) taken from material prior to inoculation; note that the three uppermost cells have recently divided, also the general absence of bacteria.

The full details of how a zoospore of a fungal parasite does reach and infect an algal cell are not known. What is clear is that the main factor in a lake is where the movements of the water take the fungal zoospores. Compared to water movements, the swimming power of a zoospore is almost insignificant until it is carried near to a host cell. Then its own swimming power is likely to be decisive, provided it can detect the presence of this algal cell. This is the situation in the cases of fungal parasites of *Asterionella* and other lacustrine planktonic algae. The smaller or more weedy a waterbody, the less important are wind-induced water movements in carrying a zoospore to its algal host. For example, in weed-filled bays of lakes, small bog pools or among mats of algal filaments a zoospore is likely to be little affected by water movements and probably will reach its algal host solely by its own swimming ability.

The small amount of evidence available to date suggests that the zoospore probably is attracted to an algal cell by substances liberated by that cell. Having reached an algal cell, it has to attach itself, encyst and produce a germ tube which can penetrate the algal cell. Even when this is possible, it may be that the host cell is not suitable. For example, there are races of *Asterionella* which are what is called hypersensitive, that is to say they are extremely reactive to parasitism. Then, in the case of *Asterionella* (fig. 611), infection of a cell by a fungal zoospore leads to its rapid death. So rapid is the death of the diatom that the fungus cannot grow into a sporangium and so multiply. Because an alga is hypersensitive to one parasite, it does not follow that it is hypersensitive to any other parasite.

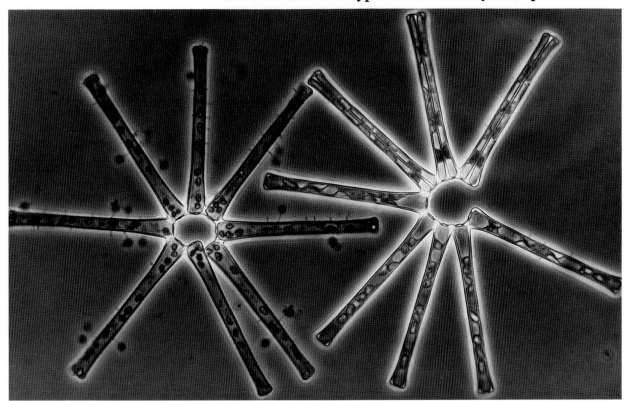

Hence, parasitism of the diatom population is severely curbed. The failure of the fungus to carry on growing into a sporangium after the death of the diatom but before its cell contents have been utilised also shows that the fungus is strictly parasitic. Otherwise, it would, like a saprophyte, continue to obtain nutriment and so grow on the dead diatom cell. *Asterionella* cells killed artificially by heat and air drying produced identical results, with no sporangial growth. In the case of *Rhizophydium planktonicum* light is also necessary for infection to take place. In very low light or darkness zoospores do not become attached to *Asterionella* cells.

Conditions determining the length of life of a zoospore are important if chance water movements and its own motions are to bring it to an algal cell before it dies. Experiments have shown that the lifetime of a zoospore of *Rhizophydium planktonicum* will be shorter in warm than in cold water. From this fact, it might seem that warm water would make fungal epidemics less likely. However, in warm water the development of sporangia on *Asterionella* will be more rapid than in cold water. Consequently, more spores can be produced in a given time. Hence, epidemics can arise at relatively high temperatures (e.g. about 20°C) as well as at lower temperatures.

The examples mentioned are only a few of the factors determining the course and severity of parasitism but do explain why the complexity of interrelations between host and parasite was emphasized in the first sentence of this section. In the case of *Asterionella formosa* and *Rhizophydium planktonicum*, these interrelationships are simpler than with some other fungal-algal parasitic associations because neither alga nor fungus has as yet been found to produce a resting spore. If either does possess a resting spore stage, then it is likely to be able to pass through a period unfavourable to growth because present evidence is that a fungus which infects ordinary vegetative cells of an alga does not infect its resting spore. Nevertheless, as we have seen, algal resting spores themselves are not immune to fungal infection and many are known to possess their own specific parasites which in turn do not infect the vegetative stage of the alga concerned (e.g. *Ceratium*, p. 284). Resting spores of a fungus will produce a potential inoculum to infect a new population of the algal host once it is large enough.

A common feature of host-parasite interrelationships is that no matter how severe the parasitism, the alga concerned reappears in abundance in a year or lesser time. For example, the periodicities of *Asterionella* and *Sphaerocystis* have been followed in certain British lakes for over 46 years. Irrespective of parasitism, they reappear year after year at much the same time. Yet parasitism can in extreme cases kill over 90% of a host population.

However, unfavourable chemical and physical conditions can cause even greater population losses, in the case of *Asterionella* these losses can exceed 99%.

The reason why populations of microscopic algae can recover so soon from such catastrophic decreases in abundance is that, like most microorganisms, they can multiply very fast. Under the best conditions which can be produced in the laboratory, cells of *Asterionella formosa* can double their number twice in one day. Of course, conditions in nature never are as favourable as these laboratory ones but, in nature, populations of *Asterionella* often double in numbers in a week and sometimes increase four-fold in that time. Let us imagine that 99% of a population of *Asterionella* is killed by parasitism, and then parasitism ends. *Asterionella* thereafter can double its numbers in each succeeding week. After one week the population is 2% of what it was, in two weeks 4% and so on. Hence, in about two months the population is as large as it was before a parasitic epidemic started. This potential ability to recover so quickly from parasitism may suggest that parasitic epidemics are not of much importance in algal ecology. In fact they play a highly significant but largely overlooked role in the rise and fall of algal populations. Very often during or shortly after an epidemic of parasitism, conditions for the growth of the parasitised alga have become or will become unfavourable. Large populations of algae favour the chances of fungal epidemics arising but large populations only arise during periods when environmental conditions favour the rapid growth of the algae concerned. If the period favourable for growth is more or less severely limited in time because of fungal parasitism, then a relatively small population of the alga concerned is likely to be produced before other unfavourable conditions restrict or stop its growth. These unfavourable conditions may be chemical, physical or biological (e.g. increased grazing by animals). However, the most unfavourable biological factor is competition, which is often linked to host specificity. At any one time there are usually a variety of algae present in a lake whose demands on the environment overlap to some or to a great extent. If the growth of any one alga is reduced by parasitism, then the others upon which the particular fungus does not grow can become dominant. Such alterations in algal diversity can take place both at the generic and the species level.

Complex though the interrelations between algae, fungal parasites and the environment may seem to be from the few matters touched on here, there is nothing intrinsically peculiar about them. All nature is governed by a complex web of interrelationships. This is why when we interfere with nature, we so often fail to foresee the full consequences of our actions.

Hyphochytrids

A striking parasite belonging to this rarely encountered group of fungi lives in cells of *Stigeoclonium* (fig. 612). *Stigeoclonium* is an attached Chlorophyte with the same kind of branched filaments and cells as *Chaetophora* (fig. 110). The plants differ from those of *Chaetophora* in that they are not embedded in mucilage. Though the plants are too small to be recognised without examination under a microscope, they often grow in groups that can be seen easily by eye (fig. 5) especially when moved about by wave action or underwater currents. The mucilaginous *Chaetophora* is not so moved. *Stigeoclonium* is common in both flowing and standing water. The part of the *Stigeoclonium* plant shown contains two large, mainly colourless cells and a third much smaller cell of the same type. These three cells contain developing sporangia of the hyphochytrid *Anisolpidium* which has caused the host cells to swell, that is to become hypertrophied. Hypertrophy can arise from other causes but if combined with the destruction of much of the cell's content (e.g. the chloroplasts), it is reasonable to suspect that it is caused by parasitism.

In many instances, the general habit of hyphochytrids living on or in algae resembles that of chytridiaceous fungi. Their zoospores are also uniflagellate. However, unlike chytrids, the flagellum is anteriorly attached and during swimming the zoospore body is carried along behind it. By special staining and viewing with a light microscope and by electron microscopy it is also seen that the flagellum is structurally different from that found in chytrids.

The sporangia of *Anisolpidium* are devoid of any special rhizoidal system and when fully developed they each produce a tube (exit tube) which penetrates the algal wall and extends outside it (like that shown for the biflagellate fungus in fig. 614). Through this tube the content of the sporangium flows out as a naked mass of protoplasm and subsequently becomes divided up into zoospores.

Biflagellate fungi

The sporangia of these fungi (figs 613-620), like *Anisolpidium*, grow inside algal cells, do not possess a separate nutrient gathering system and develop an exit tube. The fully grown fungal body may occupy a single cell or extend through many cells. It may form a single sporangium or become divided up by cross-walls into a few or numerous sporangia. Sporangial shape can be globular, tubular sac- or thread-like.

Fig. 613 shows a simple, basically sac-shaped sporangium contained within a dividing cell of the desmid *Cosmarium*. In fig. 614 a sporangium of *Myzocytium* has developed a long exit tube through the mucilage surrounding the desmid *Staurastrum*.

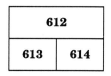

Fig. 612. *Stigeoclonium*. The three swollen, whiteish cells are infected by the hyphochytrid *Anisolpidium*. Healthy cells 6.5-14 μm br.; largest swollen (hypertrophied) cell, 34 × 24 μm.

Fig. 613. *Cosmarium* containing the sporangium of a biflagellate fungus (length 40 μm). I.

Fig. 614. *Staurastrum* parasitised by *Myzocytium*. The sporangium has produced an exit tube 33 μm long. I, N.

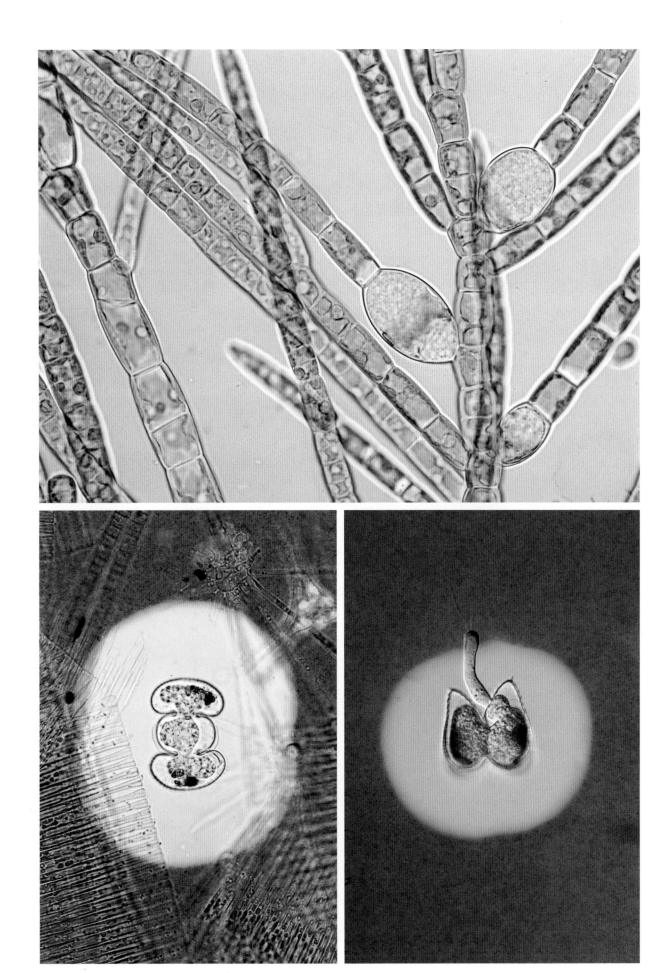

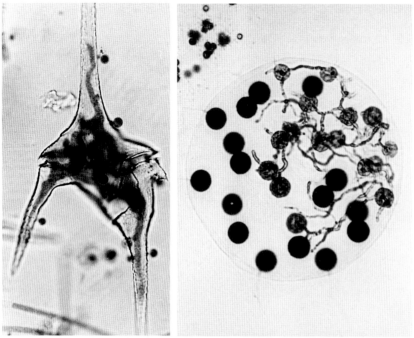

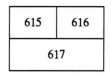

Fig. 615. A tubular sporangium of *Aphanomycopsis* inside a cell of *Ceratium*. The small, external round bodies belong to the chytrid saprophyte *Amphicypellus*. Stained with cotton blue in lactophenol (× 495).

Fig. 616. Thread-like filaments of a biflagellate fungus in a colony of *Eudorina*; host cells already killed appear grey. Stained with iodine in potassium iodide solution (× 370).

Fig. 617. An empty dehisced sporangium (central white tube) of a biflagellate fungus in a filament of *Anabaena* after squashing; far top left its exit tube (arrowed) is just visible (× 640). **P.**

The *Ceratium* cell in fig. 615 contains a fully grown, branched, tubular sporangium of *Aphanomycopsis*. The tiny round bodies attached to the alga belong to the chytrid saprophyte *Amphicypellus* (fig. 592). In fig. 616, threads of an unnamed parasite are growing among and penetrating cells of a colony of *Endorina*. When fully grown the threads will become divided by cross-walls into several sporangia. The righthand cell of *Closterium* in fig. 618 contains a large number of globose sporangia of the fungus *Myzocytium*. They have been formed by segmentation of its original cylindrical, branched fungal body.

In fig. 620 an unnamed biflagellate fungus has grown through a series of cells of *Anabaena* and an exit tube is passing through the mucilage envelope which surrounds this alga, although the mucilage cannot be seen in the picture (visible in fig. 560). This fungus is a virulent parasite and can rapidly kill entire filaments (fig. 617). Here the fungus has discharged its zoospores and is now empty.

As fig. 620 is the last picture of a Cyanophyte parasitized by a fungus, it is worth mentioning that these are the only prokaryotes parasitized by such fungi. This is another feature which they share with eukaryotic algae. On the upper side of the *Anabaena* in fig. 620, near its left hand end is an encysted zoospore with branched germ threads. It is a saprophytic chytrid which is infesting the *Anabaena* now that it has been killed by the biflagellate fungus. It is the same chytrid as was shown earlier on a dead cell of *Cosmarium* (fig. 593).

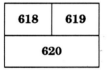

Figs 618, 619. *Closterium* infected by *Myzocytium*.

Fig. 618. Righthand cell filled with globose sporangia; lefthand cell (645 μm l.) uninfected.

Fig. 619. Formation of bean-shaped zoospores at apex of two exit tubes. The globular mass, uppermost, is still contained in a vesicle; the zoospores below are beginning to swim away.

Fig. 620. An unnamed biflagellate fungal parasite of *Anabaena*. An exit tube (5 μm br.) is developing from the, as yet, immature thread-like sporangium within the *Anabaena*. Far left: an encysted zoospore of a chytrid saprophyte (fig. 593) with rhizoids just beginning to develop. **P.**

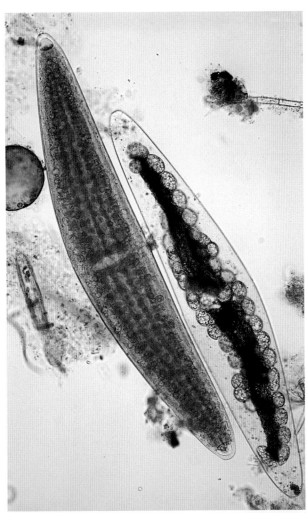
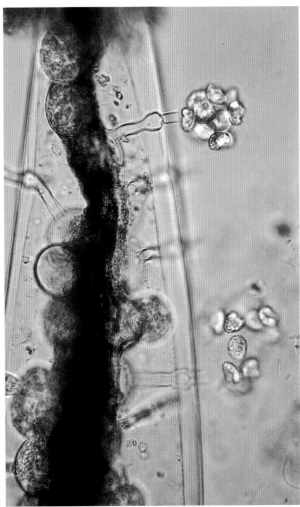
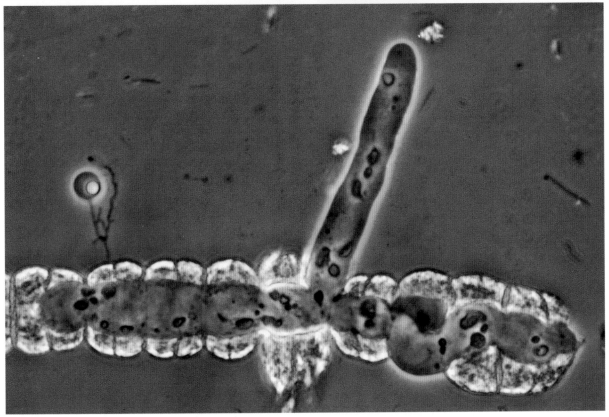

Most of the biflagellate fungi found in algae develop only one type of zoospore. However, there also exist among these biflagellate fungi those which possess a more complicated life cycle involving the production at different times of two kinds of zoospore.

In the first and simplest life cycle the zoospores are often kidney- or bean-shaped and larger (4-5 µm broad × 6-7 µm long) than the majority of chytrid zoospores. There are two laterally attached flagella which commonly are unequal in length, the shorter one being directed forwards when swimming. The body contains several small refractive globules and a conspicuous contractile vacuole (e.g. as in fig. 549) may be present near the point of attachment of the flagella. These zoospores can become fully formed within the sporangium and swim directly away via the exit tube. Alternatively, they may undergo some development within the sporangium but then are liberated into a delicate vesicle which extends from the apex of the exit tube. Here they mature and are not set free until the vesicle bursts or deliquesces. Yet other zoospores undergo their complete development within such a vesicle.

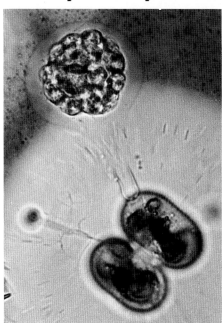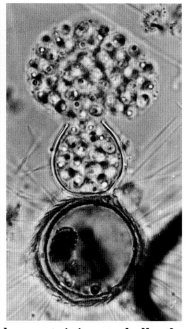

| 621 | 622 |

Fig. 621. A large vesicle within which biflagellate zoospores are undergoing development (at 11 o'clock) is continuous with the exit tube of a *Lagenidium* which projects from a cell of *Cosmarium* (× 820). I.

Fig. 622. A dehiscing sporangium of the chytrid *Pseudopileum*. The mature zoospores are flowing out into a vesicle (× 1008).

The clear outline of a vesicle, containing a ball of developing zoospores of a *Lagenidium* which has parasitized a cell of *Cosmarium*, can be seen in fig. 621. As the zoospores mature their two flagella will gradually grow out, at first causing the zoospores to oscillate slowly to and fro. Later as the flagella become longer and more active, so the zoospores may even be found trying to swim around within the confines of the vesicle. These zoospores are known as either secondary or principal type zoospores. They will seek out and infect new host cells.

In fig. 619, sporangia of the fungus *Myzocytium* (cf. fig. 618) have produced exit tubes and at the apex of two of them bunches of bean-shaped zoospores are present. In the upper bunch the zoospores are still contained in a thin vesicle; in the lower bunch zoospores are now beginning to swim away.

Fig. 629 shows a cell of the desmid *Xanthidium* which has been parasitized by a *Lagenidium* but not the same fungus as attacked *Cosmarium* (fig. 621). Abundant zoospores, awaiting release, are present within the large vesicle protruding from the host cell.

Earlier it was noted that in some chytrids the zoospores, at the time of sporangial dehiscence, can be temporarily contained within a vesicle before they swim off. This state has been reached by the *Pseudopileum* sporangium parasitic on a cyst of *Mallomonas* (fig. 622).

Although not depicted, there also exist other fungi which live entirely within algal cells, have tubular or sac-like sporangia devoid of rhizoids, and often develop a distinctive exit tube. However, the sporangia, which could be mistaken for a biflagellate or even a hyphochytrid like *Anisolpidium*, liberate posteriorly uniflagellate zoospores. Until recently these fungi were considered as chytrids.

Due to the occurrence of such parallel types of sporangial morphology it is not always easy to decide to which group a fungus belongs. When, as commonly happens in natural collections, just a few empty sporangia are found, identification may be virtually impossible. However, the addition of a solution of chlor-zinc-iodide could be helpful as it causes the walls of many biflagellate fungi to take on a blue or pinkish-mauve colouration.

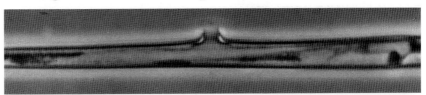

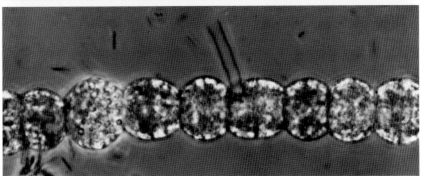

Fig. 623. Very short exit tube of a biflagellate parasite (now dehisced) in *Synedra* (× 2000). **P.**

Fig. 624. A biflagellate fungus parasitizing *Anabaena*, the empty exit tube of a sporangium projects from a central cell. The fungal body within the *Anabaena* is hidden by gas vacuoles (× 1600). **P.**

When empty, the sporangia of some biflagellate fungi are extremely difficult to see. This may be due to the delicate nature of the sporangium wall, as is shown (fig. 623) by the unnamed parasite of the diatom *Synedra*. Empty sporangia can also be hidden from view when much dead algal material is left in a cell. The *Anabaena* (fig. 624) has been infected by the same fungus as was shown in (fig. 620). The now empty exit tube indicates that within this filament there lies the remains of a long tubular sporangium. Here the fungus is masked by the gas vesicles still present in the cells of the Cyanophyte. If squashed, the outline of the fungus would become evident (cf. fig. 617).

The exit tubes of sporangia vary from extremely short (fig. 623) to long, as seen in the unnamed biflagellate which occurs in *Ceratium* spores from the bottom deposits (fig. 630). In contrast to the "bare" orifices of the empty exit tubes already shown, others may be found where a few (fig. 625) or numerous (fig. 631) round bodies are present at the apex of the tube. Such a development is indicative of the second and more complicated life cycle which some of these fungi undergo. In this case, the zoospores released from the sporangia commonly are pyriform or ovoid in shape, bear two apically attached flagella and are rather sluggish. These so-called primary or first-formed zoospores pass out of the exit tube, round off and encyst. On germination, each cyst then liberates a new zoospore which now conforms to that of the earlier described laterally biflagellate principal or secondary type. Once more, this zoospore acts as the agent of dispersal and infection.

The ungerminated cysts (fig. 625) have been produced by a second parasite which attacks spores of *Ceratium* in lake sediments. The large bunch of empty germinated cysts in fig. 631 belongs to an *Aphanomycopsis* present in a cell of the desmid *Netrium* (see fig. 80).

Fig. 629. *Lagenidium* parasitizing *Xanthidium*. Globular mass of zoospores (62 × 53 µm) enclosed in a vesicle, awaiting release. N.

Fig. 630. Cysts of *Ceratium* containing sporangia of a biflagellate fungus. Several long narrow exit tubes (4 µm br.) are present. From lake sediment. Stained with Evans blue.

Fig. 631. An exit tube of *Aphanomycopsis* projecting from a cell of *Netrium*. At its top is a group of empty zoospore cysts (cysts 6-8 µm diam.) each of which has already germinated and produced a further zoospore but of a different kind.

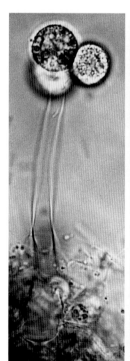
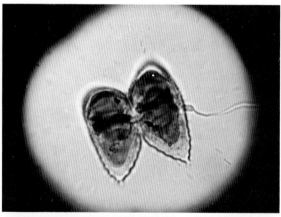
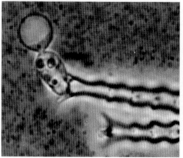
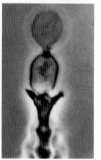

Fig. 625. The round bodies at the apex of the exit tube are the encysted (first-formed) zoospores of a biflagellate parasite which infects spores of *Ceratium* (× 1008).

Fig. 626. Live cell of *Staurastrum*. A zoospore of *Lagenidium* (hidden by Indian ink) has encysted on the edge of the mucilage envelope and produced a long germ thread. At its point of contact with the alga, a swelling (appressorium) has arisen (× 640). I.

Figs 627, 628 (× 1600). Encysted zoospore and appressorium applied to the arm of a *Staurastrum*. In fig. 627, the zoospore content has passed into the appressorium. I. In fig. 628, it has entered the desmid and both structures are empty. P.

As in chytrids, the zoospore may encyst directly on an algal cell or if there is a mucilage envelope, either traverse it until a cell is reached, settle on its margin or some distance inside and establish contact by means of a germ thread (fig. 626). The latter often becomes swollen at its

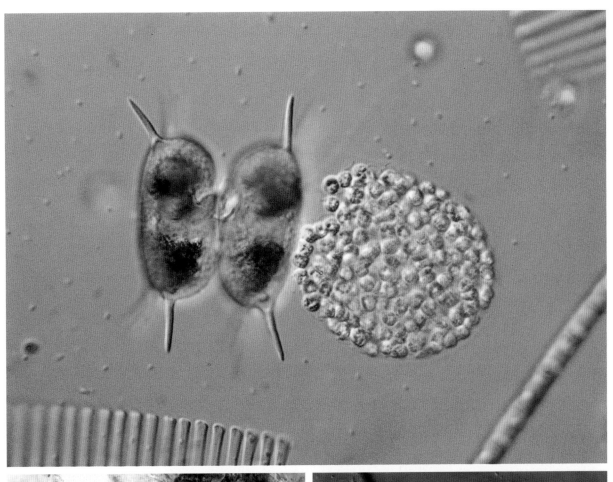
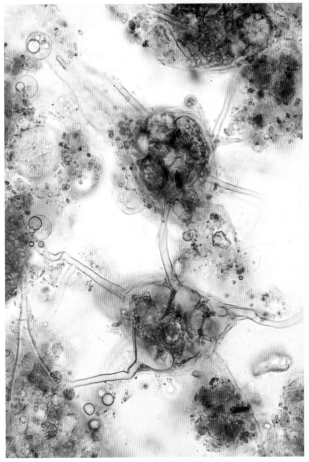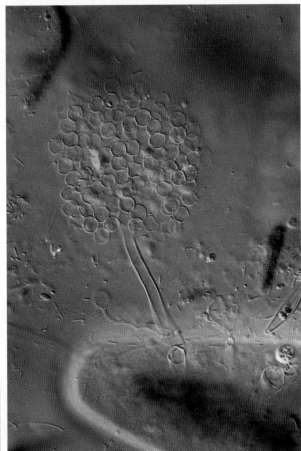

point of contact with the alga, forming a structure called an appressorium which assists in the process of fungal penetration and infection (e.g. fig. 626, *Staurastrum* and figs 627, 628 on the arm of a different *Staurastrum*). Note that in fig. 628 both the zoospore and its appressorium are devoid of content; infection has already occurred.

Biflagellate fungi also form resting spores and some can be seen in figs 633–635. Fig. 633 shows a single spore within a *Trachelomonas* (p. 100). In fig. 634 a spore of a *Myzocytium* is present in *Xanthidium*. It has been produced by a sexual process and a large empty male cell occupies the lower semi-cell of the desmid. The numerous resting spores in a cyst of *Ceratium* (fig. 635) belong to the species pictured earlier (fig. 630) from lake sediment.

Fig. 636 shows the central area of a *Ceratium* cell. Within it can be seen the empty tubular remains of the biflagellate fungus *Aphanomycopsis* (fig. 615) which killed it. Attached to the outer surface of the alga (righthand side) is a round body almost devoid of content and not in sharp focus. This was a developing sporangium of the chytrid saprophyte *Amphicypellus* (fig. 592) but it has been killed by a chytrid parasite, sporangia of which can be seen on its surface.

This one *Ceratium* cell with its three different fungi, a parasitic biflagellate, a saprophytic chytrid and its chytrid parasite, illustrates to some degree the complicated associations which can arise between algae and the fungi which depend upon them for their existence.

Three different fungi at least are known to parasitize the cyst stage of *Ceratium*, as well as two others which parasitize the free-swimming stage. When one also takes into consideration the benthic animals, including amoeboid protozoans, which can ingest cysts of *Ceratium* (e.g. fig. 632), the planktonic rotifer *Ascomorpha* which sucks out the contents of its motile stage, and the rotifer *Asplanchna* which can ingest whole cells, it is clear that there is a variety of grazing and parasitic organisms affecting its abundance. Further, it is very probable that there are yet other organisms killing *Ceratium*. Nor is this an unusual situation. It illustrates how little we know of the effects of so many organisms which prey on or parasitize algae.

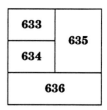

Fig. 633. Resting spore (13 μm diam.) of a biflagellate parasite in *Trachelomonas*.

Fig. 634. *Myzocytium*. Sexually produced resting spore (19 μm diam.) in the upper half of a *Xanthidium* cell, joined to the now empty male cell (39 × 15 μm) in the lower half of the desmid.

Fig. 635. Resting spores of the fungus (fig. 630) in cysts of *Ceratium* from lake sediment. Top, pear-shaped spore (29 × 23 μm).

Fig. 636. *Ceratium*. Inside the cell, tubular remains of *Aphanomycopsis* (biflagellate); outside, empty sphere (right) is a sporangium (10 μm diam.) of *Amphicypellus* (saprophytic chytrid) killed by the chytrid parasite attached to it.

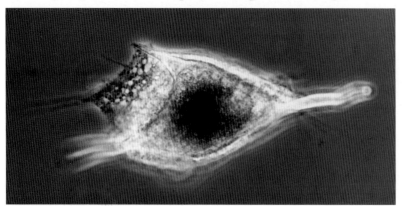

Fig. 632. An amoeboid organism which has ingested a cyst of *Ceratium* lying in lake sediment (× 640). **P**.

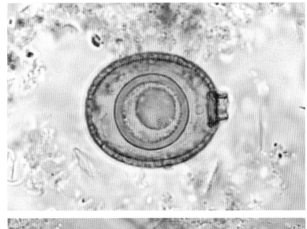
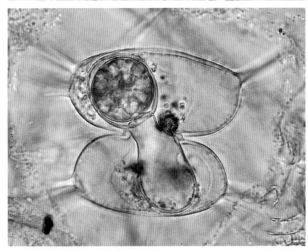
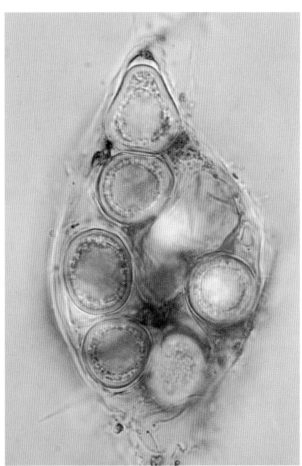
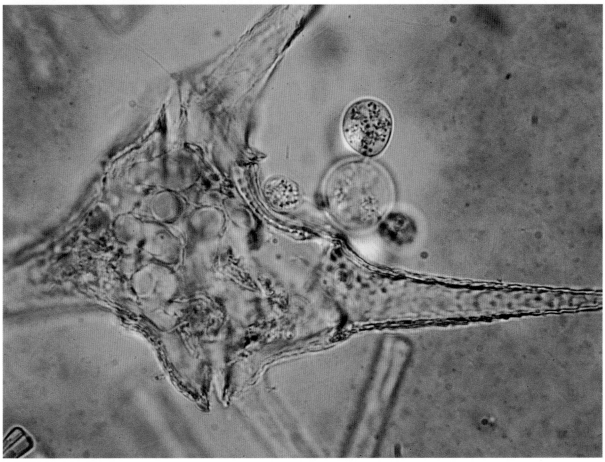

Hyperparasites

Some fungi, often belonging to the same groups or even genera as the parasitic ones on algae, can parasitize the fungi which are parasitizing the algae. As will be seen, the saying that big fleas have little fleas upon their backs to bite 'em and little fleas have lesser fleas and so on, ad infinitum, can have some relevance to fungal parasitism.

In fig. 637 a cell of *Cosmarium* has been killed by the chytrid fungus on it. This chytrid has produced a large sporangium but its content will never reach maturity and liberate zoospores. It has become infected by another parasite which is in the process of producing its own sporangium at the expense of the original parasite. This second parasite is known as a hyperparasite and the parasitism of one parasitic organism by another is called hyperparasitism.

In fig. 638 there is a row of four cells sloping from right to left. The lowest cell is a live cell of *Cyclotella*, a centric diatom. The colourless cell above it, also a *Cyclotella*, is dead. The round cell on top of this latter diatom cell is a chytrid parasite which has killed it. The topmost cell is the dehisced empty sporangium of a different chytrid fungus (hyperparasite) which has parasitised the fungus on which it sits, which fungus in turn has parasitised the dead diatom cell below it. Hence, as in the case of *Cosmarium* (fig. 637) the original parasite of the diatom cell has been prevented from completing its full life cycle and so reproducing. Under normal circumstances it would have produced a broadly triangular sporangium and liberated its zoospores via two openings as is shown by the specimen in fig. 639.

In fig. 640 there also is a row of four cells. The square cell (right) is a dead *Cyclotella* containing only the shrunken remains of its chloroplasts. The large round cell attached to it (left) is the chytrid parasite which killed it. Present on this parasite are two other fungal bodies which in turn have led to its death and so are hyperparasites. Like the hyperparasite sporangium seen at the apex in fig. 638, the one facing upwards in fig. 640 has completed its life cycle and liberated its zoospores. In contrast, the second larger hyperparasite body in fig. 640 has not been able to do so. The reason for this is the fourth body at the end of the row of cells (extreme left). This end cell also is a fungal hyperparasite, perhaps even "a hyper-hyper-parasite", which has in turn parasitized the hyperparasite of the parasite of the centric diatom. It can now be understood why earlier there was mention of the ditty beginning – big fleas have little fleas upon their backs to bite 'em!

It is noteworthy that in none of the three examples shown has the fungus been parasitized early enough in its development to prevent it killing the alga on which it is itself parasitic. Present knowledge suggests that this is the

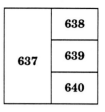

Figs 637-640. Hyperparasitism among chytrids.

Fig. 637. *Cosmarium* infected by a chytrid (24.5 × 20 μm) on top of which is another chytrid parasite (i.e. a hyperparasite) which will kill the parasite of the *Cosmarium*. **I.**

Fig. 638. In centre, chain of four cells – lowest cell, live, unparasitized cell (8.7 × 6.3 μm) of *Cyclotella*; next cell, dead *Cyclotella* parasitized by the third cell in the row which is a chytrid and, fourth (top) cell, another chytrid (hyperparasite), parasitizing the chytrid which has killed the *Cyclotella*. **P.**

Fig. 639. Lowest cell (10 μm diam.), the same *Cyclotella* species with the same parasite as in fig. 638 but here a normal empty sporangium with two lateral exit pores has developed due to the absence of any hyperparasite. **P.**

Fig. 640. Row of four cells; from right to left:-
1, dead, parasitized cell (12 × 10 μm) of *Cyclotella*;
2, its chytrid parasite;
3, hyperparasite;
4, parasite of the hyperparasite. **P.**

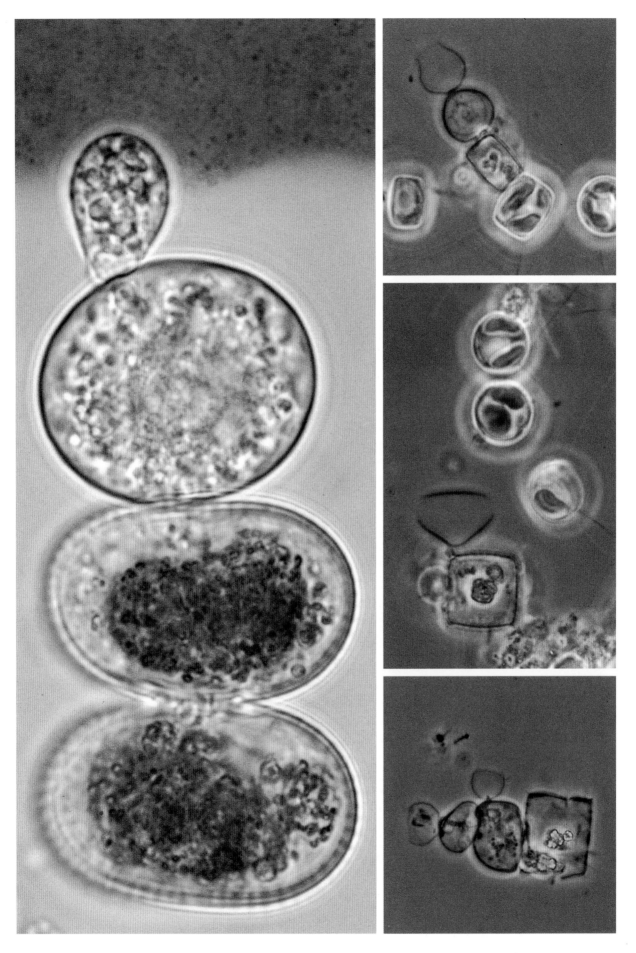

321

usual situation; namely that a parasitized alga is unlikely to be saved from death by a hyperparasite. Hence, it would seem to follow that hyperparasitism is unlikely to have any significant effect on the development or severity of a fungal epidemic. However, if one considers the effect of hyperparasitism from a population point of view, rather than that of a given alga, it can be seen that hyperparasitism can reduce the virulence of a fungal infestation. Though a parasitized alga dies, its parasite is likely to die too, as can be seen in figs 638, 640. Further, if the hyperparasite of the parasite of *Cosmarium* (fig. 637) continues to grow, as it may be expected to do, then the fungal parasite of the *Cosmarium* will die. It follows that in a population hyperparasitism reduces the output of spores by the primary parasite and so reduces the infection of further algae and the development of an epidemic. Some algae die but many are saved.

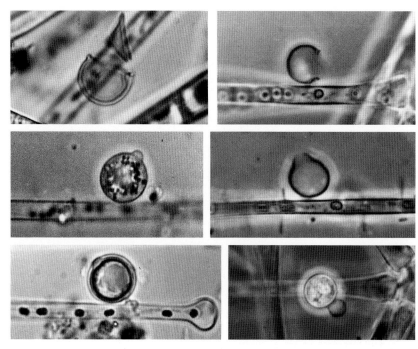

The presence of a fungus parasitizing another fungus is not always as readily visible as in the examples just shown. In some cases, instead of a hyperparasite sporangium being situated on the outer wall of the sporangium of its host fungus (e.g. fig. 637) it develops "hidden" inside the host sporangium. Since the content of both fungi concerned is colourless, the existence of the hyperparasite can be very diffficult to detect until it has reached maturity and is ready to liberate its zoospores. The method by which a hyperparasite sporangium opens (dehisces) often is quite different from that displayed by its host under normal circumstances. The chytrid *Zygorhizidium*, a parasite of *Asterionella*, possesses sporangia which at the time of dehiscence cast off a wide lid as shown in fig. 641 although here the edges of the lid have curled back giving it a triangular shape. Fig. 643 shows a specimen of the same

641	642
643	644
645	646

Figs 641, 642. Unparasitized, empty sporangia of *Zygorhizidium*; with lid (fig. 641) and wide opening (fig. 642).

Fig. 643. Sporangium of *Zygorhizidium* now containing a mature sporangium of its parasite *Rozella*; note small dehiscence papilla.

Fig. 644. Shape of empty sporangium which remains after *Rozella* has liberated its zoospores.

Fig. 645. Resting spore of *Rozella* (inner dark circle) inside a sporangium of *Zygorhizidium* (outermost wall).

Fig. 646. Unparasitized resting spore of *Zygorhizidium* with attached empty male cell at 5 o'clock.

All × 1600. Figs 642, 644, 646. P.

fungus but within it a sporangium of the hyperparasite *Rozella* has developed. In contrast to its host *Zygorhizidium*, the sporangium of *Rozella* produces a small rounded papilla which deliquesces to set free the zoospores. Because of these two differing types of dehiscence, the empty sporangia left on the diatom cells look quite different. Those of *Zygorhizidium* will possess a large, often lateral opening (fig. 642), those which have been parasitized by *Rozella* a small apical opening (fig. 644).

Although the zoospores of the two fungi involved in such a relationship may both be of the posteriorly uniflagellate type, they usually differ from one another in detail. Indeed, the presence of an internal hyperparasite may first be suspected when the "wrong" kind of zoospore emerges from the sporangium of an already well-known fungus.

The resting spores of *Rozella* also occur within the sporangia of the fungi it parasitizes. In fig. 645 the grey outer ring represents the sporangium wall of the host *Zygorhizidium*, the inner darker ring the thickened wall of a resting spore of *Rozella* (hyperparasite). For comparison, the type of resting spore the *Zygorhizidium* itself produces is shown in fig. 646. It involves a sexual process and the now empty small male cell is clearly visible adherent to the spherical spore.

Fungal parasites and man's activities

As has been mentioned, algae can be pests. Large numbers add to the difficulty and cost of producing water for domestic and industrial use, some are toxic and others interfere with amenity, conservation or fishing. Under these circumstances, epidemics of fungal parasitism are welcome and if they could be produced at will would be a valuable form of biological control.

On the other hand, there are situations where large populations of algae or certain kinds of algae are desirable and so interference with this growth by fungal parasites and the presence of other organisms is undesirable. An example is the mass cultivation of algae for food or the production of industrially valuable substances. Mass cultivation of algae under controlled factory conditions is more expensive than growing them outdoors. A problem with some outdoor algal crops, though grown under polythene or other transparent covering material, is infection by fungal parasites or other organisms such as microscopic animals. A great advantage of using algae such as *Dunaliella* (p. 83) and *Spirulina* (p. 230) is that they can grow in waters which are so salty that few other organisms can coexist with and so harm them.

Hyperparasites which prevent a parasitic fungus from reproducing occur both among chytrids and biflagellate fungi. Just as parasitism of algae by fungi is of potential use in controlling algal numbers, so hyperparasitism is a potential tool for controlling fungal epidemics. The possibility of using hyperparasites as a method of biological control in saving crops and fish (these are often attacked by fungi causing massive damage to populations) from disease is being considered more and more. However, so far as is known at present, there are formidable problems to be overcome in using fungi for algal control or the control of their parasites by hyperparasites.

Just as the effect of fungal parasitism on an algal population depends on its severity, so does the reduction of its severity by hyperparasitism depend on the abundance of hyperparasites. However, on present evidence it seems that hyperparasitism among algal parasites is of rather limited occurrence and rarely saves a significant part of a parasitized population of algae from being killed. To put it another way, it is not clear that hyperparasitism makes much effect on the course of the great majority of fungal epidemics. While this is of no advantage to the algae concerned it is good news from a waterworks point of view.

Postscript

Clearly, this is an individualistic entry into the world of freshwater algae or, to be more correct, that of inland, non-marine algae. We hope that the book will have given those unfamiliar with algae or certain aspects of their life, some idea of their variety, appearance and activities, as well as the perils they face from other microorganisms. They have predominantly been shown in the living condition with more than half of the pictures in colour.

Though the order in which the pictures appear lacks the formal planning of a textbook, it is not entirely unplanned, loose though the planning is. By starting with round green cells and progressing to thread-like green organisms, the reader unfamiliar with algae starts in a simple, plant-like world. At the end of this section (Chlorophytes) a set of free-swimming "plants" are encountered. Green colour predominates in the next two groups which, incidentally, would not follow Chlorophytes in this order in a textbook. Further "untextbook" order is seen in the Chlorophytes themselves which would start, not end, with *Chlamydomonas* and allied genera in a textbook. Though the photosynthetic Euglenophytes and Xanthophytes are green, the former are hardly plants, especially the unpigmented ones, which include some which clearly are animals. Next the reader encounters curious brown algae (diatoms), followed by various other more-or-less "unplant-like" algae, culminating in bacteria (Cyanophytes). The latter affords an easy transgression (symbiosis) to interrelations of algae with other organisms. Hence, a heterogeneous collection of pictures permits a commentary which gives some introductory insight into the algal world.

Electron microscopy, molecular biology and other modern techniques have greatly altered long-held ideas about the taxonomy, relationships and so classification of microorganisms. One effect of new knowledge has been the unprecedented number of new names or changes of old names, several of which we have had to inflict on our readers. How best to do this posed an apparently insoluble problem. If we simply used the new names, believing that they will come into general use, then the texts, indeed virtually all those mentioned in pages 343–345, apparently do not include the algae concerned. If we ignored the new names on the grounds that they have not yet been generally accepted, then we could fairly be accused of being out-of-date.

Irritated as some people may be by the apparently never-ending name changes, most of them are not the passing whims of those who enjoy digging out ancient names from obscure publications. They are made for good reasons and most are likely to get more and more widely used and finally be incorporated into textbooks.

The ferment of ideas during the last 10-20 years has led to modern classifications which ignore botany and zoology as separate disciplines. Even the simple division of living organisms into prokaryotes (bacteria) and eukaryotes (all other organisms) is doubted. From an evolutionary standpoint, two quite different groups of bacteria can be recognised. Hence, these two prokaryotic groups and the single eukaryotic one are the three great divisions among organisms.

Alga is now an almost indefinable word for a variety of plants, animals and organisms which are neither clearly plants nor animals, plus one group, bacteria. Yet it may be illogical, but alga is still a useful word. Everyone, according to degree of knowledge, knows what kinds of organisms are algae, justifiably or not.

There also are fungi which are not now considered to be fungi. These include the ones we have shown with flagellate stages in their life histories. The modern view is that they are pseudofungi (a word used to classify biflagellate forms in at least one recent classifactory system).

Though it would have been wrong to ignore these matters, obviously they could only be touched upon in such an introductory portrait gallery, nor should they be allowed to interfere with the pleasure to be had from the entry into the fascinating and beautiful world of algae.

Among several worries about the degradation of parts of the natural world and threats to the biosphere as a whole, biodiversity has become a major concern, notably in relation to tropical regions and areas about the biodiversity of which we have little knowledge and may never have much because of imminent threats, not solely to great losses of biodiversity, but even complete destruction of parts of the living world. This is not a topic raised directly in this book. The organisms shown here largely, but not wholly, came from the temperate world, in particular, from the English Lake District. It will be obvious, especially from the sections about microscopic animals and parasites, that there is still much that we do not know about the biodiversity which is, so to say, on our doorstep. Despite all the modern progress in knowledge, there is still a great need for floristic, faunistic and microbiological texts about the allegedly relatively well-known parts of the world.

Glossary

The definitions given are those pertinent to this book.

Aerobe (*Adj.* aerobic). An organism which can live and grow in an environment containing oxygen (e.g. in solution). Hence, aerobic environment, aerobic conditions, etc.

Aerotope A suggested replacement for the term gas vacuole (*q.v.*).

Agar A seaweed product from which gels can be made on which, with suitable additions of nutrients, a wide variety of microorganisms can be grown.

Akinete The name given to the spore of certain filamentous Cyanophytes and the zygospores of conjugate algae.

Algologist See phycologist.

Amoeboid Movement by means of pseudopodia (*q.v.*). A characteristic of an organism or a stage in the life history of an organism; in both cases the noun amoeba may be used (e.g. reproducing by amoebae). Named after the genus *Amoeba*. Most amoeboid organisms are protozoans (*q.v.*).

Anaerobe (*Adj*: anaerobic). An organism which can live and grow in the absence of oxygen (e.g in solution). Hence, an anaerobic environment, anaerobic conditions, etc.

Astaxanthin A carotenoid (*q.v.*) substance, red in colour and widely distributed in algae and invertebrates.

Auxospore A cell arising from the enlargement of the protoplasm of a diatom cell and its escape from the siliceous wall. Important in relation to size regulation in diatoms and usually also the product of sexual reproduction.

Axenic Applied to cultures containing only one organism (also called pure cultures). Dual cultures (e.g. parasite and host) can also be called axenic if free of all other organisms. Cultures of a single, selected organism but not free of contaminants are clonal (*q.v.*) but not axenic.

Bacterioplankton   See phytoplankton.

Basket or feeding basket   See: cytopharyngeal basket.

Benthos   The bottom and sides of a waterbody and objects arising therefrom (e.g. plants or objects on or in the bottom or sides). That part of the aquatic environment which is not the domain of planktonic organisms (e.g. the open water).

Biflagellate fungi   In the past, among the so-called Lower Fungi were included all those with a flagellate stage in their life histories. They were placed in a group called Phycomycetes ("algal fungi"; see phyco- ; phycology, the study of algae; mycology, the study of fungi). They are not included in the fungi (e.g. toadstools, rusts, yeasts, etc.) in the most recent classifications. As was suggested long ago, it is once more considered that the biflagellate fungi are related to other microorganisms with the same kinds of flagella (e.g. Chrysophytes and Xanthophytes). Chytrids, which, as the word is used here, have only one flagellum, now are neither grouped with the biflagellate fungi nor with any algal group. There seems to be uncertainty as to which major group of microorganisms they are related to.

Binary fission   The division (splitting) of a cell into two; reproduction by bipartition.

Bioassay   The use of an organism to assess the amount of a substance present (e.g. a vitamin or traces of chemical elements), the potency of a substance or mixture of substances (e.g. water, soil, human or industrial effluents) for promoting or retarding growth (e.g. nutrients or toxic substances). Algae (notably the so-called *Selenastrum capricornutum*) are often used in studies on eutrophication as well as testing for the safety of new products (e.g. herbicides and pesticides). Since algae are so diverse (e.g. plants, animals or bacteria) and since there are representatives of all the major groups in culture collections, potential populations are available for a wide variety of bioassays.

Brackish water   See fresh water.

Caesium-137 (^{137}Cs)   A radioactive isotope of the element caesium (Cs), common in fall-out from the fission products of atomic bombs and the Chernobyl disaster.

Carbohydrate	Any member of a large group of organic compounds containing carbon, hydrogen and oxygen with a basic formula of $C_xH_{2x}O_x$. Examples are starches, leucosin, paramylon, glycogen, cellulose and sugars.
Carbon-14 (^{14}C)	A radioactive isotope of the element carbon (C) produced by the action of cosmic rays.
Carotene	Yellow or orange pigments, including ß-carotene which is present in carrots and many algae. It also is a provitamin.
Carotenoid	A yellow, orange or red substance belonging to the carotene or xanthophyll (e.g. astaxanthin *q.v.*) groups.
Cell	The basic unit of which an organism is composed. Organisms can consist of a single cell (unicellular organism or unicell) or more than one cell (multicellular organism).
Chemocline	See stratification of water.
Chlorophyll	A green photosynthetic pigment. There are several kinds but all algae, including the prokaryotic (*q.v.*) Cyanophytes, contain the compound called chlorophyll α.
Chloroplast	An organelle (body) containing the photosynthetic pigments and apparatus in eukaryotic (q.v) algae and plants.
Chromatic adaptation	The alteration of the proportions in which the photosynthetic pigments are present, and so the colour of the cell, in relation to alterations in the spectral quality of the incident radiation.
Chromatophore	Any photosynthetic organelle. Here a synonym for chloroplast. Sometimes used to distinguish non-green from green coloured (e.g. in Chlorophytes) photosynthetic organelles. The word is used in other senses by bacteriologists and zoologists.
Chromosomes	Structures carrying genetic information (the genetic code). In eukaryotes, the chromosomes are enclosed in a membrane-bound body called the nucleus.
Chrysolaminarin	See leucosin.
Chytrid	See biflagellate fungi
Cilium	See flagellum. This word generally is used when large numbers of flagella are present, as in the

| | protozoan animals called ciliates. The cilia beat in unison, may be simple or compound and are used for swimming, "walking" (compound cilia) and collecting food particles. |

Clone A population derived asexually from a single ancestral cell or organism. This ancestor must not be a zygote (*q.v.*) since cells derived therefrom may not be genetically identical.

Colony A group of cells living together to form an organism of characteristic form and structure; not a mere heap of cells. The word colony also is used for an aggregation of cells growing in culture on a solid medium such as agar (*q.v.*). Such an aggregation often arises from inoculation with a single cell or organism and so is a clonal population.

Conjugation Union of two gametes within a tube (conjugation tube) connecting the cells they were formed in or after the passage of one gamete through this tube into the cell containing the other gamete.

Contractile vacuole An organelle which repeatedly fills with watery fluid, so becoming progressively larger and then contracts, so expelling this fluid.

Cryopreservation The maintenance of algae and other organisms in a viable state at temperatures well below 0° (e.g. in liquid nitrogen).

Culture As a verb: to grow or maintain populations in the laboratory or other suitable place such as a Culture Collection (or your home!). As a noun: the population so produced or maintained. Growth normally is in liquid or on solid culture media and maintenance by restricting the rate of growth (e.g. keeping in low light or in the dark and at a low temperature) or keeping the culture in a viable but non-growing state (e.g. air-dry, freeze-dried or cryo-preserved). A wide variety of algae are maintained in diverse institutes, notably in National Culture Collections.

Cyanelle A Cyanophyte-like body functioning as a chloroplast in a variety of eukaryotic microorganisms. Thought to have evolved by a symbiotic pathway.

Cyst A general word for a cell or resting spore (*q.v.*) capable of resisting unfavourable environmental conditions. Such cells usually have thick walls. See also: digestion cyst.

Cytopharyngeal basket A ribbed, funnel-shaped structure present in some ciliates, down which particles such as micro-organisms pass into the cell.

Desmids Desmids traditionally are divided into two groups, placoderms and saccoderms. Placoderm desmids have cells composed of two parts (semicells) of different age which in many genera are clearly united to one another by a narrow connecting region (isthmus) and have pores in their cell walls. They may be called "true" desmids because saccoderm desmids virtually are unicellular members of the *Spirogyra* group. Saccoderm desmids neither have cells composed of two parts nor pores in their walls.

Diatomite (*Syn.* diatomaceous earth; Kieselguhr). Deposits or geological formations rich in the remains of diatoms.

Digestion cyst The stage in certain protozoa during which food (e.g. algal cells or cell contents) is digested. In this stage a naked protozoan usually rounds off and surrounds itself with a firm wall.

DNA Deoxyribonucleic acid – see genetic code.

Dwarf male In the Oedogonium group, a structure arising in sexual reproduction. Short cells in a filament produce single zoospores, smaller but similar to asexual ones. These zoospores attach themselves to an egg-containing cell (oogonium) or nearby cell. From the zoospore arises the so-called dwarf male, consisting of a few-celled filament or single cell. In one or more cells of the dwarf male two (rarely one) sperms are produced.

Ejectosome A small organelle in Cryptophytes from which fine threads can be ejected. Old name, trichocyst (*q.v.*).

Epilimnion See stratification of water.

Eukaryote (Eucaryote). An organism, the cells of which contain a nucleus and other organelles (e.g. chloroplasts, mitochondria) which are bodies bounded by a membrane. All plants and animals are eukaryotes, all bacteria are prokaryotes (*q.v.*).

Eutrophication The state of being or the process of becoming eutrophic. The word usually is used in the media and popular language in a derogatory sense, as being in or becoming a bad state of affairs.

Eutrophy	(*Adj.* eutrophic). An environment such as soil or water which is rich in nutrients (Greek, meaning well fed) and so also rich in plants, animals and bacteria.
Eye-spot	See stigma.
Filament	Cells arranged or united together to form thread-like groups or chains.
Fine structure	Structural details of cells invisible or only partially visible by light microscopy but revealed by electron microscopy, also referred to as ultrastructure.
Flagellum	A fine fibrillar extension of the cell by the movements of which it may be propelled through the water. Most flagellate organisms (flagellates) have two or more flagella, i.e. they are biflagellate or multiflagellate. Cilia (*q.v.*) and flagella are structurally the same in all eukaryotic organisms but the flagella of bacteria have a different structure.
Freeze-drying	Maintenance in a viable condition by rapid freezing followed by drying in a vacuum.
Fresh water	Ecologists and others have erected quantitative criteria for distinguishing fresh, brackish and sea waters. So far as aquatic organisms are concerned there is no hard and fast line between them. Here by brackish we mean slightly salty, as for example in the upper reaches of estuaries. By saline water we mean salty inland waters which may be brackish or more salty than the sea.
Gamete	A sexual cell potentially capable of uniting with another such cell to produce a zygote. When gametes are markedly different in size, structure and behaviour they can be called male and female.
Gas vacuole	An aggregation of gas vesicles.
Gas vesicle	A minute gas-filled cylinder present in certain Cyanophytes and other prokaryotes but not found in any eukaryote. The cylinders are only visible using electron microscope techniques.
Genetic code	The information for protein production encoded in the DNA and the basis of heredity.
Girdle	The elements in the wall of diatoms which hold the two halves (valves) together.

Half-life	Of a radioactive material (e.g. isotope of an element), the time taken for half of the atoms in it to decay.
Hard water	Water with high concentration of salts of calcium and magnesium.
Heliozoa	Unicellular protozoans living free in water or attached to a substratum. Fine threads radiate out from the surface of their globose cells giving them the common name of sun animalcules. They can capture other microorganisms.
Helix	In ordinary parlance a synonym of spiral. Strictly, a helix is three dimensional (e.g. a spiral staircase is a helix) and a spiral is two-dimensional, like a watch spring.
Heterocyst	A special type of cell found only in certain Cyanophytes, permitting nitrogen-fixation in oxygen-containing (aerobic) environments. Heterocysts do not carry out the full process of photosynthesis and so do not liberate oxygen.
Heterocyte	Suggested new name for heterocyst, since it is not a cyst.
Holdfast	Any organ, structure or material which enables an alga to attach itself to a substratum. Examples are specialised basal parts of cells or filaments and mucilage.
Holozoic nutrition	The ingestion of live or dead organisms, parts thereof, or particles derived therefrom. Alternative name: phagotrophic nutrition.
Hormogonium	A reproductive body in filamentous Cyanophytes. Part of a filament becomes separated from the rest, often after the development of special separating structures or processes. It usually glides away. The cells of hormogonia can look just like the rest of the cells in a filament or be more-or-less markedly different, sometimes so different that they can be considered as multicellular spore bodies, and then are called hormocysts.
Hyperparasite	A parasite which parasitizes another parasite.
Hypersensitive	Here it means extremely sensitive and so reactive to infection by a parasite. The sensitivity of the host organism (e.g. an alga) may be so high that it dies before the parasite (e.g. a fungus) can reproduce.

	Hence, hypersensitivity prevents the spread of the parasite.
Hypertrophy	Here it refers to the abnormal enlargement of a cell. Commonly it means abnormal enlargement of a tissue or organ.
Hypha	A fine branched or unbranched filament of a fungus, bacterium and of certain seaweed tissues. The adjective hyphal sometimes is used informally to describe any structure or organisation reminiscent of fungal hyphae or aggregations thereof.
Hypolimnion	See stratification of water.
Ingest	To take in. Used instead of the word eat for the diverse methods by which organisms obtain and process food for digestion.
Isotope	One of two or more forms of a chemical element, differing from the other isotope or isotopes in the number of neutrons in the atomic nucleus. For example, carbon-12 (^{12}C) is a non-radioactive isotope of carbon and the standard on which all atomic weights are based; carbon-14 (^{14}C) is a radioactive isotope much used for dating objects and substances (radiocarbon dating). Nitrogen-15 (^{15}N) is a non-radioactive isotype of nitrogen much used in estimations of nitrogen fixation.
Isthmus	The narrow part between the two half-cells (semicells) of certain desmids.
Kieselguhr	See diatomite.
Leucosin	(*Syn.*: chrysolaminarin). A carbohydrate food reserve, often present as large globules in Chrysophytes. Also recorded from diatoms and Xanthophytes.
Life cycle	A series of phases through which an organism can pass during its life. The words are virtually synonymous with life history apart from the sense of uninterrupted and so successive events in a life cycle.
Life history	The various stages or phases an organism may enter or go through but not necessarily in a given order.
Light	By light is meant radiation in the visible and nearby spectral regions (e.g. near ultraviolet and infrared). The degree to which different parts of this

spectrum are used by algae depends on the "light harvesting" pigments present and their utilisation (e.g. for photosynthesis or movement of cells or their chloroplasts). The photosynthetically active part of the electromagnetic spectrum is not wholly the same as that which we call light.

Lipid — The general name for a diverse group of organic compounds which are virtually insoluble in water but largely or wholly soluble in certain organic solvents. Common examples in algae are fats and oils. Carotenoids (*q.v.*) are lipids, as are sterols (e.g. cholesterol), some vitamins and hormones.

Lyse — To cause lysis. Lysed: having undergone lysis.

Lysis — The break-up or dissolution of the cell membrane caused by internal changes or by external agents.

Macroscopic — The opposite of microscopic; easily seen by naked eye.

Magnification — Useful, that which as it increases enables ever finer details of an object to be revealed. Empty magnification enlarges the size of an object but does not reveal finer details.

Metalimnion — See stratification of water.

Molecular biology — The study of the nature and function of molecules of biological importance, especially the molecules involved in the genetic code (*q.v.*); its implementation and the proteins arising therefrom.

Monospore — A spore produced singly in a sporangium. In practice, the word usually refers to asexual, non-motile spores of Rhodophytes.

Motility — The power of movement conferred on an organism by some specific structure or cell product. Examples in algae are flagella, extrusion of mucilage, creeping or gliding mechanisms, gas vesicles.

Mucilage — Watery colloidal matter of diverse chemical composition but generally consisting largely of carbohydrates.

"Naked" cell — A cell not covered by a wall or other firm material external to the membrane enclosing the cell content.

Nanoplankton — (Nannoplankton). See plankton.

Nitrogen fixation — The utilisation of nitrogen gas (dinitrogen: N_2) to form ammonium and other nitrogenous compounds. Only prokaryotes, including many Cyanophytes, can fix nitrogen.

Nomenclature The naming of organisms, for which purpose there are international codes of practice for botany, zoology and bacteriology respectively.

Nucleolus A globose body in the nucleus with special functions.

Nucleus In eukaryotic organisms the organelle contains the main genetic material (e.g. DNA). The chloroplast also contains some genetic material, as does an organelle not shown here – the mitochondrion.

Oligotrophy (*Adj.*: oligotrophic). The opposite of eutrophy (*q.v.*). The word oligotrophication as the opposite of eutrophication is very rarely used.

Oospore A fertilized egg cell which has become thick walled and can function as a resting spore. The cell it is in is the oogonium.

Opaline silica A non-crystalline form of silica (hydrated silicon dioxide) in diatom walls which is of similar structure to the silica comprising the mineral opal.

Organelle (Diminutive of organ). Any structural component of a cell carrying out a specific function or functions. Examples are: nucleus, chloroplast, flagellum, contractile vacuole. Most organelles are bounded by a membrane. Starch, oil etc are not organelles but products of the activities of the cell and its organelles.

Palmelloid Palmelloid stage, palmelloid phase, palmella. A stage in the life history of a flagellate alga when the cells become non-motile and multiply to produce formless masses of cells embedded in mucilage (e.g. *Euglena*, fig. 161). The word palmelloid also is used loosely for any alga which normally or under certain conditions consists of cells in formless masses embedded in mucilage. *Palmella* was a genus of over 100 species of algae, the vast majority of which now are known to be stages in the life histories of a diverse collection of genera and species.

Paramylon (*Syn.*: paramylum). A carbohydrate reserve product characteristic of Euglenophytes but also recorded for certain Haptophytes. In the cells of Euglenophytes paramylon grains often are of characteristic shapes.

Parasite	An organism living on or in and at the expense of a live organism (the host organism) which it may eventually kill.
pH	A scale running from 0-14 based on the concentration of hydrogen ions present. At pH 7 the water is neither acid nor alkaline. As the pH decreases from 7, so more and more hydrogen ions are present and the water is more and more acid. The reverse takes place as the pH rises and so the water becomes more alkaline.
Photosynthesis	The conversion of radiant energy (e.g. light) into chemical energy stored in the cells (e.g. as starch, paramylon, fat) or used directly to drive some cell process. All algae can utilize carbon dioxide directly or, some of them, in the form of bicarbonate in photosynthesis and release oxygen. Certain Cyanophytes, under anaerobic conditions, can also utilise hydrogen sulphide instead of carbon dixode in a photosynthetic process not resulting in the liberation of oxygen.
Phycologist	see below.
–phyc-	This syllable is seen in the names of algal classes (Chlorophyceae, Xanthophyceae, etc.) and as a prefix in the words phycology and phycologist. It is derived from the Greek *phucos*, a seaweed. The study of seaweeds and all other kinds of algae is phycology and its devotee is a phycologist. These two words are modern replacements for the words algology and algologist, words which are still in use in some quarters. Algology is derived from the Latin name for a seaweed, alga. Linguistic purists insist on using words derived from Greek and the Greek word *algos* means pain, hence algology could be the study of pain.
–phyte-	The syllable –phyte is used here for an algal group (Chlorophyte, Xanthophyte, etc.). Such names are derived from, but not necessarily the exact equivalent of, the scientific names of major algal groups (Chlorophyta, Xanthophyta, etc.). The syllable phyte is derived from the Greek, *phuton*, a plant. Hence: phytoplankton, the so-called plant plankton (*q.v.*) and words such as phytopathology, the study of plant diseases and phytotoxin, a toxic substance produced by a plant (see also saprophyte)
Phytoplankton	See plankton.
Picoplankton	See plankton.

Placoderm desmids See desmids

Plankton The assemblage of organisms or stages in the life cycles of organisms which carry out all or most of their growth and reproduction while dispersed in the open waters of the seas, lakes, pools, rivers etc. Though they may be capable of movement, it is usually on a small scale compared to the movements of the water they are suspended in. Algal plankton is called phytoplankton (plant plankton) and includes Cyanophytes (bacteria) and photosynthetic organisms with more or less markedly animal characteristics (e.g. many flagellates). Bacterioplankton can mean planktonic bacteria other than Cyanobacteria (i.e. Cyanophytes) and the word zooplankton ("animal" plankton) may or may not include algae but usually covers non-photosynthetic organisms.

A planktonic individual is called a plankter, hence one can speak of a phytoplankter (e.g. an alga) or nanoplankter (a small plankter – see below). Size categories of planktonic organisms also are not necessarily sharply distinct. Net plankton is that retained by a net but the meshes of different kinds of net are not the same. Roughly it applies to algae less than 50-60 μm in their longest dimension. Nanoplankton (sometimes spelt nannoplankton) refers to net-passing plankton and the recent category of picoplankton to very small organisms, for example those which can pass through a membrane the pores of which are 2 μm wide and so can be less than one micrometre but, according to one proposal, are more than 0.2 μm. On this scale, nanoplankton organisms are from 0.2.-2.0 μm in longest dimension. It must be remembered that whether a cell can pass through a pore of given diameter depends on its orientation when reaching that pore. For example, a very narrow filament 1 μm broad but 1000 μm long is unlikely to pass through a pore 2 μm wide because it is unlikely to arrive perpendicular to it. Therefore, it is not a picoplankter. However, if the filament is 1 μm broad and 10 μm long it has a much greater chance of passing through a 2 μm pore. Hence absolute distinctions between such categories as picoplankton and nanoplankton or indeed any units of classification by size do not exist.

Plasmodium A multinucleate mass of protoplasm, commonly motile and often consisting of a network of pseudopodia (*q.v.*). Plasmodia are "naked".

Plastid Another word for chloroplast.

Prokaryote	(Procaryote). An organism lacking a nucleus i.e. an organelle bounded by a membrane which contains genetic and cell organising material such as chromosomes and DNA. Prokaryotes also lack chloroplasts and mitochondria (an organelle involved in respiration) and have a number of other structural and biochemical features not found in eukaryotes (*q.v.*). Bacteria are prokaryotes and in general the words bacterium and prokaryote are synonymous, hence the modern name for Cyanophytes or blue-green algae is Cyanobacteria.
Protist	See protozoan
Protoctist	An organism belonging to the Protoctista, a group composed of algae, [excluding prokaryotes (e.g. Cyanophytes)] protozoa and certain fungi, the exact definition of which involves matters not covered in this book. The whole group is called Proctoctista. This is only one of certain new classifications jostling for preeminence.
Protoplasm	The living content of a cell. Here the meaning of the word includes the membrane bounding this living matter.
Protoplast	The organized, living part of a cell (i.e. not including the cell wall, mucilage sheath or other investment).
Protozoan	(Or protozoon): A general word for lowly, microscopic organisms such as flagellate animals and algae (often collectively called flagellates) and amoeboid organisms. The classification of an organism in the kingdom of Protozoa can be considered to be obsolete. In one modern classification they are placed in the Protoctista (*q.v.*). Nevertheless, the words protozoa and protozoan are still in general use. Another name, protist, may be encountered. A protist is any small and usually unicellular organism such as an alga. The word does not have any exact scientific meaning.
Provitamin	A substance from one source which can be converted into a vitamin within another body (e.g. the conversion of β-carotene from plants into vitamin A in certain animals).
Pseudocilium	(*Syn.*: pseudoflagellum). A fine thread extending from a cell and having some of the features of a flagellum (cilium) but immobile. Pseudocilia, like flagella, often occur in pairs.

Pseudopodium	An extension of the body of an amoeboid organism, generally used for movement or capturing particles, notably other microorganisms. There are several kinds of pseudopodia, e.g. fine ones (filopodia, rhizopodia), stout, blunt-ended ones (lobopodia) and relatively rigid ones (axopodia) with a distinction between inner (axial) and outer parts.
Pure culture	See axenic.
Pyrenoid	A proteinaceous organelle associated with a chloroplast, often surrounded by starch and involved in its production.
Radon-226 (^{226}Rn)	A radioactive gaseous isotope of the element radon (Rn), which itself is a decay product of radium.
Raphe	A system of slits or fissures in the walls of one or both valves of certain pennate diatoms. Diatoms with raphes can produce a gliding motion.
Red tide	Seawater discoloured by vast numbers of reddish coloured flagellates; predominantly dinoflagellates. Sometimes the term red tide is used for mass appearance of dinoflagellates of any colour.
Reproduction	The production of a new individual or individuals. The words reproduction and multiplication (any kind of increase in the number of cells, parts or individuals) overlap in use.
Resting spore	(Cyst or resting cyst). A dormant stage in which a cell can pass through unfavourable environmental conditions. The cell usually is covered by a thick, resistant wall.
Rhizoid	In parasitic and saprophytic fungi, the fine threads involved in their nutrition. In algae, fine threads attaching the organism to a substratum.
Rhizopodium	A general word for pseudopodial (*q.v.*) extensions of the cells of a variety of protozoa, including algae such as certain Chrysophytes. Rhizopodia vary in structure and function.
Saccoderm desmids	See desmids
Saprophyte	Literally, a plant – hence the ending –phyte (*q.v.*) – deriving its nutriment from dead organic matter (i.e. organisms or their decomposition products). The word saprophyte nowadays is used for a much wider range of organisms (e.g. fungi, algae and bacteria) which absorb material produced externally

	in the breakdown of organic matter by their own enzymes or those of other organisms. Saprophytic nutrition does not include the ingestion of other organisms or parts thereof (see holozoic nutrition).
Silica	The silicon dioxide in quartz or other crystalline or amorphous mineral substance; in the latter case likely to be hydrated (chemically bonded to water molecules). See opaline silica.
Slime moulds	(Myxomycetes). Organisms which produce plasmodia and fungal-like spore bodies. They are very common in terrestrial habitats.
Soft water	Water containing low concentrations of certain calcium and magnesium salts and so not "hard".
Sporangium	The structure in which spores are formed e.g. a zoosporangium produces zoospores, an oosporangium contains and may release one or more egg cells (oospores).
Spore	A reproductive body, usually consisting of a single cell.
Stratification of water	Any process which produces a layered structure in depth because of differences in density, e.g. thermal stratification in which the temperature of the water is responsible for the density differences. In temperature stratification, the upper layer is called the epilimnion and the lower layer the hypolimnion. The intermediate layer of relatively rapid change in temperature with depth is called the metalimnion or thermocline. In density stratification caused by variations in the concentration of chemical substances, the word thermocline is replaced by chemocline.
Substratum	In ordinary parlance the word can be synonymous with substrate or surface. The word substrate is avoided here because in physiology and biochemistry it is a substance which an organism utilises for nutrition or as a source of energy.
Taxon	Any taxonomic category. Thus, species, genera, families etc. are taxa of different taxonomic status. Sometimes the word is used loosely to refer to a given species, genus etc.
Taxonomy	The process of grading, comparing and grouping organisms. The basis of classification.

Tetraspore One of the four spores in a sporangium (tetrasporangium) found in many Rhodophytes. During the process of the development of tetraspores the number of chromosomes per cell is halved.

Thermal stratification See stratification of water.

Toxin (*Adj.*: toxic). Here, the word toxin means a substance produced by one organism which harms or poisons another organism. In common parlance, unless stated otherwise, a toxin usually is such a substance produced by a microorganism (e.g. Cyanophyte or other bacterium) which is toxic to man.

Trichocyst General name for a variety of structures from which fine threads can be ejected (cf. ejectosomes) in various algae and animals.

Ultrastructure See fine structure.

Undulipodium Name suggested for the cilia and flagella of eukaryotes. Undulipodia differ in structure from prokaryote (bacterial) flagella.

Unicellular (*Noun*: unicell). An organism or stage thereof consisting of a single cell.

Uniseriate Arranged in a single row, e.g. cells in a filament when the filament itself can be referred to as uniseriate.

Vacuole An organelle containing watery matter (e.g. sap) and various kinds of solid matter (e.g. barium sulphate crystals in desmids, or organisms or parts thereof in digestion vacuoles). Vacuoles are bounded by a membrane. Gas vacuoles (*q.v.*) are not true vacuoles.

Valve One of the two halves of which the wall of a diatom is composed. A cell seen from above or below is said to be in valve view and from the side to be in girdle view.

Vorticellid A group of ciliate protozoans (protoctists), the most famous genus of which is *Vorticella*.

Wall The non-living integument round a plant or plant-like (e.g. algal) cell, commonly composed largely or wholly of carbohydrates (e.g. cellulose). Though itself nonliving it may be penetrated by protoplasm.

Waterbloom	(Water bloom). A superficial aggregation of planktonic algae, commonly of gas-vacuolate Cyanophytes. Algae containing large amounts of fat and oil also can form waterblooms or flagellate algae can if water movements are slight (e.g. under ice). The word bloom can be used as a synonym of waterbloom but usually refers to a large development or population of algae, not restricted to superficial layers of a waterbody, which may make the water turbid or discoloured.
Xanthophylls	A group of carotenoid compounds (astaxanthin is a xanthophyll).
Zooplankton	See plankton.
Zoospore	A flagellate (ciliate) motile spore, so called because an ability to swim suggests an animal characteristic (*Greek*: zoon, an animal). Zoospores are formed in zoosporangia.
Zygote	A cell produced by the fusion of two gametes (strictly, of their nuclei).
Zygospore	A zygote which passes into a dormant stage, like a resting spore (*q.v.*).

Suggestions for further reading

Since the freshwater algal flora is much the same all over the world, works on a given area or about a given aspect of algal life are likely to be much the same in any language.

Beginners' guides or richly illustrated surveys of genera and species.

Barber, H. G. & Haworth, E. Y. (1981). *A Guide to the Morphology of the Diatom Frustule*. Scientific Publication No. 44, pp. 112. The Freshwater Biological Association, Ambleside, U.K.

Belcher, H. & Swale, E. (1978). *A Beginner's Guide to Freshwater Algae.* pp. 47. H. M. Stationery Office, London, U.K.

Belcher, H. & Swale, E. (1979). *An Illustrated Guide to River Phytoplankton.* pp. 64. H. M. Stationery Office, London, U.K.

Belcher, H. & Swale, E. (1988). *Culturing Algae. A Guide for Schools and Colleges.* pp. 25. Culture Collection of Algae and Protozoa, Ambleside, U.K.

Bellinger, E. G. (1992). *A Key to Common Algae.* 4th ed., pp. v + 138. Institution of Water and Environmental Management, London, U.K.

Hausmann, K. & Patterson, D. J. (1983). *Taschenatlas der Einzeller.* pp. 71. Kosmos, Stuttgart, Germany.

Lee, J. J., Hutner, S. H. & Bovee, E. C. (eds) (1985). *An Illustrated Guide to the Protozoa* pp. ix + 629. Society of Protozoologists, Lawrence, Kansas, U.S.A.

Lind, E. M. & Brook, A. J. (1980). *Desmids of the English Lake District.* pp. 123. Scientific Publication No. 42. Freshwater Biological Association, Ambleside, U.K. (All the desmids covered are of worldwide distribution).

Moore, J. A. (1986). *Charophytes of Great Britain and Ireland.* pp. 140. Handbook No. 5, Botanical Society of the British Isles, London, U.K. (Though not included here, these are large, attractive Chlorophytes, easily seen by eye).

Patterson, D. J. & Hedley, S. (1992). *Free-living Freshwater Protozoa.* pp. 282. Wolfe Publishing Ltd., London, U.K.

Prescott, G. W. (1978). *How to Know Freshwater Algae.* 3rd edit. pp. 348. W.C.Brown, Dubuque, Iowa, U.S.A.

Rieth, A. (1961). *Jochalgen (Konjugaten).* pp. 87. Kosmos-Verlag, Stuttgart, Germany.

Sandhall, A. & Berggren, H. (1985). *Planktonkunde. Bilder aus der Microwelt von Teich und See.* pp. 107. Kosmos-Verlag, Stuttgart, Germany.

Strebler, H. & Krauter, D. (1988). *Das Leben im Wassertropfen.* pp. 400. Kosmos-Verlag, Stuttgart, Germany.

Vinyard, W. C. (1979). *Diatoms of North America.* pp. 119. Mad River Press Inc., Eureka, California, U.S.A.

For an oversight of all genera up to the time the books were published the following can be consulted.

Bourrelly, P. (1966-1988). *Les Algues d' Eau Douce. 1. Les Algues Vertes; 2. Les Agues Jaunes et Brunes; 3. Les Algues Bleues et Rouge, Les Egléniens, Peridiniens et Cryptomonadines.* Societé Nouvelle des Editions Boubée, Paris-Vle, France.

There are many national or regional floras. A recent one now being published which gives figures, taken from the world's literature and covering a wide variety of algae is:-

Dillard, G. E. (1989-1993). *Freshwater Algae of the Southeastern United States.* Parts 1-6. Bibliotheca Phycologia 81, 83, 85, 89, 90, 93. Cramer, Berlin, Germany.

Four textbooks for graduate and postgraduate students, university lectures, etc. are:-

Bold, H. C. & Wynne M. J. (1985). *Introduction to the Algae.* 2nd ed., pp. xv + 920. Prentice-Hall, Inc., Englewood Cliffs, New Jersey, U.S.A.
Hoek, C. van den, Mann, D. G. & Jahns, H. M. (1994). Algae: An Introduction to Phycology, pp. 576. Cambridge University Press, Cambridge, U.K.
Lee, R. E. (1989). *Phycology.* 2nd ed. pp. xv + 645. Cambridge University Press, Cambridge, U.K.
South, G. R. & Whittick, K. A. (1987). *Introduction to Phycology.* pp. viii + 339. Blackwell Scientific Publications, Oxford, U.K.

There are many advanced texts on algae which are beyond the scope of this book. However, they are worth looking at for the illustrations alone. Many illustrations are electron microscope micrographs revealing the fine structure of algae and often also of great beauty.

Fogg, G. E., Stewart, W. D. P., Fay, P. and Walsby, A. E. (1973). *The Blue-green Algae.* pp. vii + 459. Academic Press, London & New York.
Still the best introduction to the Cyanophytes (Cyanobacteria).

Margulis, L., Corliss, J. O., Melkonian, M. and Chapman, D. V., eds (1989). *Handbook of Protoctista.* pp. xli + 914. Jones & Bartlett, Boston, U.S.A.
A major recent classification and oversight of algae, protozoa, aquatic "fungi", etc.

Round, F. E., Crawford, R. M. and Mann, D. G. (1990). The Diatoms. Biology and morphology of the genera. pp. viii + 747. Cambridge University Press, Cambridge, U.K.

Concerning uses to which algae can be put, eutrophication and toxic algae, the following can be recommended:-

Anon. (1990): *Toxic Blue-green Algae.* pp. 125. Water Quality Series No. 2. National Rivers Authority, Peterborough, Northants., U.K.
Borowitzka, M. A. & Borowitzka, L. J. (eds) (1988). *Microalgal Biotechnology.* pp. x + 477. Cambridge University Press, Cambridge, U.K.
Cresswell, R. C., Rees, T. A. V. & Shah, N. (eds) (1989). *Algal and Cyanobacterial Biotechnology.* pp. xvi + 341. Longman Scientific & Technical, Harlow, Essex, U.K.
Codd, G. A. & Roberts, C. (eds) (1991). *Public Health Aspects of Cyanobacteria (Blue-green Algae).* pp. 20 Public Health Laboratory Service, Association of Medical Microbiologists, London. U.K. (PHLS Microbiology Digest 8 (3) 80-100).
Lembi, C. A. & Waaland, J. R. (eds) (1988). *Algae and Human Affairs.* pp. viii +

590. Cambridge University Press, Cambridge, U.K.

Pickup, J. & Meddle, C. (1992). *Trouble with Algae.* pp. 11. Issues, Hobson Publishing, Cambridge, U.K.

Premazzi, G. & Volterra, L. (1993). Microphyte Toxins. pp. 336. Commission of the European Communities, Luxembourg.

Richmond, A. (1990). *Large scale microalgal culture and applications.* In: Round, F. E. & Chapman, D. J. (eds). *Progress in Phycological Research*, 7, 269-330. Biopress Ltd., Bristol, U.K.

Vallentyne, J. R. (1974). *The Algal Bowl. Lakes and Man.* pp. 185. Department of the Environment, Fisheries and Marine Service, Ottawa, Canada.

For information on symbiosis see:-

Smith, D. C. & Douglas, A. E. (1987). *The Biology of Symbiosis.* pp. xi + 302. Edward Arnold. London. U.K.

Other organisms. Suggestions for further reading about non-algal organisms is outside the scope of this book. However the following works already cited also include a variety of invertebrates: Hausmann & Patterson (1983), Lee *et al.* (1985), Patterson & Hedley (1992), Sandhall & Berggren (1985) and Strebler & Krauter (1988). For a comprehensive coverage of freshwater invertebrates, together with some algae, the following can be recommended.

Edmondson, W. T. (ed.) (1959). Ward and Whipple's *Freshwater Biology.* pp. xx + 1248. John Wiley & Sons, Inc., New York, U.S.A.

For an introduction to lichens see:-

Ahmadjian, V. (1993). The Lichen Symbiosis. pp. xv + 249. John Wiley & Sons Inc. New York, U.S.A.

Dobson, F. (1992). *Lichens; an Illustrated Guide to British and Irish Species.* 3rd ed., pp. xv + 249. Richmond Publishing Co., Slough, Berks., U.K.

There are no beginners' guides to fungi parasitizing algae but the following are richly illustrated accounts with a wide coverage of the groups of fungi concerned.

Batko, A. (1975). *Zarys Hydromikologii.* pp. 478. Panstwowe Wydawnictwo Naukowe, Warsaw, Poland.

Karling, J. S. (1977). *Chytridiomycetarum Iconographia.* pp. vii + 414. J. Cramer, Vaduz, Liechtenstein.

Karling, J. S. (1981). *Predominantly Holocarpic and Eucarpic Simple Biflagellate Phycomycetes.* 2nd. ed., pp. 252. J. Cramer, Vaduz, Liechtenstein.

Sparrow, F. K. (1960). *Aquatic Phycomycetes.* 2nd. ed., pp. vii + 1187. University of Michigan Press, Ann Arbor, Michigan, U.S.A.

The following is due for publication late in 1994 (number of pages not yet known).

Dick, M. W. *Straminopilous Fungi: a new classification for the biflagellate fungi and their uniflagellate relatives with particular reference to Lagenidiaceous fungi.* CAB International (Mycological Papers Series), Wallingford, Oxon, U.K.

Index of Algal Genera

This index refers to algae as algae, irrespective of whether any of them can also be called animals and that some are bacteria. The following indexes cover associations with animals and fungi. A few of the pictures in these latter indexes also appear here because they show certain features particularly well (e.g. the colonies of *Aphanizomenon* in fig. 447). On the other hand, the figures of a few genera, only referred to in relation to grazing or parasitism, show so little of the algae concerned that they are omitted from this index (e.g. *Pseudocarteria*, fig. 334 and *Chlorogonium*, fig. 574). Figure numbers are in bold type.

Achnanthes 128: **232**
Aegagropila 74
Amphora 126: **222, 379**
Amscottia 44: **55**
Anabaena 208, 210, 240, (241, *Trichormus*): **397-405, 407**, (**453, 454**, *Trichormus*)
Anabaenopsis 214: **406**
Ankistrodesmus 30: **31**
Apatococcus 250, 252: **468**
Aphanizomenon 216: **409-412, 477**
Aphanochaete 70: **112-114**
Aphanothece 196: **215, 216, 371, 372**
Apiocystis 24: **16**
Arthrospira 230
Astasia 104: **177-183, 187, 188**
Asterionella 132, 136, 138: **240, 246-249**
Asterococcus 24: **15**
Audouinella – see *Rhodochorton*
Aulacoseira – see *Melosira*

Batrachospermum 184-190: **344-358**
Botrydiopsis 114: **204**
Botrydium 112, 114: **197-200**
Botryococcus 12, 36, 38: **11, 39-41**
Brachiomonas 82, 84: **133**
Bulbochaete 62, 64: **97, 99-101**

Cephaleuros 246
Ceratium 172, 174, 176, 178: **317-325, 332**
Chaetophora 68: **108, 110**
Chaetosphaeridium 70: **115-117**
Chamaesiphon 200: **380-382**
Chantransia 186, 188, 193
Characium 288: **566**
Chilomonas 170: **316**
Chlamydocapsa 282: **556**
Chlamydomonas 80, 82: **8, 19, 131, 134-137**
Chlorella 242-245: **455-459**
Chlorococcum 22: **13**
Chlorogonium 292
Chromulina 163
Chroococcus 196, 198: **375-377**
Chroomonas 170

Chrysamoeba 163, 165: **301-304**
Chrysochromulina 166, 168: **305, 308**
Chrysopyxis 160: **296**
Cladophora 172, 174, 176: **6, 118-123**
Closterium 38, 40, 46, 52: **42, 43, 71**
Coelastrum 26, 28: **21, 23**
Coelosphaerium 208
Colacium 98, 100: **168, 169**
Conferva 218
Cosmarium 40, 48: **44, 45, 72, 73, 637**
Craticula 130: **233, 234**
Crucigenia 28: **24, 25**
Cryptomonas 170: **313-315**
Cyanothece 195, 196: **373, 374**
Cyclonexis 158: **295**
Cyclotella 121, 122: **209, 211, 213, 214**
Cylindrocystis 54: **75-77**
Cylindrospermum 216: **408**
Cymbella 126: **226,** (**227**, *Encyonema*), **253, 254**
Cystodinium 180: **337**

Desmococcus 251
Dictyosphaerium 25, 26: **19, 20, 31**
Didymosphenia 126, 140: **219, 220, 257**
Dinobryon 154, 156: **276-279, 282, 283**
Distigma 104: **184, 185**
Draparnaldia 68: **107, 109, 111**
Dunaliella 83: **132**

Ellerbeckia 122: **210, 212**
Encyonema see *Cymbella*
Entermorpha 78: **126-130**
Epithemia 124: **217**
Eremosphaera 21, 22: **12**
Euastrum 40: **46, 49, 74**
Eudorina 88: **141, 144, 147**
Euglena 94, 96, 98: **7, 158-161, 163, 165-167**
Eutetramorus – see *Sphaerocystis*

Fragilaria 134: **243**

Gemellicystis – see *Pseudosphaerocystis*
Geminella 66: **106**
Glaucocystis 252, 253: **469**
Gloeocystis – see *Chlamydocapsa*
Gloeotrichia 218: **416-420**
Gomphonema 128: **228, 230, 231**
Gomphosphaeria 206, 208: **384, 389-396**
Gonium 86: **140, 143, 144**
Gonyostomum 168, 169: **309-312**
Gymnodinium 180: **338, 339**
Gyrosigma 126: **225**

Haematococcus 84-86: **9, 138, 139**
Hapalosiphon 228: **431, 432**
Hildenbrandia 182: **342**
Hydrodictyon 32, 34, 36: **32-34, 38**

Hydrurus 160, 162: **291-294, 297**

Kephyrion 158: **286**
Kephyriopsis 158
Kirchneriella 30, 32: **30, 31**

Lagerheimia 24: **17**
Lemanea 182, 184: **343**
Lepocinclis 98: **162, 164**
Limnothrix — see *Oscillatoria*
Lyngbya 226

Mallomonas 148, 156: **261-266, 269**
Mallomonopsis 148
Melosira 138, (140, *Aulacoseira*): **250-252**, (**255, 256**, *Aulacoseira*)
Meridion 130: **236**
Merismopedia 198: **378, 379**
Mesotaenium 54: **78**
Micractinium 24: **18**
Micrasterias 38, 42: **42, 52, 54, 65, 67, 68**
Microcystis 202: **383-385**
Mischococcus 110, 112: **196**
Monoraphidium 30
Mougeotia 58: **81, 89, 201**

Navicula 130: **237**
Netrium 54: **80**
Nitzschia 130: **222-234**
Nostoc 234, 236-238, 240, 242: **445-447, 449, 451, 452**

Ochromonas 163: **299, 300**
Oedogonium 62, 64: **93-96, 98, 102**
Ophiocytium 114, 116: **201-203, 205, 206**
Oscillatoria 222, 224: (**422**, *Planktothrix*), (**423**, *Tychonema*), (**424**, *Limnothrix*), **430**

Palmella 334
Pandorina 88: **146**
Pediastrum 32: **31, 35-37**
Penium 42: **48**
Peranema 107
Peranemopsis 106, 107: **189-191**
Peridinium 176, 179: **326-331, 333, 335, 336, 340**
Petalonema 228: **436**
Phacus 102: **174-176, 186**
Phaeodermatium 162: **298**
Phormidium 224, 226: **425, 426**
Pinnularia 120, 124: **207, 208, 218**
Pithophora 76, 77: **124, 125**
Planktothrix — see *Oscillatoria*
Pleodorina 88
Pleurococcus 250, 252
Pleurotaenium 42: **50, 53**
Porphyridium 182: **341**
Prymnesium 166: **306, 307**
Pseudocarteria 179
Pseudokephyrion 158
Pseudosphaerocystis 286: **565**

Pseudotrebouxia 246

Rhizochrysis 165
Rhizosolenia 134: **241, 242, 245**
Rhodochorton 190, 192, 193: **359-370**
Rhodomonas 170

Scenedesmus 28, 30: **26-28**
Scytonema 228: **435**, (**436**, *Petalonema* stage)
Selenastrum 30, 32: **29**
Snowella 205: **386-388**
Sphaerocystis 22, 24: **14**
Spirogyra 56, 58, 60: **81, 82, 84-88, 90-92**
Spirotaenia 54: **79**
Spirulina 230, 232: **437-439, 441-444**
Spondylosium 48: **63**
Spumella 260: **485**
Staurastrum 42, 44, 46, 52: **56-58, 60, 62, 300, 301**
Staurodesmus 44: **61, 64**
Stauroneis 126: **221**
Stephanodiscus 120, 122: **215, 216**
Stichococcus 246, 248: **467**
Stigeoclonium 310: **5, 612**
Stylosphaeridium 208: **396**
Surirella 126: **223, 224**
Synechococcus 196
Synechocystis 196: **393, 394**
Syncrypta 152
Synedra 134: **229, 244**
Synura 150, 152, 156: **267, 268, 270-275, 281, 284**

Tabellaria 132: **238, 239, 599, 603**
Tetmemorus 46: **59**
Tetrastrum 28: **22**
Tolypothrix 228: **433, 434**
Trachelomonas 100, 102: **170-173**
Trebouxia 246
Trentepohlia 246: **465, 466**
Tribonema 110: **195**
Trichormus – see *Anabaena*
Triploceras 42: **51**
Tychonema – see *Oscillatoria*

Ulothrix 66: **103-105**
Ulva 78
Uroglena 158: **280, 287-290**
Urosolenia – see *Rhizosolenia*

Vaucheria 108, 110: **192-194**
Volvox 88, 90, 92: **145, 148-157**

Woronichinia – see *Gomphosphaeria*

Xanthidium 50: **66, 69, 70**

Zoochlorella 242
Zygnema 56: **12, 81, 83**

Index of Animal Genera

"Animal" (e.g. *Peranemopsis*, certain species of *Peridinium*) or animal-like (e.g. some Chrysophytes) algae are in the algal index. Unnamed animals are listed after the generic names. Figure numbers are in bold type.

Ascomorpha 276: **539, 542**
Asplanchna 276: **541**
Asterocaelum 260, 262: **487-498**

Bicosoeca 302: **602**
Bosmina 274: **535**

Cephalodella 276, 278: **543-546**
Codosiga 302: **599, 603**
Cyclidium 306: **608**

Daphnia 274: **536**

Enteromyxa 270-272: **520-532**
Euplotes 242: **456**

Frontonia 242, 254: **457, 472**

Hydra 244: **459-461**

Keratella 100, 276: **169**

Lembadion 254: **471**

Monas 260

Nassula 256, 258: **474-478, 481, 482**

Paulinella 252, 253: **470**

Salpingoeca 302: **600**

Vampyrella 264-270: **449-501, 503-519**

Unnamed animals

Amoeboid organisms 260, 264: **486, 502**
Choanoflagellates 302
Ciliate 254: **473**
Heliozoa 306: **604-607**
Plasmodial organisms 272: **533, 534**
Rotifer 276: **540**
Tintinnid 258, 260: **479, 480, 483, 484**
Vorticellid ciliates 212, 214, 242: **407, 455**
Worm 274: **538**

Index of Fungal Genera

The fungi are not described as discrete genera in the way the algae are but shown in relation to stages in their life histories, to parasitism or, in a few cases, saprophytism. Illustrations of unnamed fungi are listed after the generic names. Figure numbers are in bold type.

Amphicypellus 298, 318: **592, 615, 636**
Anisolpidium 310: **612**
Aphanomycopsis 298, 312, 316, 318: **615, 631, 636**

Blastocladiella 300: **597**

Chytridium 288, 292, 296: **566, 575-577, 584-587**

Endocoenobium 284: **561**

Lagenidium 314-316: **621, 626, 629**

Micromyces 297, 298: **582, 583, 588-591**
Myzocytium 300, 310, 312, 314, 318: **595, 614, 618, 619, 634**

Podochytrium 290: **569**
Pseudopileum 294, 296, 315: **581, 622**

Rhizidium 286: **565**
Rhizophydium 156, 282, 286, 288, 294, 297, 298, 308, 309: **282, 555, 558, 564, 568, 580, 593**
Rhizosiphon 284, 292: **560, 578**
Rozella 322: **643, 644**

Scherffeliomyces 288: **567**

Zygorhizidium 282, 286, 300, 322: **556, 562, 563, 596, 598, 641-646**

Unnamed fungi:

Chytrids **557, 559, 570, 574, 579, 637-640**

Biflagellate fungi **613, 616, 617, 620, 623-625, 630, 633, 635**

Index of Algal Genera associated with Animals or Fungi

References to algae associated with animals (e.g. as food or a substratum to live on) are preceded by an (A) and those associated with a fungus (e.g. parasitized by, or symbiont with) by (F). A few references to unnamed algae are given at the end of the index. Figure numbers are in bold type.

Algae associated with other algae are covered by the algal index. The general index covers associations with plants, man, etc.

Anabaena
  (A) 212, 214, 256, 258: **407, 476, 481**
  (F) 284, 288, 292, 300, 312, 315: **560, 567, 578, 597, 617, 620, 624**

Aphanizomenon
  (A) 256: **475, 477, 478**

Asterionella
  (A) 276, 302: **541, 600, 602**
  (F) 288, 300, 304, 305, 307-309, 322, 323: **568, 598, 601, 611, 641-646**

Batrachospermum
  (F) 282: **557**

Botryococcus
  (F) 284: **559**

Ceratium
free-living
  (A) 276: **539, 542**
  (F) 298, 312, 318: **592, 615, 636**
spore
  (A) 318: **632**
  (F) 284, 294, 316, 318: **558, 579, 625, 630, 635**

Characium
  (F) 288: **566**

Chlamydocapsa
  (F) 282: **556**

Chlamydomonas
  (F) 290: **570**

Chlorella
  (A) 242, 244: **455-459**

Chlorogonium
  (F) 292: **574**

Chroococcus
  (A) 274: **538**

Closterium
  (A) 268: **516, 517**
  (F) 297, 312: **554, 582, 618, 619**

Colacium
  (A) 98, 100: **169**

Cosmarium
  (F) 298, 310, 314, 320: **593, 613, 621, 637**

Cryptomonas
   (A) 254: **471**
Cyclotella
   (F) 320: **638-640**

Dinobryon
   (A) 156: **282**
   (F) 156: **282**

Eudorina
   (A) 276: **541**
   (F) 282, 284, 312: **555, 561, 616**

Euastrum
   (A) 274: **538**

Geminella
   (A) 264-268: **449-501, 504-515**

Kirchneriella
   (F) 286: **562, 563**

Mallomonas (spore)
   (F) 294, 315: **581, 622**

Microcystis
   (F) 292: **577**

Netrium
   (F) 316: **631**

Ochromonas (spore)
   (F) 296: **584-587**

Oedogonium
   (A) 264: **503**

Oscillatoria
   (A) 258, 270, 271: **482, 520-532**

Oscillatoria (Planktothrix)
   (A) 254, 256: **472, 474**
   (F) 286: **564**

Pinnularia
   (F) 290: **569**

Pseudosphaerocystis
   (F) 286: **565**

Pseudotrebouxia
   (F) 246

Sphaerocystis
   (F) 300: **596**

Spirogyra
   (F) 297, 298: **583, 588-591**

Staurastrum
   (F) 300, 310, 318: **594, 595, 614, 626-628**

Staurodesmus
   (F) 292: **576**

Stephanodiscus
   (A) **491**

Stichococcus
   (F) 246, 248

Stigeoclonium
   (F) 310: **612**

Synedra
   (F) 315: **623**

Tabellaria
   (A) 302: **599, 603**
   (F) 292: **575**

Tetmemorus
   (A) 274: **538**

Trachelomonas
   (F) 318: **633**

Trebouxia
   (F) 246

Trentepohlia
   (F) 246

Uroglena
   (A) 276, 277: **543-546**

Volvox
   (A) 272: **533, 534**

Xanthidium
   (F) 315, 318: **629, 634**

Unnamed algae

Diatoms
   centric (A) 258, 260, 262: **483, 486-488, 491-498**
   pennate (A) 254, 264, 274, 276: **473, 502, 538, 540**
   mixed (A) **480**

Kephyrioid Chrysophytes
   (A) 258: **484**

General Index

acid rain, acidification, air pollution, 146, 168, 169, 252
akinete (Cyanophytes) 210, 218: **399-402, 405, 408, 410, 416, 417**
algae living in the mucilage of other algae 26, 206, 208: **19, 393, 396**
astaxanthins 84, 86, 98
atom bombs 147
auxospore — see diatoms
Azolla 240, 241: **448**

bacteria
   in the mucilage of algae 26, 48, 208: **20, 63, 395**
   ingested by algae 154, 158: **276, 277**
barium sulphate (barytes: barite) 40
β-carotene 83
biflagellate fungi — see chapter 13
   appressoria 318: **626-628**
   epidemic 300: **594, 595**
   exit tubes 310, 312, 315: **614, 619, 620, 623, 624, 630, 631**
   male cell 318: **634**
   resting spores 318: **633-635**
   sporangia, dehiscing 314, 315: **619, 629**
      developing 310, 312: **613-616, 618, 620**
      empty 312, 315: **617, 623, 630, 636**
   zoospores
      primary (encysted) 316: **625, 631**
      secondary
         contained in a vesicle 314, 315: **619, 621, 629**
         free swimming 280: **549, 550**
         infecting alga 316: **626-628**
bilharzia 76
bioassay 30, 31, 328
bird bath, a garden ornament 8, 83: **9**
blanket weed 74
boghead coal 36
botulism 222
burgundy blood alga 220, 258

calcium carbonate (limestone, chalk) 112, 168: **198**
calcium sulphate (gypsum) 40
Chernobyl 147
chloroplast movements 58, 108
choanoflagellates 302: **599, 603**
chromatic adaptation 198, 200
chytrid fungi — see chapter 13
   "asterosphere" 297: **583**
   ecology 300-309
   epidemics 300: **596-598**
   hyperparasites 320, 322, 323: **637, 638, 640, 643-645**
   hypersensitivity 307: **611**
   male cell 294, 322: **580, 581, 646**
   parasitizing a saprophyte 318: **636**
   resting spores 294: **567, 579-581, 646**
   rhizoids 286, 288, 296: **563-568, 570, 585, 587**

saprophytes 298, 312, 318: **592, 593, 620, 636**
　　sporangia,　dehiscing 290, 292: **571-573, 622**
　　　　"　　　　development of 281, 282, 284, 296, 297: **555-562, 588, 589**
　　　　"　　　　mature 288, 289: **568-570, 585, 591**
　　　　"　　　　empty 292: **574-578, 586, 639, 641, 642**
　　zoospores,　creeping 281: **552, 553**
　　　　"　　　　encysted 281: **554, 584, 601**
　　　　"　　　　free swimming 280, 281: **547, 548, 551**
　　　　"　　　　ingestion of 306: **604-610**
　　　　"　　　　liberation of 290, 314: **571-573, 622**
　　　　"　　　　with contractile vacuole 281: **553**
cilia — see flagella
ciliates — see chapter 12
　　attached to Cyanophyte 212, 214: **407**
　　ingestion of chytrid zoospores 306: **608**
　　symbiosis with *Chlorella* 242, 244: **455-458**
Cladophora balls 72, 74: **122, 123**
coccoliths 168
contractile vacuoles — see vacuoles
cooragonite 36, 38
crustaceans — see chapter 12
cryopreservation 305
cyanelles 252: **469, 470**
Cyanophytes — see chapter 10.
　　See also akinetes, eutrophication, folklore, gas vacuoles, grazing, heterocysts, human food, nitrogen fixation, symbiosis, toxic algae, waterblooms
cyst — see spore

dating the past 146, 147
deserts 10, 236
diatomite 142, 144: **258, 259**
diatoms — see chapter 5
　　artistic arrangements 144: **260**
　　auxospores 138, 140: **250, 253, 254, 256**
　　cell size 136, 138: **246-249**
　　centric and pennate 120
　　drowning 144
　　environmental indicators 146, 147
　　food source 276
　　fossil 142, 146
　　hypersensitivity 307: **611**
　　industrial use 142, 144
　　lake history 146, 147
　　movement 124, 126
　　sex — see auxospores
digestion cysts — see chapter 12
dwarf male 64: **100**
dynamite 144

ecology
　　of algae 8-20
　　of fungi 300-309
environmental indicators
　　Cyanophytes — see eutrophication and water blooms
　　diatoms — see lake history
eutrophication — see also waterblooms 12, 14, 34, 36, 74, 76, 220-222
evolution — see symbiosis theory

experimental use of algae
   Chlamydomonas 82
   Chorella 245
   Euglena 98
   Selenastrum (*Kirchneriella*) 30
eye-spot 88: **105, 147, 148, 158-160, 171-176, 282, 290**

fat – see lipid
fertilizers 241
fish
   flesh colour – see astaxanthin
   poisoning – see toxic algae
flagellum 80
flotation – see gas vacuoles and oil
folklore 74, 220, 237
fossils
   Botryococcus 36
   Chrysophytes 165
   Cyanophytes 195
   diatoms 142, 146
   Dinophytes 178
   Haptophytes 168
freeze-drying – see cryopreservation
frogspawn alga 184
fungi – see chapter 13

gametes – see sex
gas-vacuoles, gas vesicles 202, 204
global warming and "greenhouse" gases 17, 38, 221
glycerol 83
grazing
   by algae on bacteria 154, 158: **276, 277**
   by algae on fungal zoospores 306: **609, 610**
   by algae on other algae 180: **339**
   by algae on protozoan spores 179: **335, 336**
   of algae by protozoa, rotifers and crustaceans – see chapter 12
greenhouse effect 38
Gunnera 238, 240: **450**
gypsum – see calcium sulphate

haptonema 166: **305, 307, 308**
Heliozoa 306
heterocyst 210, 238: **399, 406, 408, 433, 3434**
hormogonium 226: **427-429**
human food
   Chlorella 245
   Nostoc 236: **447**
   Spirogyra 58
   Spirulina 230, 232: **438, 440**
Hydra 244: **460, 461**
hypertrophy 298, 310: **612**
hyphochytrids 310: **612**

ice and snow 16, 20: **8**
industrial uses
   Botryococcus 36, 38: **41**
   Cladophora 74

    diatoms 144
    Dunaliella 83
    Haematococcus 84, 86
iron and manganese-colouring
    cases of *Trachelomonas* 100: **171-173**
    holdfasts of *Ophiocytium* and *Characium* 116, 288: **202, 203, 566**
    warts of *Euglena* 96: **159**

Kieselguhr – see diatomite

lake history
    Chrysophytes 165
    diatoms 146, 147
leucosin 156: **264, 281**
lichens 246, 248, 250: **462-464**
little "hedgehog alga" 218: **418**

manganese – see iron

naming of organisms (nomenclature) 4-7, 24, 107, 130, 165
nitrogen fixation 210, 212
nucleus and nucleolus 40, 58, 174, 176, 260: **47, 86, 321, 333, 335, 485**

odours
    Asterionella 132
    Dinobryon, Synura and *Uroglena* 152
    Hydrurus 160
oil
    Botrydiopsis 114
    Botryococcus 36, 38: **41**
ozone layer 20, 84

palmella or palmelloid stage 96, 170, 335
    Euglena 96: **161**
    Cryptomonas 170: **315**
paramylon 96, 102: **165, 174-176, 186**
parasites, fungal – see chapter 13
plankton 14-20, 337; see also fungi and waterblooms
plasmodium 272: **533, 534**
prokaryotes 1, 194, 337, 338
protozoa – see chapter 12 and 338

red tides 172
rice cultivation 212
rotifers – see chapter 12

saprophytes, fungal – see chapter 13
schistosomiasis – see bilharzia
sex
    Chlorophytes 52, 54, 58, 60, 64, 66, 82, 92: **62, 71-75, 90, 99-101, 134-137, 155, 156**
    diatoms 138, 140: **257**
    fungi 294, 318, 322: **580, 581, 634, 646**
    Rhodophytes 186, 188, 190, 192, 193: **347-355**
    Xanthophytes 108: **194**
shellfish poisoning 172
silica
    Chrysophytes 148

diatoms 118
　　　Paulinella 148
skin rashes – see swimmer's itch
slime mould 272
space travel 245
stratification of water 16, 17
stigma – see eye-spot
swimmer's itch and skin irritations 76, 168, 169, 222
symbiosis – see chapter 11
symbiosis evolutionary theory 170, 180, 252, 253

taxonomy and classification 5, 6, 8, 32, 194, 195, 230
tintinnid – see chapter 12
torbanite 36
toxic algae
　　　Cyanophytes 205
　　　Dinophytes 172
　　　Haptophytes 166
trichocysts (including ejectosomes) 163, 168, 170

ultraviolet radiation 20, 84, 98, 204

vacuoles
　　containing crystals 40: **47**
　　containing food 260: **486**
　　containing gas – see gas vacuoles
　　contractile – in
　　　　Chlamydomonas 80, 82: **131**
　　　　Eudorina 141
　　　　fungal zoospores 281, 314: **549, 552**
　　Gonium 140
　　Gonyostomum 312
　　Ochromonas 300
　　Rhizosolenia 134: **241**
　　Volvox 90: **152**

waterblooms and scums – see also eutrophication 98, 114, 202, 204, 220, 222: **7, 8, 10, 11, 166, 204, 403, 413, 421, 422**
water net alga 5, 32, 34, 35: **38**